Perspectives on Marriage

Perspectives on Marriage

A READER

EDITED BY
Kieran Scott
Michael Warren

New York Oxford
OXFORD UNIVERSITY PRESS
1993

Oxford University Press

Oxford New York Toronto
Delhi Bombay Calcutta Madras Karachi
Kuala Lumpur Singapore Hong Kong Tokyo
Nairobi Dar es Salaam Cape Town
Melbourne Auckland Madrid

and associated companies in
Berlin Ibadan

Published by Oxford University Press, Inc.
200 Madison Avenue, New York, New York 10016

Library of Congress Cataloging-in-Publication Data
Perspectives on marriage : a reader / edited by Kieran
Scott, Michael Warren
p. cm.
ISBN 0-19-507804-7
1. Marriage—Religious aspects—Christianity. I. Warren,
Michael. II. Scott, Kieran.
BV835.P47 1993
248.4—dc20 92-14595

9 8 7 6 5 4 3 2 1
Printed in the United States of America
on acid-free paper

To Connie and Ellen,
who taught us the meaning of marriage

Acknowledgments

We give our deepest thanks to the administrators, Theology Department faculty, and chairs at St. Bonaventure and St. John's universities, who made available research grants, teaching reductions, and yearlong sabbaticals during the period of preparation, organization, and construction of this volume. Without their generous help, this book would never have been published.

Contents

V. Communication, Conflict, and Change

VI. Getting Beyond Stereotypes

VII. Divorce/Annulment

Perspectives on Marriage

Introduction

Kieran Scott and Michael Warren

This book is the result of a need recognized for several years by many teachers of the marriage course. The idea of putting together this book emerged from a casual conversation between the two editors at a meeting of professors. Each expressed a zest for the marriage course, while deploring the lack of any single "reader"-type book, full of rich resources. There were fine books about aspects of marriage, but no single resource approaching marriage from a variety of angles. In that conversation we each spoke of our ideal table of contents in a marriage reader. Several years and twelve revisions later, this book is the fruit of that casual conversation.

In the slow process of assembling the materials here, we asked colleagues about their marriage courses and found a variety of approaches to the topic: some more theological, some more sociological, some psychological, but all seeking to provide a tone to catch the serious attention of students. Almost all combined a theological perspective with a social science one. In asking these persons about the sort of reader that might be helpful, we concluded that variety would be a key component of such helpfulness.

Born of such background, this book could not be intended as a course outline, and we who have edited it warn against letting the book's contents design the marriage course. Instead, the book should be used the other way around: the vision of the course should determine what here will or might be useful in a particular approach to marriage. This reader offers a series of field trips or excursions into aspects of marriage. We suspect what gives special life to the course in marriage is the artistry worked out by particular teachers examining marriage from insightful angles based on their own study and gifts.

Most teachers of the university-level course in the theology of marriage see their efforts as an academic enterprise, not a ministerial or church one. They seek to disclose historical traditions, including religious ones, affecting marriage, while attending to recent social and cultural shifts affecting attitudes toward the practice of marriage. Still, an academic course dealing with the religious dimensions of marriage will have some things in common with church efforts in ministry to marriage. One of these is concern for marriage preparation.

In a society where at the end of the 1980s 25 percent of all births were out of wedlock and where the number of children living with a single parent rose from 7 percent in 1960 to a current 25 percent, adequate preparation for marriage will be a major concern of all institutions, including church and academy.[1] Students themselves crowd into course sections in the theology of marriage apparently well aware that the stakes for them in this enterprise are quite high. Our goal in this reader is to press them with radical questions and humanizing perspectives with important consequences for their futures.

Any course that puts marriage in a religious perspective has the possibility of contesting some currents in comtemporary culture that are hostile to enduring commitments, hostile to the kind of love ethic needed for successful parenting over the long haul, and hostile to the wider networks of support needed for families in the future. While some religious assumptions about marriage themselves need critique, many religious convictions offer a radically humanizing view of the marital relationship, quite different from some of the approaches to marriage found in current film and television depictions.

The readings in this collection contest various assumptions about marriage, including many recently created by a consumerist culture and some being challenged from within religious traditions themselves. Careful reading will spark earnest, if not heated, discussion and some conflicts over basic points of view. Many of the underlying, larger issues cannot be settled in a semester. We expect many students will leave their courses with lingering questions needing continuing reflection: the seeds of emancipatory practice.

The editors recognize the variety of academic levels in what they have gathered here. Some readings are historically "dense" and will need careful preteaching before being read. Others make use of theologically sophisticated concepts that will also need explanation before becoming accessible to many students. Other fare is less weighty but not less important. We have included several popularizations of first-rate social science scholarship. The ideas and issues in such essays should not be dismissed as "magazine pop psychology," since they make accessible for serious and informed discussion sophisticated research conducted, in some cases, over several years.

The organization and selection of material for this reader was a process of mutual reflection, dialogue, evaluation, and consensus. It was a genuine collaborative endeavor. We found ourselves sharing some common resources, proposing diverse materials and partners in writing the introductions to the essays. Each essay is contextualized, given its appropriate frame of reference, and its reading guided by a set of questions to facilitate and direct classroom discussions.

Part I, "Marriage in Historical Perspective," brings together a rich set of materials on marriage in the Jewish and Christian traditions. This historical perspective is indispensable to acquiring a developmental sense of ecclesial teaching.

Part II, "Contemporary Perspectives on the Theology of Marriage, Commitment, and Fidelity," presents marriage as a basic Christian sacrament and an act of worship. Essays by James Nelson and Margaret Farley explore the varied meanings of fidelity and the various forms of commitment.

Part III, "Marriage: Change, Character, and Context," is a challenging and provocative section. It presents important materials on the changing meanings of marriage, the family as a school for character, and the church as a nurturing and supportive context for marital relations and family life.

Part IV, ''Attitudes Toward Sexuality,'' contains a treasure of materials for study and pedagogy. The history of romantic love in the West, sexual revolution, conflicting sexual attitudes, the nature and role of intimacy in marriage, and a plea for a new moral discourse in sexual morality are offered in imaginative and stimulating readings.

Part V, ''Communication, Conflict, and Change,'' has its eye in the practical. It offers, in a more popular form, social research on the role and value of communication and conflict in marriage. Students should find the material interesting and empowering.

Part VI, ''Getting Beyond Stereotypes,'' is the fruit of the contemporary women's and men's liberation movements. The shifting work roles in marriages, the demise of the good-provider role, and the emergence of the quest for new male identity are issues central to this material . . . and to contemporary marriage in the West.

Part VII, ''Divorce/Annulment,'' acknowledges the failure, at times, of the marital ideal and its effects on people's lives. Canonical and theological perspectives on divorce and remarriage are offered, and the dilemma remarriage poses for Catholics seeking ecclesial belonging is raised.

Part VIII, ''Interreligious Perspectives on Marriage,'' widens the context for exploration and interpretation. Most of our resources and materials, up to this point, have been from a Christian perspective. This viewpoint is enriched in this section by comparison and contact with Jewish, Islamic, and Hindu perspectives on marriage. The common ground is fascinating and the distinctiveness of each inspiring.

Finally, Part IX, ''Appendix: Church Documents,'' presents a sample of church documents that couples may find themselves working with and being guided by in preparation for marriage or in filing for an annulment. These documentary guidelines are now common in Roman Catholic dioceses in the United States.

As a concluding word, we offer this volume with hope and expectation. Our hope is that students and teachers will find here a ready source of literature to challenge and stretch the imagination. Our expectation is that this volume will make a contribution to an emerging field and study of marriage.

NOTE

1. See Don S. and Carol Browning, ''The Church and the Family Crisis: A New Love Ethic,'' *The Christian Century* (August 7–14, 1991): 746–749.

I

Marriage in Historical Perspective

1.

Marriage in the Bible

Michael G. Lawler

We all have attitudes toward marriage that have been socially shaped for us within our own lifetime. We can easily tend to think of marriage as it is today—or even as it is depicted for us today in movies and television. Marriage has been imagined for us by our own parents, by the behavior of other married people we know, and by the many literary, TV, and film depictions of it.

The following essay is an account of the religious imagination of marriage, particularly the Jewish and the Christian imagination of what it could and should be. Some will find the account unbelievable, perhaps almost shocking. Not all those who call themselves Jews or Christians are aware of the quite radical approach to the marriage relationship their faith calls for.

This essay calls for careful reading. The best way to read it is to have read immediately beforehand the passages from the Old and New Testaments the author refers to and then to have them open before you while reading this important account of the profound Jewish and Christian understanding of marriage. Some helpful questions follow:

1. *The author stresses the differences between Israel and its neighbors in their understanding of sexuality and marriage. What differences are highlighted here, and exactly why are they so different?*
2. *Of the Israelite understanding of marriage, do you find any features that are still not widely accepted today—say, in the way marriage is presented in movies and on television?*
3. *Is the author claiming that marriage is the analogy for understanding our relation to God or that our relation to God is the analogy for understanding marriage?*
4. *Were Hosea living today, some would say he was "hung up" on Gomer or, in more pyschological terms, "obsessed" by her. Obession is not a beautiful thing. How would you explain the beauty of Hosea's commitment to Gomer, and what is its religious significance?*
5. *In more than one place, the author stresses that covenant love in the*

Chapter 1 of *Secular Marriage/Christian Sacrament* (Mystic, CT: Twenty-third Publications, 1985), pp. 5–22.

Bible is "not the same as a passionate affection for a person of the opposite sex." Does the author's approach make covenant love seem dull or unexciting? What is his point in this emphasis?

6. *In dealing with New Testament passages about marriage, the author spends the most time dealing with several verses in Ephesians. Why does he see these verses as so significant? If the ideas of these verses were actually lived out, what difference would they make in the way people "love" one another?*
7. *Mutual giving way is an important concept here. What might be some examples of this mutual giving way in an actual marriage? Could you describe five situations in which it might take place?*
8. *What did you make of the author's claim "If marital love exists only inchoately on a wedding day, as it surely does, indissolubility also exists only inchoately. Marital love is . . . a task to be undertaken"?*

As in all other matters, the biblical teaching on marriage should be seen in the context of the Near Eastern cultures with which the people of the Bible had intimate links, specifically the Mesopotamian, Syrian, and Canaanite. It is not my intention here to dwell on these cultures and their teachings on marriage and sexuality. They were all quite syncretistic, and a general overview sufficiently gives both a sense of the context and their specific distinctions from the Jewish Bible.

Underlying the themes of sexuality, fertility, and marriage in these cultures are the archetypal figures of the god-father and the goddess-mother, the sources of universal life in the divine, the natural, and the human spheres. Myths celebrated the marriage, the sexual intercourse, and the fertility of this divine pair, legitimating the marriage, the intercourse, and the fertility of every earthly pair. Rituals acted out the myths, establishing a concrete link between the divine and the earthly worlds and enabling men and women to share not only in the divine action but also in the efficacy of that action. This is especially true of sexual rituals, which bless sexual intercourse and ensure that the unfailing divine fertility is shared by man's plants and animals and wives, all important elements of his struggle for survival in those cultures.[1] In Mesopotamia, the divine couple is Ishtar and Tammuz; in Egypt, Isis and Osiris; in Canaan, Ashtarte (or Asherah) and, sometimes, Eshmun. After the Hellenization of Canaan, Eshmun is given the title of Adonis.

Marriage in the Old Testament

The Biblical view of sexuality, marriage, and fertility makes a radical break with this polytheistic perspective. The Old Testament, whose view of marriage I do not intend to treat fully here but only as it provides the basis for the New Testament view of Christian marriage, does not portray a god-goddess couple, but only Yahweh who led Israel out of Egypt and is unique (Deuteronomy 6:4). There is no goddess associated with him. He needs no goddess, for he creates by his word alone. This God created man and woman, "male and female he created them and he named *them 'adam*" (Genesis 5:2). This fact alone, that God names male and female together *'adam* (that is, earthling or humankind), founds the equality of man and woman as human beings, whatever be their distinction in functions. It establishes them as "bone of bone and flesh of flesh" (Genesis 2:23), and enables them "therefore" to marry and to become "one body" (Genesis 2:24). These details are taken from the early Yahwist creation account. But the much later priestly account which we find in Genesis I also records the creation of male and female as *'adam* and the injunction given them to "be fruitful and multiply and fill the earth" (Genesis 1:28).

Equal man and woman, with their separate sexualities and fertilities, do not derive from a divine pair whom they are to imitate. They are called into being by the ceative action of the sovereign God. Man and woman *'adam,* their sexuality, their marriage, their fertility are all good, because they are the good gifts of the Creator. Later Christian history, as we shall see, will have recurring doubts about the goodness of sexuality and its use in marriage, but the Jewish biblical tradition had none. As gifts of the Creator God, who "saw everything that he had made and behold it was very good" (Genesis 1:31), sexuality, marriage, and fertility were all good, and belonged to man and woman as their own, not as something derived from some divine pair. When looked at within this context of creation-gift, all acquired a deeply religious significance in Israel. That is not to say that they were sacred in the sense in which the fertility cults interpreted them as sacred, namely, as participation in the sexuality and sexual activity of the divine pair. In that sense they were not sacred, but quite secular. But in another sense, the sense that they were from God and linked man and woman to God, they were both sacred and religious. "It was not the sacred rites that surrounded marriage that made it a holy thing. The great rite which sanctified marriage was God's act of creation itself."[2] It was God alone, unaided by any partner, who not only created *'adam* with sexuality and for marriage but also blessed him and them, thus making them inevitably good.

Man and woman together are named *'adam.* They are equal in human dignity and complementary to one another; there is no full humanity without both together. Human creation, indeed, is not complete until they stand together. It is precisely because man and woman are equal, because they are *'adam,* because they are "bone of bone and flesh of flesh," that is, because they share human strengths and weaknesses, that they may marry and become "one body" (Genesis 2:24). Among the birds of the air and the animals of the field there "was not found a helper fit" for the man (Genesis 2:20), and it is not difficult to imagine man's cry of delight when confronted by woman. Here, finally, was one who was his equal, one whom he could marry and with whom he could become one body.

That man and woman become one body in marriage has been much too exclusively linked in the Western tradition to one facet of marriage, namely, the genital. That facet is included in becoming one body, but it is not all there is. For *body* here implies the entire person. "One personality would translate it better, for 'flesh' in the Jewish idiom means 'real human life.'"[3] In marriage a man and a woman enter into a fully personal union, not just a sexual or genital one. In such a union they become one person, one life, and so complement one another that they become *'adam*. They enter into a union which establishes not just a legal relationship, but a blood relationship which makes them one person. Rabbis go so far as to teach that it is only after marriage and the union of man and woman into one person that the image of God may be discerned in them. An unmarried man, in their eyes, is not a whole man.[4] And the mythic stories,[5] interested as always in aetiology, the origin of things, proclaim that it was so "in the beginning," and that it was so by the express design of God. There could be for a Jew, and for a Christian, no greater foundation for the human and religious goodness of sexuality, marriage, and fertility. Nor could there be a secular reality better than marriage for pointing to God and his steadfastly loving relationship with Israel. That was the next step in the development of the religious character of marriage.

Marriage as Covenant Symbol

Central to the Israelite notion of their special relationship with God was the idea of the covenant. The Deuteronomist reminded the assembled people: "You have declared this day concerning Yahweh that he is your God and Yahweh has declared this day concerning you that you are a people for his own possession" (Deuteronomy 26: 17–19). Yahweh is the God of Israel; Israel is the people of Yahweh. Together Yahweh and Israel form a community of grace, a community of salvation, a community, one could say, of one body. It was probably only a matter of time until the people began to imagine this covenant relationship in terms drawn from marriage, and it was the prophet Hosea who first did so. He preached about the covenant relationship of Yahweh and Israel within the biographical context of his own marriage to a harlot wife, Gomer. To understand his preaching, about marriage and about the covenant, we must first understand the times in which Hosea lived.

Hosea preached around the middle of the eighth century B.C. at a time when Israel was well established in Canaan. Many Israelites thought that the former nomads had become too well established in their promised land, for as they learned their new art of agriculture they learned also the cult of the Canaanite fertility god, Baal. This cult, which seriously challenged their worship of Yahweh, was situated in the classic mold presented earlier, that of the god-goddess pair, with Baal as the Lord of the earth, and Anat as his wife (and sister). The sexual intercourse and fertility of these two were looked upon as establishing the pattern both of creation and of the fertile intercourse of every human pair. The relationship of human intercourse and its fertility to that of the divine couple was acted out in temple prostitution, which required both *kedushim* and *kedushoth,* that is, male and female prostitutes. These were prohibited in the cult of Yahweh (Deuteronomy 23:18), and any Jewish maiden participating in temple prostitution was regarded as a harlot. It was such a harlot, Gomer, that Hosea says Yahweh instructed him to take for his wife (1:2–3).

It is quite irrelevant to the present discussion whether the book of Hosea tells us what Hosea did in historical reality, namely, took a harlot-wife and remained faithful to her despite her infidelity to him, or whether it offers a parable about marriage as steadfast covenant. What is relevant is that Hosea found in marriage, either in his own marriage or in marriage in general, an image in which to show his people the steadfastness of Yahweh's convenantal love for them. On a superficial level, the marriage of Hosea and Gomer is just like any other marriage. But on a more profound level, it serves as prophetic symbol, proclaiming and realizing and celebrating in representation the covenant relationship between Yahweh and Israel.

The names of Hosea's two younger children reflect the sad state of that relationship: a daughter is Not Pitified (1:6), and a son is Not My People (1:9). As Gomer left Hosea for another, so too did Israel abandon Yahweh in favor of Baal and become Not Pitied and Not My People. But Hosea's remarkable reaction proclaims and makes real in representation the remarkable reaction of Yahweh. He buys Gomer back (3:2); that is, he redeems her. He loves her "even as Yahweh loves the people of Israel, though they turn to other gods" (3:1). Hosea's action towards Gomer reveals and makes real in representation the action of Yahweh's unfailing love for Israel. In both cases, that of the human marriage symbol and of the divine covenant symbolized, the one body relationship had been placed in jeopardy. But Hosea's posture both is modeled upon and models that of Yahweh. As Hosea has pity on Gomer, so Yahweh "will have pity on Not Pitied," and will "say to Not My People 'you are my people,'" and they will say to him, "Thou art my God" (2:23). The covenant union, that between Hosea and Gomer as well as that between Yahweh and Israel, is restored. A sundering of the marital covenant relationship is not possible for Hosea because he recognized that his God is not a God who can abide the dissolution of covenant, no matter what the provocation. He believed what the prophet Malachi would later proclaim: "I hate divorce, says Yahweh, the God of Israel . . . so take heed to yourselves and do not be faithless" (2:16).

There are two possibilities of anachronism to be avoided here. The first is that overworked word *love*. In its contemporary usage, it always means a strong affection for another person, frequently a passionate affection for another person of the opposite sex. When we find the word in our Bible it is easy to assume that it means the same thing. But it does not. Covenant Love, of which Hosea speaks and which we read of first in Deuteronomy 6:5, is not a love of interpersonal affection but a love that is "defined in terms of loyalty, service and obedience."[6] When we read, therefore, of Hosea's steadfast love for Gomer and of Yahweh's faithful love for Israel, we ought to understand loyalty, service and obedience, and not interpersonal affection. The second possibility of anachronism rests in the hatred of divorce proclaimed by Malachi. "In the circumstances addressed by Malachi, what God hates is the divorce of Jew and Jew; there is silence about the divorce of Jew and non-Jew."[7] The post-exilic reforms of Ezra and Nehemiah require the divorce of all non-Jewish wives and marriage to Jewish ones. Malachi speaks for this period. The divorce of Jewish wives is hated, but the divorce of all non-jewish ones is obligatory. As we shall see, Paul will adapt this strategy to the needs of his Corinthian church, and it continues to be a crucial factor in the Catholic strategy toward divorce in our day.

What ought we to make of the story of his marriage that Hosea leaves to us? There is a first, and very clear, meaning about Yahweh: he is faithful. But there is also a second, and somewhat more mysterious, meaning about human marriage: not only is it the loving union of a man and a woman, but it is also a prophetic symbol, proclaiming and making real in representative image the steadfast love of Yahweh for Israel. First articulated by the prophet Hosea, such a view of marriage recurs again in the prophets Jeremiah and Ezekiel. Ultimately, it yields the view of Christian marriage that we find in the New Testament.

Both Jeremiah and Ezekiel present Yahweh as having two wives. Israel and Judah (Jeremiah 3:6–14). Oholah-Samaria and Oholibah-Jerusalem (Ezekiel 23:4). Faithless Israel is first "sent away with a decree of divorce" (Jeremiah 3:8), but that does not deter an even more faithless Judah from "committing adultery with stone and tree" (Jeremiah 3:9). Israel and Judah are as much the harlots as Gomer but Yahweh's faithfulness is as unending as Hosea's. He offers a declaration of undying love: "I have loved you with an everlasting love; therefore, I have continued my faithfulness to you" (Jeremiah 31:3; cf. Ezekiel 16:63; Isaiah 54:7–8). The flow of meaning, as in Hosea, is not from human marriage to divine covenant, but from divine covenant to human marriage. The belief in and experience of covenant fidelity creates the belief in and the possibility of fidelity in marriage, which then and only then becomes a prophetic symbol of the covenant. Yahweh's covenant fidelity becomes a characteristic to be imitated, a challenge to be accepted, in every Jewish marriage. Malachi, as we saw already, puts it in a nutshell: "I hate divorce, says Yahweh . . . so do not be faithless" (2:16).

Marriage in the New Testament

The conception of marriage as a prophetic symbol, a representative image of a mutually faithful covenant relationship is continued in the New Testament. But there is a change of *dramatis personae,* from Yahweh-Israel to Christ-Church. Rather than presenting marriage in the then-classical Jewish way as an image of the covenant union between Yahweh and Israel, the writer of the letter to the Ephesians[8] presents it as an image of the relationship between the Christ and the new Israel, his church. This presentation is of such central importance to the development of a Christian view of marriage and unfortunately has been used to sustain such a diminished Christian view that we shall have to consider it here in some detail.

The passage in which the writer offers his view of marriage (EPH.5:21–33) is situated within a larger context (EPH.5:21–6:9) which sets forth a list of household duties that exist within a family at that time. This list is addressed to wives (EPH.5:22), husbands (EPH.5:25), children (EPH.6:1), fathers (EPH.6:4), slaves (EPH.6:5), and masters (EPH.6:9). All that concerns us here is, of course, what is said of the pair, wife/husband. There are two similar lists in the New Testament, one in the letter to the Colossians (3:18–4:1), the other in the first letter of Peter (2:13–3:7). But the Ephesians' list is the only one to open with a strange injunction. "Because you fear Christ subordinate yourselves to one another";[9] or "give way to one another in obedience to Christ";[10] or, in the weaker translation of the Revised Standard Version, "be subject to one another out of reverence for Christ."[11] This junction, most commentators agree, is an essential element of what follows.

The writer takes over the household list from traditional material, but critiques it in 5:21. His critique challenges the absolute authority of any one Christian group over any other, of husbands, for instance, over wives, of fathers over children, of masters over slaves. It establishes a basic attitude required of all Christians, an attitude of giving way or of mutual obedience, an attitude which covers all he has to say not only to wives, children, and slaves, but also to husbands, fathers, and masters.[12] Mutual submission is an attitude of all Christians, because their basic attitude is that they "fear Christ." That phrase probably will ring strangely in many ears, clashing with the deeply rooted Augustinian-Lutheran claim that the basic attitude toward the Lord is not one of fear, but of love. It is probably for this reason that the Revised Standard Version rounds off the rough edge of the Greek *phobos* and renders it as *reverence*. But *phobos* does not mean reverence. It means fear, as in the Old Testament aphorism: the fear of the Lord is the beginning of wisdom (Proverbs 1:5; 9:10;15:33; Psalms 111:10).

The apostle Paul is quite comfortable with this Old Testament perspective. Twice in his second letter to the Corinthians (2 Cor. 5:11 and 7:1) he uses the phrase *fear of God*. In his commentary on Ephesians, Schlier finds the former text more illuminating of Ephesians 5:21.[13] But I am persuaded, with Sampley, that the latter is a better parallel.[14] Second Corinthians 6:14–18 recalls the initiatives of God in the covenant with Israel and applies these initiatives to Christians, who are invited to respond with holiness "in the fear of God" (2 Cor. 7:1). The fear of God that is the beginning of wisdom is a radical awe and reverence that grasps the mighty acts of God and responds to them with holiness. In 2 Corinthians 6:14–17 that holiness is specified as avoiding marriage with unbelievers; in Ephesians 5:21 it is specified as giving way to one another. That mutual giving way is required of all Christians, even of husbands and wives as they seek holiness together in marriage, and even in spite of traditional family relationships which permitted husbands to lord it over their wives.

As Christians have all been admonished to give away to one another, it comes as no surprise that a Christian wife is to give way to her husband, "as to the Lord" (EPH.5:22). What does come as a surprise, at least to the ingrained male attitude that sees the husband as supreme lord and master of his wife and appeals ro Ephesians 5:22–23 to ground and sustain that un-Christian (superior) attitude, is that a husband is to give way to his wife. That follows from the general instruction that Christians are to give way to one another. It follows also from the specific instruction about husbands. That instruction is not that "the husband is the head of the wife" (which is the way in which males prefer to read and cite it), but rather that "in the same way that the Messiah is the head of the church is the husband the head of the wife."[15] A Christian husband's headship over his wife is in image of, and totally exemplified by, Christ's leadership over the Church. When a Christian husband understands this, he will understand the Christian responsibility he assumes toward the woman-gift he receives in marriage as his wife. In a Christian marriage, spouses are required to give way mutually, not because of any inequality between them, not because of any subordination of one to the other, not because of fear, but only because they have such a personal unity that they live only for the good of that one person. Mutual giving way, mutual subordination, and mutual obedience are nothing other than total availability and responsiveness to one another so that both spouses can become one body.

The way Christ exercises headship over the church is set forth unequivocally in Mark 10:45: "The Son of Man came not to be served but to serve, and to give his life as a ransom (redemption) for many." *Diakonia,* service, is the Christ way of exercising authority, and our author testifies that it was thus that "Christ loved the church and gave himself up for her" (Ephesians 5:25). A Christian husband, therefore, is instructed to be head over his wife by serving, giving way to, and giving himself up for her. Headship and authority modeled on those of Christ does not mean control, giving orders, making unreasonable demands, reducing another human person to the status of servant or, worse, slave to one's every whim. It means service. The Christian husband-head, as Markus Barth puts it so beautifully, becomes "the first servant of his wife."[16] It is such a husband-head, and only such a one, that a wife is to fear (v. 33b) as all Christians fear Christ (v. 21b).

(In this section of Ephesians) the reversal of verses 22 and 25 in verse 33 is interesting and significant. Verse 22 enjoined wives to be subject to their husbands and verse 25 enjoined husbands to love their wives. Verse 33 reverses that order, first commanding that husbands love their wives and then warmly wishing that wives fear their husbands. This fear is not fear of a master. Rather it is awe and reverence for loving service, and response to it in a love-as-giving way. Such love cannot be commanded by a tyrant. It is won only by a lover, as the church's love and giving way to Christ is won by a lover who gave, and continues to give, himself for her. This is the author's recipe for becoming one body, joyous giving way in response to, and for the sake of, love. It is a recipe echoed unwittingly by many a modern marriage counselor, though we need to keep in mind that the love the Bible urges upon spouses is not interpersonal affection but loyalty, service, and obedience. That such love is to be mutual is clear from v. 21, "Be subject to one another," though it is not stated that a wife is to love her husband. The reasons that the writer adduces for husbands to love their wives apply to all Christians, even those called wives!

Three reasons are offered to husbands for loving their wives, all of them basically the same. First of all, "husbands should love their wives as [for they are] their own bodies" (v. 28a); secondly, the husband "who loves his wife loves himself" (v. 28b); thirdly, "the two shall become one body" (v. 31b), a reading which is obscured by the Revised Standard Version's "the two shall become one." There is abundant evidence in the Jewish tradition for equating a man's wife to his body.[17] But even if there was no such evidence, the sustained comparison throughout Ephesians 5:21–33 between Christ-Church and husband-wife, coupled with the frequent equation in Ephesians of church and body of Christ (EPH.1:22–23; 2:14–16; 3:6; 4:4–16; 5:22–30), clarifies both the meaning of the term *body* and the fact that it is a title of honor rather than of debasement.

Love is always essentially creative. The love of Christ brought into existence the Church and made its believers "members of his body" (v. 30). In the same way, the mutual love of a husband and a wife brings such a unity between them that, in image of Christ and Church, she may be called his body and his love for her, therefore, may be called love for his body or for himself. But it is only within the creative love of marriage that, in the Genesis phrase, "the two shall become one body" (Gen. 2:24). Prior to marriage, a man did not have this body, nor did a woman have this head. Each receives a gift in marriage, a complement neither had before, which so fulfills each of them that they are no longer two separate persons but one blood person. For each to love the other, therefore, is for each to love herself or himself.

The second reason offered to a husband for loving his wife is that "he who loves his wife loves himself" (EPH. 5:28B, V. 33A). Viewed within the perspective I have just elaborated, such reasoning makes sense. It makes even more Christian sense when one realizes that it is a paraphrase of the great commandment of Leviticus 19:18, cited by Jesus in Mark 12:31: "You shall love your neighbor as yourself." Ephesians, of course, does not say that a husband should love his neighbor as himself, but that he should so love his wife. Where, then, is the link to the great commandment? It is provided through that most beautiful and most sexual of Jewish love songs, the Song of Songs, where in the Septuagint version the lover addresses his bride nine separate times as *plesion,* neighbor (1:9, 15; 2:2, 10, 13; 4:1, 7; 5:2; 6:4). "The context of the occurrence of *plesion* in the Song of Songs confirms that *plesion* is used as a term of endearment for the bride."[18] Other Jewish usage further confirms that conclusion, leaving little doubt that the author of Ephesians had Leviticus 19:18 in mind when instructing a husband to love his wife as himself.

The great Torah and Gospel injunction applies also in marriage: "you shall love your neighbor as yourself." As all Christians are to give way to one another, so also each is to love the other as himself or herself, including husband and wife in marriage. The paraphrase of Leviticus 19:18 repeats in another form what had already been said before in the own-body and the one-body images [of Ephesians]. What the writer [of Ephesians] concludes about the Genesis one-body image, namely. "This is a great mystery, and I mean in reference to Christ and the church" (v. 32), will conclude our analysis of this central teaching of the New Testament on marriage.

"*This* is a great mystery," namely, as most scholars agree, the Genesis 2:24 text just cited, "the two shall become one body." The mystery, as the Anchor Bible translation seeks to show, is that "this (passage) has an eminent secret meaning," which is that it refers to Christ and the Church. All that has gone before about Christ and the Church comes to the forefront here: that Christ chose the Church to be united to him, as body to head; that he loved the Church and gave himself up for her; that the Church responds to this love of Christ in fear and giving way. Christ who loves the Church, and the Church who responds in love, thus constitute one body, the Body of Christ (Ephesians 1:22–23; 2:14–16; 3:6; 4:4–16; 5:22–30), just as Genesis 2:24 said they would. The writer is well aware that this meaning is not the meaning traditionally given to the text in Judaism, and he states this forthrightly. Just as in the great anithesis of the Sermon on the Mount, Jesus puts forward his interpretations of biblical texts in opposition to traditional interpretations ("You have heard that it was said to the men of old . . . but I say to you"), so also here the writer asserts clearly that it is his own reading of the text ("*I* mean in reference to Christ and the church," v. 32b).

Genesis 2:24: "That is why a man leaves his father and mother and clings to his wife, and the two of them become one body," was an excellent text for the purpose the writer had in mind, for it was a central Old Testament text traditionally employed to ordain and legitimate marriage. He acknowledges the meaning that husband and wife become one body in marriage; indeed, in v. 33, he returns to and demands that husband and wife live up to this very meaning. But he chooses to go beyond this meaning and insinuate another. Not only does the text refer to the union of husband and wife in marriage, but it refers also to that union of Christ and his church which he has underscored throughout Ephesians 5:1–33. On one level, Genesis 2:24 refers to human marriage; on another level, it refers to the covenant union between Christ and his Church. It is a small step to see human marriage as prophetically representing the covenant between Christ and his Church. In its turn, the union between Christ and Church provides an ideal model for human marriage and for the mutual conduct of the spouses within it.

Ephesians is not, of course, the only New Testament passage to speak of marriage and of the relationship between husband and wife. Paul does so in 1 Corinthians 7, apparently in response to a question which the Corinthians had submitted to him. The question was: "Is it better for a man to have no relations with a woman?" (7:1). The answer is an implied yes, but not an absolute yes. "Because of the temptation to sexual immorality, each man should have his own wife and each woman her own husband" (7.2). Marriage is good, even for Christians, he seems to say, as a safeguard against sexual sins, a point he underscores again in vv. 5–9. I do not wish to dwell, however, on this unenthusiastic affirmation of marriage. I wish only to highlight the equal relationship Paul assumes in marriage between a husband and a wife, a relationship he makes explicit in vv. 3–4. "The husband should give to the wife her conjugal rights, and likewise the wife to her husband. For the wife does not rule over her own body, but the husband does; likewise the husband does not rule over his own body, but the wife does."

A modern Christian might seize (as did medieval canonists seeking a precise legal definition of marriage) on Paul's dealing with marital sexual intercourse as an obligation owed mutually by the spouses to one another. But his contemporaries would have seized on something else, something much more surprising to them, namely, his assertion of strict equality between husband and wife in this matter. As Mackin puts it, correctly: "A modern Christian may wince at finding the apostle writing of sexual intercourse as an obligation, or even a debt, owed by spouses to one another, and writing of husbands' and wives' marital relationship as containing authority over one another's bodies. But Paul's contemporaries—at least those bred in the tradition of Torah and of its rabbinic interpreters—would have winced for another reason. This was Paul's assertion of equality between husbands and wives, and equality exactly on the juridical ground of authority and obligations owed."[19]

The author of 1 Timothy 2:8–15 also has something to say about the attitudes of men and women, laying down somewhat disproportionately what is expected of men (v. 8) and women (vv. 9–15). Of great interest in this text are the two traditional reasons he advances for the authority of men over women and the submission of women to men. The first is that Adam was created before Eve, and the other that it was Eve, not Adam, who was deceived by the serpent. Here the submission of women to men, and therefore of wives to husbands, is legitimated by collected stories of the mythical first human pair. For his part, the author of 1 Peter 3:1–6 requires that wives be submissive to their husbands "as Sarah obeyed Abraham" (v. 6). Such widespread views on such Old Testament bases were common in the Jewish world in which the Christian church originated, which makes the attitude of the writer to the Ephesians all the more surprising.

The Old Testament passage that the writer chooses to comment on is one which emphasizes the unity in marriage of the first pair, and therefore of all subsequent pairs, rather than their distinction. He embellishes it not with Old Testament references to creation and to fall, but with New Testament references to the Messiah and to his love. This leads him to a positive appraisal of marriage in the Lord that was not at all customary in the Jewish and Christian milieu of his time. While he echoes the customary *no* to any form of sexual immorality (5:3–5), he offers a more-than-customary *yes* to marriage and sexual intercourse. For him marriage means the union of two people in one body, the formation of a new covenant pair, which is the gift of both God who created it and his Christ who established it in the love he has for the church. So much so that the Christian marriage between a man and a woman becomes the prophetic symbol of the union that exists between Christ and the Church.

This doctrine does not mythicize marriage as an imitation of the marriage of some divine pair, nor does it idealize it so that men and women will not recognize it. Rather it leaves marriage what it is, a secular reality in which a man and a woman seek to become one person in love. What is added is only this, simple and yet mysteriously complex. As they become one body-person in love, they provide through their marriage a prophetic symbol of a similar oneness that exists between their Christ and their Church.

Qualities of Christian Marriage

The qualities of Christian marriage already appear from our biblical analysis. The root quality, the one that irradiates all the others, is the fulfillment of the great Torah and Gospel injunction: "You shall love your neighbor as yourself" (Leviticus 19:18; Mark 12:31; Matthew 19:19).The Apostle Paul instructed the Romans that every other commandment was "summed up in this sentence, 'You shall love your neighbor as yourself' " (13:9). It is an instruction that holds true even, perhaps especially, in marriage, *Love,* of course, is a reality that is not easy to specify in words. It has a variety of different meanings. In Christian marriage love between the spouses, in fulfillment of the great commandment, is so radically necessary that in our time the Roman Rota, the Supreme Marriage Tribunal of the Roman Catholic Church, has ruled that where it is lacking from the beginning a Christian marriage is invalid.[20] That is how important Christian love is between Christian spouses.

We recall here that covenant love in the Bible is a love that is defined in terms of loyalty, service, and obedience, not in terms of interpersonal affection. The Letter to the Ephesians specifies that the love that is demanded in a Christian marriage is that kind of love. It is, first, love as mutual giving way, love as mutual obedience. The love of the spouses in a Christian marriage is a love that "does not insist on its own way" (1 Corinthians 13:4), a love that does not seek to dominate and control the other spouse. Rather is it a love that seeks to give way to the other whenever possible, so that two persons might become one body. There are individuals whose goal in life appears to be to get their own way always. The New Testament message proclaims that there is no place for such individuals in a marriage, least of all in a Christian marriage. That is not to say that there is no place in a marriage for individual differences. It is to say only that spouses who value getting their own way always, who value the domination of their spouses, who never dream of giving way, will never become one person with anyone, perhaps not even with themselves. In a Christian marriage, love requires not insisting on one's own way, but a mutual empathy with and compassion for the needs, feelings, and desires of one's spouse, and a mutual giving way to those needs, feelings, and desires when the occasion demands for the sake of, and in response to, love. Love that is exclusively *eros* is not the kind of love that is apt to ensure that two persons shoud become one body.

Love in a Christian marriage is, secondly, love as mutual service. All Christians are called to, and are sealed in baptism for, the imitation of Christ, who came "not to be served but to serve" (Mark 10:45). It cannot be otherwise in Christian marriage. In such a marriage there is no master, no mistress, no lady, no lord, but only mutual servants, seeking to be of service to the other, so that each may become one in herself/himself and one also with the other. This is required not just because it is good general counsel for marriage, but specifically because these Christian spouses are called in their marriage both to be imitators of Christ their Lord and to provide a prophetic symbol of his mutual servant-covenant with his church. For Christian spouses their married life is where they are to encounter Christ daily, and thereby come to holiness.

The love that constitutes Christian marriage is, finally, steadfast and faithful. The writer to the Ephesians instructs a husband to love his wife "as Christ loved the church." We can be sure that he intends the same instruction also for a wife. Now Christ loves the Church as Hosea loves Gomer, steadfastly and faithfully. A Christian husband and wife, therefore, are to love each other faithfully. This mutually faithful love, traditionally called fidelity, makes Christian marriage exclusive and permanent, and therefore an indissoluble community of love. On the question of indissolubility I want my position to be clear. Christian marriage is indissoluble because Christian love is steadfast and faithful. Indissolubility is a quality of Christian marriage because it is, first, a quality of Christian love. If marital love exists only inchoately on a wedding day, as it surely does, indissolubility also exists only inchoately. Marital love, as mutual giving way, as mutual service, as mutual fidelity, as mainspring of indissoluble community, is not a given in a Christian marriage but a task to be undertaken. It has an essentially eschatological dimension. *Eschatological* is a grand theological word for simple and constant human reality, namely the experience of having to admit "already, but not yet." Already mutual love, but not yet steadfast; already mutual service, but not yet without the desire to control; already one body, but not yet one person; already indissoluble in hope and expectation, but not yet in full human reality; already prophetic representation of the covenant union between Christ and his church, but not yet totally adequate representation. For authentic Christian spouses, Christian marriage is always a challenge to which they are called to respond as followers of the Christ who is for them the prophetic symbol of God.

Summary

Four things we have seen in this chapter need to be underscored. First, human marriage is not an imitation of the eternal marriage of some divine couple, but a truly human, and therefore truly secular, reality which man and woman, *'adam,* hold as their own as gift from their Creator-God. In the giving and receiving of this gift, the Giver, the gift and the recipient are essentially and forever bound together. Secondly, this bond is explicated by the prophet Hosea, who brings into conscious focus the fact that marriage between a man and a woman is also the prophetic symbol of the covenant union between Yahweh and his people. Thirdly, the author of the letter to the Ephesians further clarifies the symbolic nature of marriage by proclaiming "a great mystery." The great mystery is that as a man and a woman become one body-person in marriage, so also are Christ and his Church one body-person, and that the one reflects the other. From such thinking Roman Catholic theologians will be led slowly to declare that *human* marriage, on occasion, may be also *Christian* sacrament. Fourthly, Christian marriage is both a covenant and a community of love between a man and a woman, love that does not seek its own, love that gives way, love that serves, love that is steadfastly faithful. Because it is a covenant and a community of steadfast love, it is a permanent and exclusive state and a prophetic symbol of the steadfast covenant and community of love between Christ and his Church. That Christian marriage is such a reality, though, is not something that is to be taken blindly as being so. Rather it is something that in steadfast continuity is to be made so. Permanence is not a static, ontological quality of a marriage, but a dynamic, living quality of human love on which marriage, both human and Christian, thrives.

Question for Reflection and Discussion

1. In your judgment, what is the radical distinction between the ancient Jewish mythology about sexuality, marriage, and fertility and that of the peoples surrounding them in the ancient Near East? Does that distinction make any contribution to the mythology you hold about those same realities?
2. If you believed that sexuality and marriage were gifts of God, would that be enough for you to say that they related you to God? If you believed they were gifts of God, would that be enough for you to say that they were sacramental? If yes, in what sense?
3. Do you look upon marriage as sacramental? What does *sacramental* mean to you?
4. The two great commandments in Judaism and Christianity prescribe the love of God and the love of neighbor. According to the letter to the Ephesians, how are these commandments to be lived in a Christian marriage?
5. What does it mean to you to say that a man and a woman become one body in marriage? Do you understand their one-body relationship to be a legal or a kind of blood relationship? If it were a kind of blood relationship, how would you go about getting a divorce?

NOTES

1. For more detailed information see M. Eliade, *Patterns in Comparative Religion* (London: Sheed and Ward, 1979); E. O. James, *The Cult of the Mother-Goddess* (London: Thames and Hudson, 1959).

2. E. Schillebeeckx, *Marriage: Secular Reality and Saving Mystery,* Vol. 1 (London: Sheed and Ward, 1965), 39.

3. F. R. Barry, *A Philosophy from Prison* (London: SCM, 1926), 151. Cp. Schillebeeckx, *Marriage,* 43; Markus Barth, *Ephesians: Translation and Commentary on Chapters Four to Six. The Anchor Bible* (New York: Doubleday, 1974), 734–738; X. Leon—Dufour (ed.), *Vocabulaire de Theologie Biblique,* 2nd ed. rev. (Paris: Cerf, 1970), 146–152.

4. See Richard Batey, "The *mid sarx* Union of Christ and the Church," *New Testament Studies* 13 (1966–67), 272.

5. For a discussion of whether the term *myth* should be applied to any biblical passage, and for a suggestion of alternative language, see John McKenzie, "Myth and the Old Testament," CBQ 21 (1959), 265–282.

6. William L. Moran, "The Ancient Near Eastern Background of the Love of God in Deuteronomy," CBQ 25 (1963), 82.

7. Bruce J. Malina, *The Testament World: Insights from Cultural Anthropology* (Atlanta: John Knox, 1981), 110.

8. It is of no interest to any thesis in this book whether the Apostle Paul was or was not the author of Ephesians, and so I do not deal with that disputed question, referring only to *the writer*. Those who require information on the question may consult any of the modern commentaries.

9. Markus Barth, *Ephesians,* 607.

10. *The Jerusalem Bible* (London: Darton, Longman and Todd, 1966).

11. *The Holy Bible: Revised Standard Version* (London: Nelson, 1959).
12. Barth, *Ephesians,* 609.
13. Heinrich Schlier, *Der Brief an die Epheser* (Dusseldorf: Patmos, 1962), 252.
14. J. Paul Sampley, *And the Two Shall Become One Flesh: A Study of Traditions in Ephesians* 5:21–33 (Cambridge: University Press, 1971), 119–121.
15. Barth, *Ephesians,* 607.
16. Ibid., 618.
17. Cf. Sampley, *The Two Shall Become One Flesh,* 33.
18. Cf. Ibid., 30. See 30–34; cp. Barth, *Ephesians,* 704–708.
19. Theodore Mackin, *What is Marriage?* (New York: Paulist, 1982), 56.
20. Cited in Paul F. Palmer, "Christian Marriage: Contract or Covenant?" TS 33 (1972): 647–648.

2.

The Primitive Christian Understanding of Marriage

Theodore Mackin, S.J.

Most people understand marriage only within the customs of their own day and are thus unable to consider the arbitrariness of many marriage customs. Thus, our own socially determined customs take on a kind of sacredness unable to be questioned. Theodore Mackin's description of the Jewish marriage customs of Jesus' day helps us see how differently various societies can go about the tasks and duties of marriage. Even more important, this account helps us understand the way Jesus himself would have thought of marriage and then suggests how unusual his contemporaries would have found the fact that he himself did not marry.

The specific roles of women and men in marriage, the role of families in determining whom one would marry, and the centrality of having children are all matters that make the following essay fascinating. This essay could open up discussion of the subtle ways some of these ideas, attitudes, and customs are still with us in a time when they seem so foreign.

In Mackin's book, readers will find an equally fascinating description of a later period: "Christian Marriage in the Roman Empire," pp. 69–79.

Since it is accurate enough to date the establishing of the community of Jesus' followers as a church within the decade 27–37 A.D., we may put the end of the second Christian generation at the turn of the first century. By that time two major traditions, the Jewish and the Hellenistic-Roman, had formed the character of early Christian life in community. By the end of these sixty years the confluence of these two traditions in Christianity was complete. Each had had time to make its contribution to Christian life in full strength. Each had conditioned the other. And even though the first communities of Jesus' followers in Palestine had been destroyed, or at least dispersed, during the devastation of Judea and the siege of

From *What Is Marriage?* (New Jersey: Paulist Press, 1982) pp. 38–41, 47–50. Reprinted without footnotes by permission.

Jerusalem by the Roman army under Vespasian and then Titus in 68–70, the Jewish influence continued past that date because the first Christian communities outside Palestine had been formed mainly from the synagogue communities scattered around the eastern half of the empire. Consequently, though the Hellenistic-Roman character eventually overwhelmed at least Jewish custom as the Jewish portion of the Christian population dwindled, the Jewish vision of marriage left its effect in the early Christian consciousness. It is this vision that we must seek to understand first.

Palestinian Jewish Marriage Custom

Among Jesus' people marriage was regarded as obligatory. The male who had reached eighteen years but had not yet married could be compelled by a court to do so. While girls, as also boys, could not be married before puberty, the former were ordinarily married as soon as they reached it, which was set at twelve years and six months. Thirteen was the earliest age at which a boy might marry. These early ages for marrying hint clearly at the meaning of Jewish marriage in Palestine at that time. It was for family decisively and in multiple ways. The principal motive for marrying was to provide children so as to preserve the husband's family, to keep his and his father's name from dying out, to keep the tribe, the nation itself as the people of God in existence—withal to honor the ancient covenant commitment with Yahweh to be a light to the nations. Since infant mortality was high and life-expectancy low even for those who survived until adulthood, it was important that the childbearer marry at the onset of puberty so as to bear as many children as possible.

Hesitant belief in personal survival beyond death reinforced this tradition. If a man had no heirs, even if he had daughters but no sons, he risked having his name die out; the memory of him might vanish and be lost to his people forever and in every way. Hence marriage was before all else what Christian tradition would later call an *officium,* a dutiful vocation, a vocation motivated by *pietas,* loyal loving concern for one's parents and one's people. The Jewish people itself had already been given and had accepted the *officium* mentioned earlier, to be a light to the nations. Obviously it could be faithful to this vocation only if its sons and daughters bore and nurtured children.

The Jewish people of Jesus' time were not a defining people in the Western sense of assigning an entity to generic and specific categories of existence. But if one were to ask in terms of these categories about the essence of marriage among this people, no more exact answer could be given than that it was an abiding relationship between a man and a woman intended mainly although not exclusively for children.

Although the negotiations between the two families to settle the dowry and the *mohar* (the bride-price given by the husband-to-be to the girl's father) were contractual in form, the marriage itself was not thought of as a contract. For one thing an element essential to a contract was missing. The mutual consent by the contracting parties themselves to the exchange of a contractual good was deemed functionally unnecessary. Jewish fathers in consultation with their wives negotiated their children's marriages; the wills of the fathers held good for those of their children. The latters' marital consent was more an obedient acceptance of the partners chosen for them, although loving parents took their children's desires into consideration, especially if the desire were against the tentatively chosen partner. Thus the marriage was a covenant between two families. It was made orally by the parents, but by Jesus' time the covenant was confirmed legally in writing.

The preferred source wherein to search for spouses for one's children was one's own relatives, within those forms and degrees of kindred permitted marriage by Torah and the traditions of the fathers. Marriage between uncle and niece was not uncommon; the girl's father found a husband for her among his own brothers. Marriage of cousins was common. This marrying within the family gained two understandable advantages. One could be less uncertain about the quality of a spouse in the uncertain human enterprise of marrying. And there was less likelihood of divorce if a dissatisfied husband had to answer to his brother for dismissing the latter's daughter from his house. And if she did fall victim to divorce, there was considerable comfort for her in having to fall back only to a family she had never really left.

When the girl married she became a member of her husband's family. Since he owed obedience to his father as long as the latter lived, so too did his wife. She became in effect her father-in-law's newest daughter. Often enough the newly married pair, especially if they were quite young, lived in the paternal household. When they did, the young wife took her place as an added daughter in the household managed by her mother-in-law under her father-in-law's authority.

The Religious Reality of Jewish Marriage

Was this marriage a religious reality or was it secular? The question would have been meaningless for a Jew of Jesus' time, for he simply recognized no division in life between these two realms. All of life for him was religious because all the world came from the hand of God. Getting married was one form of faithful response to God's covenant invitation to his people Israel. Marriage in turn was God's gift to men and women as part of his creation. According to the creation poem of Genesis his work of creating came to its climax in the sexually differentiated pair who were given the *officium* of using their sexuality in order to populate the earth and rule it. According to the Genesis parable of the garden the sexually different and therefore desirable creature, the woman, was given to the man as God's gift to him to relieve his loneliness. He made them two and sexually different expressly so that they could become one—even "one flesh," one person before God and the people. The sexual uniting was holy because it was God's creation and given as his gift. Yet it was of this world and worldly, not sacred in the sense of the ancient fertility cults' sexuality interpreted as participation in genital activity of the gods. It was not sacred precisely because it was created and therefore creaturely in substance. It was created by a supra-sexual deity, therefore by one who could not be imitated sexually and in this way bribed by sexually magic ritual.

Because in the parable of the garden God "brought the woman to him" (the man) as a gift to relieve his loneliness, marriage at least in this ancient tradition could only with difficulty be construed as a contract. An added difficulty against this was the fact that the most excellent religious model for the husband-wife relationship proposed by the prophetic tradition could hardly be contractual. This was the husband-bride relationship of God to Israel pictured by the prophets Hosea (1:1–3:35), Jeremiah (2:1–2), and Ezekiel (16). Common acceptance of this religious metaphor as a normative model, where it was done, overweighted the husband's authority in the marriage severely. The model presumed the wife's infidelity and need to be disciplined; it gave the husband a divine warrant to discipline her. But for this reason alone the model blocked the possibility of seeing Jewish marriages as contracts. As the bride Israel could not negotiate with God, could not in justice hold him to contractual conditions, neither could the Jewish wife with her husband. She could expect and demand that he care for her, protect her, forgive her, but none of these as obligations in justice. As God had promised these to Israel only out of love and in sovereign freedom, so, according to this model, the wife could expect these from her husband only as a gift of his love freely given.

I have said that the marriage itself was not understood as a contract, and that where this prophetic model was in the forefront of awareness, any temptation to understand a marriage contractually was suppressed. But given the contractual nature of the marriage arrangements, especially as these were detailed in the *ketuba,* the inclination to think of the relationship as contractual must have been strong. But since the pre-marital agreement had been struck not between the spouses but between the families, the inclination was to think of these as the contractors.

The solemnization of the Jewish marriage was, in the senses I have explained above, both religious and nonsacral. The *kiddushin,* the year-long trial period (literally "the sanctification"), was sealed and begun with the handing over of the *mohar* to the bride's father. The boy and girl were dedicated to one another and held to sexual fidelity. The bride's intercourse with a third person during that year, but only hers, was considered adultery. In the northern Israel the two were forbidden intercourse with one another until after the wedding feast. In the south, in Judea, betrothal and marriage were in effect the same in permitting intercourse.

The wedding feast lasted a week. All of it was deemed religious, as I have said, but the core of the "liturgical" part of the celebration was a series of seven benedictions read by the father of the groom. No part of the feast took place in either temple or synagogue. No priest, levite or or rabbi had a part in it as an exercise of his office. It was a family affair, supervised and conducted by parents. It was private in that sense but thoroughly public in the sense that the entire village might join in the celebration. Since the groom's father presided over the celebration, it ordinarily took place in his home. The bride's father gave her away, and the groom's father took her for his son. The ceremony ended and climaxed with the groom's leading his bride in procession to his home. If he was quite young, this would be his parents' home.

Joachim Jeremias supplies more detail when describing the condition of women and wives in Jesus' time:

> The wife's first *duties* were household duties. She had to grind meal, bake, wash, cook, suckle the children, prepare her husband's bed and, as repayment for her keep . . . to work the wool by spinning and weaving. . . . Other duties were that of preparing her husband's cup, and of washing his face, hands and feet. . . . These duties express her servile relationship with her husband; but his rights over her went even further. He laid claim to anything his wife found . . . as well as any earnings from her manual work, and he had the right . . . to annul her vows. . . . The wife was obliged to obey her husband as she would a master—the husband was called *rab*—indeed this obedience was a religious duty. . . . This duty of obedience went so far that the husband could force a vow upon his wife, but any vows which put the wife in a discreditable position gave her the right to demand divorce before the court. . . . Relationships between children and parents were also determined by the woman's duty of obedience to her husband; the children had to put respect for their father before respect for their mother, for she was obliged to give a similar respect to the father of her children. . . .
>
> Two facts are particularly significant of the degree of the wife's dependence on her husband:
>
> (a) Polygamy was permissible; the wife had therefore to tolerate concubines living with her. Of course, we must add that for economic reasons the possession of several wives was not very frequent. Mostly we hear of a husband taking a second wife if there was dissension with the first, but because of the high price fixed in the marriage contract, he could not afford to divorce her. . . .
>
> (b) The right to divorce was exclusively the husband's. . . . In Jesus' time (Matt. 19:3) the Shammaites and Hillelites were in dispute over the exegesis of Deut. 24:1, which gives, as a reason for a man divorcing his wife, a case where he finds in her "some unseemly thing," *'erwat dabar*. The Shammaites' exegesis was in accord with the meaning of the phrase, but the Hillelites explained it as, first, the wife's unchastity (*'erwat*) and, secondly, something (*dabar*) displeasing to the husband; either gave him the right to put away his wife. . . . In this way the Hillelite view made the unilateral right of divorce entirely dependent on the husband's caprice. From Philo (*De spec. leg.* III, 30) and from Josephus (*Ant.* 4.253), both of whom knew only the Hillelite point of view and championed it, it appears that this must already have been the prevailing view in the first half of the first century A.D. However, reunion of the separated parties could take place; also by reason of divorce there was a public stigma on the husband as well as on the wife and daughters . . . ; then, too, when he divorced his wife, the husband had to give her the sum of money prescribed in the marriage contract; so in practice these last two facts must often have been obstacles to any hasty divorce of his wife. As for his wife, she could occasionally take things into her own hands and go back to her father's house, e.g., in the case of injury received. . . . But in spite of all this, the Hillelite view represented a considerable degradation of women.
>
> . . . to have children, particularly sons, was extremely important for a woman. The absence of children was considered a great misfortune, even a divine punish-

ment. . . . As the mother of a son the wife was respected; she had given her husband the most precious gift of all.

As a widow too a woman was still bound to her husband, that is, if he died without leaving a son. . . . In this case she had to wait, unable to make any move on her side, until the brother or brothers of her dead husband should contract a levirate marriage with her or publish a refusal to do so; without this refusal she could not remarry.

The conditions we have just described were also reflected in the prescriptions of religious legislation of the period. So from a *religious* point of view too, especially with regard to the Torah, a woman was inferior to a man. She was subject to all the *prohibitions* of the Torah (except for the three concerning only men . . .), and to the whole force of civil and penal legislation, including the penalty of death. . . . However as to the *commandments* of the Torah, here is what was said: "The observance of all the positive ordinances that depend on the time of the year is incumbent on men but not on women."

As a woman's religious *duties* were limited, so were her religious *rights*. According to Josephus, women could go no further in the Temple than into the Courts of the Gentiles and of the Women. . . . During the time of their monthly purification, and also for forty days after the birth of a son . . . and eighty days after the birth of a daughter . . . they were not allowed even into the Court of Gentiles. . . . By virtue of Deut. 31:12 women, like men and children, could participate in the synagogue service . . . but barriers of lattice separated the women's section. . . . In the liturgical service, women were there simply to listen . . . Women were forbidden to teach. . . . In the house, the wife was not reckoned among the number of persons summoned to pronounce a benediction after a meal. . . . Finally we must record that a woman had no right to bear witness, because it was concluded from Gen. 18:15 that she was a liar. Her witness was acceptable only in a few very exceptional cases, and that of a Gentile slave was also acceptable in the same cases . . . e.g., on the remarriage of a widow, the witness of a woman as to the death of the husband was accepted.

On the whole, the position of women in religious legislation is best expressed in the constantly repeated formula: "Women, (Gentile) slaves and children (minors)." . . . like a non-Jewish slave and a child under age, a woman has over her a man who is her master . . . and this likewise limits her participation in divine service, which is why from a religious point of view she is inferior to a man.

From a distance of nineteen and a half centuries it is impossible to know with certainty how much this subservant condition of wives in Jesus' time among his people was due simply to the customs of a patriarchal culture, and how much was the consequence of husbands' asserting the lordship they had found granted them in Genesis 3:16b. No doubt the two causes converged and reinforced one another; custom and inclination were justified by what was taken to be a divine decree. And insofar as the decree was understood to be a warrant to punish, the punishment was thought in turn to be for the violation of a covenant. That the violation was also thought to have the first husband's manipulable need as one of its contributing causes could not have diminished the inclination to punish.

3.

Marriage

Joseph Martos

The history of marriage in the Christian church is as complicated as it is important. Marriage was influenced by shifts in society, in philosophy, in theology, and in church organization. Joseph Martos's detailed examination of this history is worth careful study for the light it sheds on the questions religious people still ask about marriage today. Martos shows us where certain ways of thinking about marriage came from, helping us reflect on them in the light of these origins.

Here, this historical survey has been broken into three sections: I: Early Roman, Jewish, and Christian Understandings of Marriage; II: Marriage in the Western Church; III: Marriage from the Reformation to the Present. Headings have been added to make the sections easier to follow. In some ways, each of these sections would be best explained in detail before being read and further reflected upon.

Section I, "Early Roman, Jewish, and Christian Understandings of Marriage," *shows how marriage could be at the same time religious but not under ecclesiastical control. This section highlights shifts in marriage celebrations. It also shows Jesus' teaching on marriage as in opposition to the major religious teachers of his day. Finally, it surveys marriage in the first eight centuries of the church.*

Section II, "Marriage in the Western Church," *explains the gradual shift from secular to ecclesiastically governed marriages. In part, the conflict between Roman and Germanic marriage traditions led to more church control. As differing understandings of what constituted a marriage had to be resolved, so did various understandings of what the sacrament of marriage meant. These matters were settled only to an extent, since we find these controversies still with us today. Understanding these origins helps us reexamine marriage today.*

Section III, "Marriage from the Reformation to the Present," *explains how the Reformation's attack on abuses of authority in the Church, including arbitrary annulments for the wealthy, led to a questioning of marriage as a sacrament under clerical control. Luther, Calvin, and other reformers saw*

Chapter XI of *Doors to the Sacred* (Garden City: Doubleday, 1981), pp. 399–451.

marriage as a civil, not a church, matter, with divorce permissible under certain conditions. In response, the Roman Catholic Church reaffirmed its power over marriage as a sacrament and took special action against secret marriages. Marriage was a public act, able to be affirmed by witnesses. Forthcoming marriages were to be publicly announced and then officially entered into church records.

However, questions persisted about exactly what made up the sacrament. These differing Protestant and Catholic positions on marriage were later complicated by civil laws requiring all persons to go through nonchurch weddings and by the question of marriages between Catholics and non-Catholics.

Closer to our own day, changes in social attitudes toward marriage have continued both to challenge the church and to call for reexamining the older understandings of marriage. These challenges and resulting reexaminations have led to new insights into the reality of marriage but also to new questions posed by the complexity of human personality and behavior.

I: Early Roman, Jewish, and Christian Understandings of Marriage

Relatively early in the history of Christianity, marriage was regarded as a sacrament in the broad sense, but it was only in the twelfth century that it came to be regarded as a sacrament in the same sense as baptism and the other official sacraments. In fact, before the eleventh century there was no such thing as a Christian wedding ceremony, and throughout the Middle Ages there was no single church ritual for solemnizing marriages between Christians. It was only after the Council of Trent, because of the need to eliminate abuses in the practice of private marriages, that a standard Catholic wedding rite came into existence.

Parallel to the absence of any church ceremony for uniting Christians in marriage was the absence of any uniform ecclesiastical regulations regarding marriage during the early centuries of Christianity. As long as the Roman empire lasted—and it lasted longer in the east than in the west—church leaders relied primarily on the civil government to regulate marriage and divorce between Christians and non-Christians alike. It was only when the imperial government was no longer able to enforce its own statutes that Christian bishops began to take legal control over marriage and make it an official church function. In the west, church leaders eventually adopted the position that marriages between Christians could not be dissolved by anything but death; in the east they followed the civil practice of allowing the dissolution of marriages in certain cases. To safeguard the permanence of marriage, Roman Catholicism gradually developed an elaborate system of church laws and ecclesiastical courts, which was challenged by the Protestant reformers as being unscriptural and unnecessary. Today some Catholic theologians and canon lawyers are themselves asking whether it might be better to let the legal regulation of marriage revert back to civil control, without denying that church weddings are important communal celebrations or that Christian marriages are sacramental.

Parallels and Precedents

The origins of marriage are as obscure as the origins of the human race itself. Certainly they both started sometime and somewhere since they are both here today. But whether humankind began with an original pair or with a widely scattered population is not known and perhaps never will be known with certainty. And whether marriage began with promiscuity or fidelity, monogamy or polygamy, matriarchy or patriarchy is a historical question that likewise may never be answered.

In prehistoric and ancient cultures, however, marriage in some form or other was already well established as part of the network of relationships that bound people together in kinship and friendship, by occupation and social position. As an accepted custom it had a variety of forms in different parts of the world at different times, and marriage practices through the ages seem to have been just as diverse as the cultures in which they were found. But whether marriages were permanent or temporary, headed by men or women, joining clans or individuals, marriage was always a socially institutionalized way of defining relationships between the sexes, of establishing rights and responsibilities for parents and offspring, of providing for cohesiveness and continuity in society. Such things were important in every culture whether it was nomadic or sedentary in its lifestyle, hunting or farming for its existence, tribal or urbanized in its organization. And since social relationships were so important, the marriage customs that surrounded and supported them were usually revered and sacred, and in that broad sense, religious.

In early Rome, for example, marriage was a religious affair, but the religion was that of the family; there was as yet no official state religion uniting the various clans that lived in central Italy. The father of each family acted as priest and preserver of his household religion, which included reverence for the gods of the home and respect for the spirits of departed ancestors. Besides leading his family in worship, one of the Roman father's chief duties was to provide children (primarily sons) to continue the family religion. In primitive times girls had been captured from neighboring clans to continue the family line, but eventually this practice gave way to a less violent one of arranged marriages: fathers obtained wives for their sons in exchange for a "bride-price" which compensated the girl's family for the loss of a skilled and fertile member. After she was escorted by her father to her new home, the bride was now only ritually abducted by the husband, who carried her over the threshold of his home and fed her a piece of sacred cake, which inducted her into her new religion and established her communion with her husband's family gods.

At the beginning the Roman family was absolutely patriarchal: the father was not only the head of the household but he also possessed all of his family's legal rights; his wife and children had none. He could beat and punish them as he saw fit; he could sell his children into slavery and even put them to death; his wife's personal property became his, and he could divorce her if she did not live up to his expectations, especially if she failed to provide him with male heirs. And although such extreme exercises of paternal power may have been rare because of social pressure, they were legally permissible and they were in fact practiced.

What changed the social status of women and children as well as the institution of marriage was war. When the Romans began to extend their republic throughout Italy and build their empire in the Mediterranean, men were often away for long periods of time, and sometimes they did not return home. Women learned to manage their family's affairs and children began to make decisions that used to be made for them. The older family values were replaced by more nationalistic ones, and the individual religion of hearth and home gave way to a state religion of glory and gods. Many of the traditional wedding customs were kept, like handing over the bride and eating the cake, but they no longer had the religious meaning that they had in the past. The wedding was still primarily a family affair, but now a betrothal ring rook the place of the bride-price, and the marriage was based on the mutual consent of the partners themselves and not on the consent of their parents. At the wedding ceremony the bride was usually dressed in white with a red veil and perhaps a garland of flowers in the fashion of the Greeks, and when she and her husband gave their consent to each other it was customary for them to join their right hands, but there was no legal formula they had to use and there was no special religious significance attached to this action. If they invited a Roman priest to their wedding it was only to offer a sacrifice to the gods or to divine their prospects for a happy future, and if they got married without a family celebration Roman law presumed they were husband and wife if they lived together for a year. But marriage by consent also implied divorce by consent, and by the second century B.C. divorces were not uncommon at least among the upper classes. And like marriage, divorce was a private affair which could be initiated and carried out by either partner; it did not require the approval of any civil authority or the judgment of any court.

But women in the Roman world were never entirely equal to men before the law, nor were they socially equal to men in the middle east, which was also largely patriarchal. In ancient Israel marriage was a family matter that was arranged by fathers for their children usually when they were adolescents. The Jewish scriptures say little about marriage customs and nothing at all about wedding ceremonies, since marriages were private agreements and weddings were not public religious functions. Most Israelite men had only one wife, but those who could afford the bride-price and maintenance of more than one sometimes had more. Women had few legal rights, and when they were married off they in effect were transferred from the property of their father to that of their husband. Adultery was forbidden by the Torah, for example, not because it was sexually immoral but because it violated the property rights of the woman's father or husband, and even the "ten commandments" placed coveting a neighbor's wife on the same footing as coveting his goods (Exodus 20:17).

At the height of the Israelite kingdom in the tenth century B.C., much of the early Hebrew folklore was collected into books, including the Genesis story of the first man and woman. At the time, however, this story was not taken as a divine endorsement of monogamy since Solomon and other kings had many wives and concubines. But after the conquest of the kingdom by Assyria and Babylonia, the later Israelite prophets pondered the lesson to be learned from that tragedy, and they began to propose that Yahweh had punished his people for not living up to their calling as a chosen people. They were supposed to be a holy people with high moral standards, being just and merciful to all, and not chasing after the false gods of wealth and power. And as part of the new morality that they preached, some of the prophets began to propose that the moral ideal in marriage was faithful love between a husband and one wife. Ezekiel 16 described the relationship between Yahweh and Israel as the marriage of a man who had given his wife everything, only to be deserted by her, and Hosea 1–3 portrayed Israel as a faithless prostitute married to Yahweh, who was still faithful to her and longed to take her back. The Song of Songs extolled the ecstasy of love between a bride and her beloved, and Tobit 6–8 presented the perfect marriage as one bound by the love of one man for one woman. The wisdom literature of Sirach 25–26 described the dangers and rewards of domestic life, and the book of Proverbs 5–7, 31 praised the virtues of the perfect wife and advised husbands against adultery.

The prophet Malachi (2:16) denounced the men who divorced their Jewish wives to marry daughters of the conquerors, but apart from this divorce was an accepted even if regretted way to end an unhappy marriage. If a wife had sexual relations with another man her husband could divorce her, and if he caught her in the act he could have her stoned to death (Deuteronomy 20:22–24). Jewish law was fair-minded inasmuch as it also called for the death of the male accomplice in adultery, but in the case of divorce only the husband had a legal right to demand it. A woman who wanted to be freed from her husband had to ask him to grant her a divorce, and if her refused she was obliged to stay with him. Divorce in the ancient world always included the possibility of remarriage, but in Jewish law there was an exception to this: a woman who was divorced a second time could not remarry her first husband (Deuteronomy 24:1–4).

The Torah allowed a man to give his wife a writ of dismissal if she was guilty of "impropriety," and at the beginning of the Christian era there were two schools of thought about how serious the misconduct had to be for her husband to justifiably divorce her. The followers of Rabbi Shammai authorized divorce only for blatantly shameful behavior, such as adultery; the followers of Rabbi Hillel permitted divorce for almost anything she did that displeased her husband.

Early Christian Marriage

Jesus said very little about marriage and divorce, but what he did say put him squarely in opposition to both rabbinical schools. In Luke's gospel Jesus' denunciation of divorce and remarriage is apodictic: "Everyone who divorces his wife and marries another is guilty of adultery, and the man who marries a woman divorced by her husband commits adultery" (16:18). Mark's gospel contains the same short statement by Jesus, but attaches it to an argument between Jesus and Jewish religious leaders on the question of divorce (10:1–12). Against their insistence that the Torah permitted divorce, Jesus replied that Moses had allowed it only because the Israelites in his day were so hard-hearted. Using the book of Genesis as the basis of his own teaching, Jesus went on: "But from the beginning of creation God made them male and female. This is why a man must leave his father and mother, and the two become one body. They are no longer two, therefore, but one body. So what God has united, man must not divide." In the opinion of most scripture scholars today, what Jesus taught about the permanence of marriage was a radical departure from the traditional Jewish acceptance of divorce.

But then how do the scholars explain the fact that Matthew's gospel contains both the short statement by Jesus (5:31–32) and his argument with the Jewish leaders (19:3–9), but that in both of these places Jesus seems to allow divorce and remarriage (at least for the husband) in certain cases? In Matthew's version of the argument the religious leaders question Jesus about the possible grounds for divorce, apparently trying to get him to side with either Shammai or Hillel. Jesus' first answer is that God never intended the separation of men and women, but then he goes on to say that remarriage after divorce is tantamount to adultery except in the case of immorality.

The texts in Matthew raise a number of questions. Are the passages written by Luke and Mark closer to the words of Jesus, or the passages written by Matthew? What is the reason for the difference between the texts? And what is the immorality referred to?

Many if not all scholars, both Catholic and Protestant, agree that the original teaching of Jesus is found in Luke and Mark. That is to say, Jesus taught that divorce was wrong, that God did not intend it to happen, and that he himself saw it as falling far short of moral perfection. Jesus' standards of morality were high, his call to perfection was revolutionary, and he often presented his teachings in radical, absolute statements. In his "sermon on the mount," for example, Jesus proclaimed that anger is a capital offense, that lust is equivalent to adultery, that swearing oaths is wrong, and that loving one's enemies is right. No less forthrightly he commanded his followers to do good only in secret, to renounce wealth, to avoid judging prople, and to cut off any part of their body that sins (Matthew 5–7). In this manner, then, Jesus also preached an ideal of lasting fidelity in marriage, and he proposed it as a norm for all those who heeded his call to moral perfection.

The book of Matthew was written in its present form around the year 80, and when it was composed its final author (who was not the apostle Matthew) modified the earlier, stark saying of Jesus by adding the words, "except for immorality," to Christ's apodictic condemnation of divorce. Many scholars theorize that the author was a Jewish Christian writing for other Jewish converts, and that he simply wrote down the teaching of Jesus as it was understood at that time in his Jewish Christian community, which allowed divorce in certain cases and believed that it was in conformity with Christ's teaching.

But what were those cases? The Greek word *porneia* can be tramslated as "immorality" or "indecency," but it can also be translated as "adultery" or "fornication," and some scholars believe that the Jewish Christians continued in the tradition of Rabbi Shammai and allowed husbands to divorce their wives for serious sexual misconduct. Other scholars suggest that the early community continued the Jewish practice of not allowing a pagan convert to Judaism to keep his wife if she had been related to him before they were married. The Torah regarded marriages between certain close relatives as indecent (Leviticus 18:6–18), and according to this interpretation the author of Matthew is indicating that he did not believe that Jesus prohibited divorce in this situation, which might arise if a pagan wished to join a community of Christians who still followed the Law of Moses. No matter which interpretation is accepted, however, it seems that at least some Christians allowed divorce for certain reasons at the time that the gospels were written.

Paul, too, stressed the ideal of marital fidelity, but he also allowed divorce in certain situations. On the one hand he acknowledged that according to Christ, "a wife must not leave her husband—or if she does leave him she must remain unmarried or else make it up with her husband—nor must a husband send his wife away" (I Corinthians 7:10–11; Romans 7:2–3). On the other hand, he viewed marriages to non-Christians as fraught with spiritual dangers, and he ventured the opinion that if a Christian brother or sister was married to an unbeliever who wanted a divorce, he or she might grant it and be free to marry again (I Corinthians 7:12–16; II Corinthians 6:14–18). Ideally the Christian spouses should be a source of salvation for their marriage partners; but Paul reluctantly admitted that this might not always be possible.

Much of what Paul had to say about marriage is found in I Corinthians 7, but it has to be read in the light of his belief that "the world as we know it is passing away" (7:31) and that the second coming of Christ would happen soon. Thus Paul advised the Christians in Corinth not to make any great changes in their lives and to devote their full attention to the things of the Lord. Those who were married should stay married; and if they abstained from intercourse it should be only for a time, to devote themselves to prayer, and only by mutual consent. Those who were single should stay single; it was no sin to get married, but it was better to remain celibate. The same advice applied to widows: it was better for them to remain free, but if they could not get along without a man they should marry again.

About five years later Paul wrote to the community at Ephesus, and this time he devoted a section of his letter to the way husbands and wives should behave toward one another (Ephesians 5:21–33). He accepted the patriarchal marriage system of his own day, including the notion that the relationship of wife to husband was not the same as the relationship of husband to wife. But he saw no reason why those relationships could not be lived "in the Lord," the way Christ gave himself in love to the church and the way the church submitted in love to Christ. Wives should regard their husbands as their head, and be obedient to them in all things; husbands should regard their wives as their own body, caring for them and looking after them. So in the marriage relationship between husband and wife Paul saw an image of the spiritual relationship between Christ and the church. It was, he said, "a great mystery," for it was a reflection of an even greater mystery.

Paul's standard of Christian marriage was echoed in I Peter 3:1–7, and apart from this there is not much more in the New Testament explicitly about marriage. In one of his later letters Paul spoke again about widows, but this time he advised that the younger ones should remarry (I Timothy 5:14). He no longer urged celibacy for those who were engaged in the Lord's work; instead he insisted that the leaders in the church should be men of character, who were faithful to their wives and able to manage their households (I Timothy 3:1–13).

Little is known about Christian wedding and marriage customs in the decades that immediately followed the writing of the New Testament. Most Christians were adult converts from Judaism and other religions, and presumably many of them were married according to their own customs before they were baptized. Some of them were not, however, and around the year 110 Ignatius of Antioch wrote in a letter that "those who are married should be united with the consent of their bishop, to be sure that they are marrying according to the Lord and not to satisfy their lust" (*To Polycarp* 5). But there is no indication that this was a widespread practice; probably most young people needed only the consent of their parents, and Roman law allowed those who were old enough to give their own consent. Still, the idealism of the first generations of Christians was high, and evidently Jesus' statement against divorce continued to be taken as a moral norm.

The First Eight Centuries

During the first three centuries of Christianity, then, the fathers of the church did not say much about marriage, but when they did they talked about it as an important aspect of Christian life, not as an ecclesiastical institution. When Christians married they did so according to the civil laws of the time, in a traditional family ceremony, and often without any special church blessing on their union. The early Christian writers implicitly accepted the government's right to regulate marriage and divorce, and when they spoke about marriage they usually limited themselves to pastoral matters, affirming the goodness of marriage, urging Christians to marry within their own community, and warning them not to get drunk and unruly at wedding feasts. The bishops did not approve of divorce but they did not absolutely prohibit it, and they even allowed remarriage in some places that we know of and perhaps others.

Even after the edict of toleration by the emperor Constantine in 313, the great patristic writers said little about marriage compared to the amounts they wrote about other liturgical and doctrinal matters. For one thing, there was no liturgical ceremony for marriage as there was for baptism and the eucharist, and so nothing had to be said to explain or defend it. For another, the Christian teaching on marriage was not complicated, and apart from the periodic bouts with gnostic sects, there were no great doctrinal controversies over marriage to call forth a flood of literature on the subject. All of the bishops agreed that sexual licentiousness and easy divorce were wrong, and they sometimes spoke out against what they perceived as the pagan immorality of the Roman world. They all agreed that marriage was a divine institution sanctioned by Christ, and they sometimes cited the biblical account of creation and Jesus' miracle at the wedding at Cana to prove their point. But by and large they continued to regard marriage as a mundane matter in which Christians were expected to follow the norms of the gospels and epistles no less than in their other daily affairs.

This meant that the legal regulation of marriage and divorce was left to the government, and even though Constantine gave bishops the authority to act as civil magistrates, there is little indication that they were given any marriage cases to decide. Marriage under Roman law was still by the mutual consent of the parties involved (which in rural areas often meant the consent of the parents), and divorces came into court only when they were contested or involved property litigation. But the Roman government itself was concerned about marriage for its own reasons, and since the time of the emperor Augustus it had passed laws to discourage the childless marriages which were becoming more frequent among the aristocracy, and to encourage larger families by means of financial and other legal incentives. Now the emperor Constantine tried to eliminate the injustice of many one-sided divorces by making it illegal for men or women to reject their spouses for trivial reasons. In 331 he passed a law that allowed a woman to be left only if she were an adulteress, a procuress, or a dealer in medicines and poisons, and that allowed a man to be left only if he were a murderer, a dealer in medicines and poisons, or a violator of tombs. Those who abandoned their spouses for other reasons could lose their property and their right to remarry, but the law said nothing and imposed no sanctions against the traditional practice of divorce by agreement.

Remarriage and the Double Standard

The Christian bishops, on the other hand, recognized only adultery as grounds for divorce. The local council of Elvira in Spain in the early 300s prohibited a woman from remarrying if she left an unfaithful spouse but said nothing prohibiting a man from doing so. A similar council at Arles in France in 314 declared that young men who caught their wives in adultery should be counseled not to remarry, though it was not forbidden. Later in the century an unknown Christian author in Italy wrote that it was clear from the scriptures that a woman who was divorced for adultery could not marry again, but that nothing prevented the husband from doing so. Thus in the west, many churchmen recognized infidelity as grounds for divorce, and some even allowed injured husbands to remarry.

The same situation—and the same double standard—existed for Christians in the eastern part of the Roman empire. Writing around 375, Basil of Caesaria summarized the current church regulations on marriage and divorce that he was aware of, and offered some of his own suggestions to a fellow bishop. According to Basil, "The Lord's statement that married persons may not leave their spouses except on account of immorality should, according to logic, apply equally to both men and women. However, the custom is different, and women are treated with greater severity" (*Letters* 188). For example, the wife of an unfaithful husband was obliged to remain with him, but the husband of an unfaithful wife could leave her. It was true that a man who commited adultery had to enroll in the order of the penitents, but she could not leave him and remarry even if he physically abused her. If she did so she was considered an adulteress and had to do penance for her sin, but the man she left could remarry without any further ecclesiastical penalty. Even a man who abandoned his wife just to marry another woman could be readmitted to the ranks of the faithful after doing penance for seven years. Furthermore, a woman who was unjustly deserted by her husband would be regarded as an adulteress if she remarried, but a man who was unjustly deserted by his wife could be forgiven if he remarried. Noted Basil, "It is not easy to find a reason for this difference in treatment, but such is the prevailing custom" (*Letters* 199).

Even if divorce and remarriage in the case of adultery was not universally taught during the patristic era, many bishops accepted it particularly in the eastern empire, although they sometimes imposed a period of penance on the persons involved. Those who took advantage of their civil right to divorce by consent and then remarry, for example, were often treated as adulterers in the penitential system, but their second marriage was generally recognized as legal and binding. This is not to say that divorces were frequent among Christians, but it does show that they were possible and that there was no universally recognized prohibition against divorce at this time. In any case, churchmen during this period had no legal say in the matter of marriages, divorces, and remarriages, and so they had to deal with these happenings as pastoral rather than legal matters.

Many of the church fathers were against second marriages of any sort even by widows and widowers, not only because of Paul's advice against it but also because the gnostic movements sometimes tried to prove their moral superiority by rigorously forbidding them. Promiscuity, prostitution, and lewd public entertainments in the cities even led Roman philosophers of the period to insist that the only legitimate purpose of sex was to found a family, and so Christian writers were sometimes hard put to defend why people should remarry if they already had children. Gradually, sexual abstinence became increasingly associated with moral perfection, but most bishops tended to follow Paul's lead by only discouraging and not denouncing second marriages. The sole exception to this trend was divorce and remarriage by converts to Christianity. Paul had sanctioned separation by mutual consent if pagan husbands or wives strongly opposed their spouse's new religion, and bishops still allowed Christians who were divorced under these circumstances to marry again.

Even after Christianity was made the official religion of the Roman empire in 380, there was no great change in the civil marriage laws, though some Christian writers continued to denounce the ease with which people could obtain divorces even under a Christian emperor. It was not until 449 that Theodosius II passed a law prohibiting consensual divorce, but at the same time he expanded the list of legitimate grounds for leaving one's spouse to include robbery, kidnapping, treason, and other serious crimes, and for the first time it became legal for a woman to divorce her husband for adultery. This change in the law reflected a growing tendency among bishops in the east to interpret the *porneia* of Matthew

19:9 as any kind of gross immorality, as well as their concern that innocent wives should not suffer because of the sins of their husbands.

The outlawing of divorce by mutual consent did not last, however, and later Christian emperors made it possible again. Then, in the sixth century, Justinian brought about a sweeping reform in the laws of the eastern empire, including the laws on marriage and divorce. According to the Code of Justinian the basis of marriage was mutual affection between the sexes, but like any other human contracts marriages should be able to be made and unmade according to law. The code provided for nearly equal grounds for divorce by either husbands or wives, and for the first time in Roman law the children of dissolved marriages were specifically provided for. The traditional grounds for one-sided divorce were reaffirmed, and to them was added impotence, absence due to slavery, and "the renunciation of marriage" by entering a monastery. Couples could also divorce by mutual consent if one of them wanted to pursue the spiritual perfection of the monastic life. In short, permanent marriage was still regarded as the ideal, but the law provided practical norms for those who could not or would not live up to it.

This late Roman attitude was generally accepted by the churches of Greece, Asia Minor, Syria, Palestine, and Egypt, which had already become more involved in the marriage practices of those regions. On the whole marriage was still primarily a family and secular affair with the bride's father playing the chief role in the wedding ceremony. Though there were local variations, the usual custom was that on the wedding day the father handed over his daughter to the groom in her own family's house, after which the bridal party walked in procession to her new husband's house for concluding ceremonies and a wedding feast. The principal part of the ceremony was the handing over of the bride, during which her right hand was placed in the groom's, and the draping of a garland of flowers over the couple to symbolize their happy union. There were no official words that had to be spoken, and there was no ecclesiastical blessing that had to be given to make the marriage legal and binding.

Late in the fourth century, however, it became customary in some places in the east for a priest or bishop to give his blessing to the newly wedded couple either during the wedding feast or even on the day beforehand. This blessing was usually considered something of an honor, showing the clergy's approval of the marriage, and it was not given if the bride or groom had been married before. Then, in the fifth century, especially in Greece and Asia Minor, the clergy began to take a more active role in the main ceremony itself, in some places joining the couple's hands together, in other places putting the garland over them, and in others doing both. Gradually the wedding ceremony developed into a liturgical action in which the priest joined the couple in marriage and blessed their union, but still this ceremony was not mandatory and through the seventh century Christians could still get married in a purely secular ceremony.

By the eighth century, liturgical weddings had become quite common, and they were usually performed in a church rather than in a home as before. New civil legislation was passed recognizing this new form of wedding ceremony as legally valid, and in later centuries other laws required that a priest officiate at all weddings. Marriage in the Greek church (for by this time the rest of the eastern empire had been conquered by the Moslems) thus became an ecclesiastical ceremony, and in the view of eastern theologians the priest's blessing was essential for the joining of two people in a Christian, sacramental marriage.

This same view persists in the Orthodox churches today, for which marriage is a sacrament of Christian transformation, a transition ritual from one state in life to another which both effects that transition and symbolizes its spiritual goal. Just as baptism both initiates a person into the kingdom of God and symbolizes the sinless way that life in the kingdom should be lived, so marriage unites and consecrates two persons in fidelity to each other and symbolizes the love and respect that married people should always have for each other. That relationship, as Paul said, is like the relationship between Christ and his church, and so just as by baptism Christians enter into and participate in the mystery of the redemption, so also by marriage Christians enter into and participate in the mystery of union with Christ. It is a true earthly union of one man with one woman, which both symbolizes and takes place within the spiritual union of the one Lord with the one church.

It sometimes happens, however, that Christians do not live up to their spiritual calling, that they do not live up to what was symbolized and initiated in their baptism and marriage, and it sometimes happens drastically and publicly as in the case of apostasy or divorce. When it does happen the church does not approve of it but it does recognize it as a fact. Christians should not renounce their faith, and they should not be divorced, but when these things happen they must be recognized as realities, and ways must be found to deal with the persons involved justly and mercifully. In the case of lapsed Christians, the way is always left open to reconciliation with the church, and in the case of divorced Christians the way is left open to reconciliation with each other or, if that is not possible, to remarriage.

Orthodox churches, therefore, do not grant divorces but they do allow for the possibility that Christians might receive a civil divorce and later want to remarry. Rather than exclude the innocent and repentant from communion in the church, then, they allow divorced Christians to remarry in the church as a concession to human needs and imperfections even though this second marriage cannot symbolize, as the first one did, the union between the one head and the one body of the church. Thus this second marriage is recognized as a real marriage but it is not regarded as a sacramental marriage, for it is not a full sign of the unique and eternal commitment that the first one promised to be. It is an approach to marriage which focuses on the liturgical aspects of the sacrament rather than on its juridical aspects, for that is the early church tradition out of which it grew.

II: Marriage in the Western Churches

There was, however, another church tradition regarding marriage. It developed not in the east but in the western half of the Roman empire, and it developed along quite different lines.

Initially in the west, as in the east, marriage between Christians involved no distinct wedding ceremony. A Christian marriage was simply one that was contracted between two Christians by their mutual consent and lived "in the Lord." In the fourth century, however, once Christians could practice their religion openly, bishops and priests were sometimes invited to wedding feasts and to bestow their blessing on the newly wedded couple after the family marriage ceremonies were concluded. Sometimes this blessing was given instead during a eucharistic liturgy a day or more after the wedding itself. As in the east, the priest's blessing was usually given as a favor to the family or as a sign of his approval on the marriage; it was not a standard or universal practice. In some places the bishop or priest occasionally participated in the wedding ceremony itself by draping a veil over the newly united couple, a custom that was parallel to the practice of garlanding in the east. Bishop Ambrose of Milan even insisted that "marriage should be sanctified by the priestly veil and blessing" (*Letters* 19), but there is no other evidence that the veiling was ordinarily done by a cleric during this period.

Shortly before the end of the fourth century, however, Pope Siricius ordered that all clerics under his jurisdiction must henceforth have their marriages solemnized by a priest, and Innocent I at the beginning of the next century issued another decree to the same effect. Around the year 400, then, the only Christians who had to receive an ecclesiastical blessing on their marriage were priests and deacons.

Ambrose and Augustine

Ambrose was also the first Christian churchman to write that no marriage should be dissolved for any reason, and to insist that not even men had the right to remarry as long as their wives were alive. "You dismiss your wife as though you had a right to do this because the human law does not forbid it. But the law of God does forbid it. You should be standing in fear of God, but instead you obey human rulers. Listen to the word of God, whom those who make the laws are supposed to obey: 'What God has joined together, let no man put asunder'" (*Commentary on Luke* VIII, 5). Ambrose would not even allow divorce and remarriage in the case of adultery. And his firm stand on the permanence of marriage was taken up by one of his converts to Christianity, Augustine, later the bishop of Hippo in North Africa and one of the greatest influences on the early schoolmen in the Middle Ages.

Augustine's attitude toward marriage was an ambivalent one. On the one hand he viewed it as a beneficial social institution, necessary for the preservation of society and the continuation of the human race, and sanctioned by God since the creation of the first man and woman. On the other hand he saw sexual desire as a dangerous and destructive human energy which could tear society apart if it were not kept within bounds. As a young man he had had two mistresses and a child by one of them, but he had also been attracted by the moral asceticism of the Manichean religion, to which he had belonged for a time. He was the first and only patristic author to write extensively about sex and marriage, and in the end he affirmed that marriage was good even though sex was not.

It was an attitude that was common among the intellectuals of his day. The stoic philosophers taught that strong impulses should be controlled in order to have peace of soul and harmony in society, and that the only justification for intercourse was to produce offspring. The Manicheans and other gnostics were often sexual puritans, condemning sensuality as evil and even forbidding marriage among the devout members of their sects. Christian ascetics like the hermits and monks of the desert sought to quench the desires of the flesh in order to free their minds for prayer and meditation. A number of the fathers of the church, including Gregory of Nyssa and John Chrysostom, taught that intercourse and childbearing were the result of Adam and Eve's fall from grace, and that if they had not sinned God would have populated the earth in some other way. Virginity for both men and women was extolled as the way of perfection for those who sought first the kingdom of God and wished to devote themselves to the things of the Lord. In Augustine's mind sexual desire was evil, a result of original sin, so those who gave in to it cooperated with evil and committed a further sin, even in marriage: "A man who is too ardent a lover of his wife is an adulterer, if the pleasure he finds in her is sought for its own sake" (*Against Julian* II, 7).

According to Augustine, therefore, only those who remained unmarried could successfully avoid the sin that almost always accompanied the use of sex. Those who were married usually committed at least a slight sin when they engaged in sexual intercourse, but they could be excused if they did it for the right reason. "Those who use the shameful sex appetite in a legitimate way make good use of evil, but those who use it in other ways make evil use of evil" (*On Marriage and Concupiscence* II, 21). And for Augustine, as for the stoics, the only fully legitimate reason for having sexual relations was to produce children.

Children were thus the first of the good things in marriage which counterbalanced the necessary use of sex. Another was the faithfulness that it fostered between a man and a woman, so that they did not seek sex for pleasure with other partners. These two benefits could even be found in pagan marriages, said Augustine, but Christians also received a third benefit, which was mentioned by Paul in one of his epistles: it was a sacred sign, a *sacramentum,* of the union between Christ and the church. Augustine read the New Testament in Latin, not Greek, and in Ephesians 5:32 Paul's word *mysterion* had been translated into Latin as *sacramentum*. Augustine took it to mean that marriage was a visible sign of the invisible union between Christ and his spouse, the church. But he also saw a deeper meaning in the word by understanding it in the sense that a soldier's pledge of loyalty was called a *sacramentum:* it was a sacred pledge of fidelity. In this sense it was something similar to the sacramental character that Christians received in baptism. Just as baptism formed the soul in the image of Christ's death and resurrection, so marriage formed the soul in the image of Christ's fidelity to the church; and just as Christians could not be rebaptized and receive another image of Christ, so spouses could not remarry and receive another image of his fidelity. The *sacramentum* of marriage was therefore not only a sacred sign of a divine reality but it was also a sacred bond between the husband and wife. And like the *sacramentum* of baptism, it was something permanent, or nearly so:

"The marriage bond is dissolved only by the death of one of the partners" (*On the Good of Marriage* 24).

It was this invisible *sacramentum,* argued Augustine, that reminded Christians that they should be faithful even to a partner who was not. Marriage between Christians therefore should not be disrupted even in the case of adultery; instead, like the prophet Hosea in the Old Testament, Christians should try to win back their erring spouses. If it happened that the visible sign of this union was broken by their being separated, they still had no right to remarry, for the invisible sign of their union, the bond that was formed in the image of Christ's union with the church, remained. And if they did take another partner while their first spouse was still living, the *sacramentum* of their first marriage marked them as adulterers.

Augustine presented a strong theological case for the prohibition of divorce and remarriage, and the council of Carthage which he attended in 407 reflected his position by forbidding divorced men as well as divorced women to remarry. It would be some centuries, however, before the Latin church would turn to Augustine's writings to justify an absolute prohibition against remarriage by either spouse while the other remained alive.

Jerome, a scripture scholar and contemporary of Augustine, spoke out strongly against remarriage after divorce, but he spoke out much more strongly in the case of women than men. He took Christ's words in the Bible as an absolute prohibition for women: "A husband may commit adultery and sodomy, he may be stained with every conceivable crime, and his wife may even have left him because of his vices. Yet he is still her husband, and she may not remarry anyone else as long as he lives" (*Letters* 55). But for men who had divorced unfaithful wives he had a different warning: "If you have been made miserable by your first marriage, why do you expose yourself to the peril of a new one?" (*Commentary on Matthew* 19). He even insisted on scriptural grounds that a man should not continue to live with an adulterous woman, although he was equally insistent that nothing but adultery was a valid reason for separation.

In the opinion of Innocent I, those who divorced by mutual consent as the civil law allowed were both adulterers if they remarried, regardless if they were men or women. He was equally clear in affirming that a woman who had been legally dismissed for adultery could not remarry as long as her husband was alive, but he apparently held that a husband in that case could remarry. Around the year 415 he wrote to a Roman magistrate about a woman who had been carried off by barbarian invaders sometime before and had returned later, only to find her husband remarried. Claiming to be his rightful wife, she appealed to her bishop and Innocent agreed with her: "The arrangement with the second woman cannot be legitimate since the first wife is still alive and she was never dismissed by means of a divorce" (*Letters* 36). Presumably if her husband had divorced her (since women in captivity were usually violated) Innocent would have considered the second marriage legitmate.

Later in the century bishops began to get even more involved in marriage cases as the Germanic invasions led to a breakdown of Roman civil authority. The council of Vannes in France decided that husbands who left their wives and remarried without providing proof of adultery should be barred from communion. And a similar council at Adge in 506 imposed the same penalty on men who failed to justify their divorce before an ecclesiastical court before they remarried. At the close of the patrisitic period in the west, then, Christian bishops were becoming legally involved with marriage, and there was still no universally recognized prohibition against divorce and remarriage for both sexes. And apart from Augustine, no one spoke of marriage as a sacrament.

From Secular to Ecclesiastical Marriage

With the coming of the dark ages in Europe after the fall of the Roman empire, churchmen were called upon more and more to decide marriage cases. Centuries before, Constantine had given them authority to act as judges in certain civil matters, and now that authority grew as the regular judicial system collapsed. Bishops also began to issue canonical regulations about persons who should not marry because they were too closely related. Initially the churchmen simply adopted the prevailing Roman customs, although they sometimes added prohibitions that were found in the Old Testament. Later they incorporated the customs of the invading Germanic peoples into the church's laws. These customs varied somewhat from tribe to tribe, but generally speaking persons who were more closely related than the seventh degree of kinship (for example, second cousins) were not allowed to marry legally.

Moreover, just as the bishops had earlier accepted Roman wedding customs, so they now also accepted the marriage practices of the Germanic peoples who settled within the old Roman provinces. Again these varied from tribe to tribe, although they, too, followed a general cultural pattern. Marriages were basically property arrangements by which a man purchased a wife from her father or some other family guardian to be his wife. The arrangement involved a mutual exchange of gifts, spoken and sometimes written agreements between the groom and the bride's guardian. In many places brides were betrothed ahead of time in return for a token of earnestness such as a small sum of money or a ring from the prospective husband, which would be forfeited if the marriage did not take place as agreed. On the wedding day the guardian handed over the woman and her dowry of personal possessions to her new husband, and received the bride-price as compensation for the loss her family incurred by allowing her to leave it. After the wedding feast that was celebrated by the relatives and other witnesses to the marriage, the bride and groom entered a specially prepared wedding chamber for their first act of intercourse, which formally sealed the arrangement.

Throughout this early period, then, marriage was still a family matter similar to what it had been in the Roman empire, and the clergy were not involved in wedding ceremonies except as guests. Bishops in their sermons and letters tried to impress their people with the Christian ideal of marriage found in the New Testament, and they sometimes urged them to have their marriages blessed by the clergy, but again this blessing was not essential to the marriage itself. In some places it was given during the wedding feast, in others it was a blessing of the wedding chamber, and in others it was a blessing during a mass after the wedding. Some bishops in southern Europe also suggested that the Roman custom of veiling the bridal couple should be done by a priest, but it was not a very common practice.

Just as churchmen were not officially involved in weddings, so also they were not officially involved in divorces when they occurred. However, some divorces ran counter to accepted Christian practices, and when they occurred those who were responsible for them had to confess their sin and do penance for it. The penitential books from the early Middle Ages show that divorce was more accepted in some places than others, but almost all allowed husbands to dismiss unfaithful wives and marry again. An Irish penitential book written in the seventh century instructed that if one spouse allowed the other to enter the service of God in a monastery or convent, he or she was free to remarry. The penitential of Theodore, archbishop of Canterbury in the same century, gave the following prescriptions: a husband could divorce an adulterous wife and marry again; the wife in that case could remarry after doing penance for five years; a man who was deserted by his wife could remarry after two years, provided he had his bishop's consent; a woman whose husband was imprisoned for a serious crime could remarry, but again only with the bishop's consent; a man whose wife was abducted by an enemy could remarry, and if she later returned she could also remarry; freed slaves who could not purchase either spouse's freedom were allowed to marry free persons. Other penitential books on the continent contained similar provisions.

The penitential books contained only unofficial guidelines to be followed in the administration of private penance, but conciliar and other church documents contained more official regulations. Again here these were not uniform, and ecclesiastical practices during this period ranged from extreme strictness to extreme laxness, but at least they show that there was no universal prohibition against divorce. In Spain the third and fourth councils of Toledo in 589 and 633 invoked the "Pauline privilege" in allowing Christian converts from Judaism to remarry. Irish councils in the seventh century allowed husbands of unfaithful wives to remarry, and although the council of Hereford in England advised against remarriage it did not forbid it. In eighth-century France the council of Compiègne allowed men whose wives committed adultery to remarry, and it allowed women whose husbands contracted leprosy to remarry with their husband's consent. In 752 the council of Verberie enacted legislation which allowed both men and women to remarry if their spouses committed adultery with a relative, and it prohibited those who committed the sin from marrying each other or anyone else. It also permitted a man to divorce and remarry if his wife plotted to kill him, or if he had to leave his homeland permanently and his wife refused to go with him. Pope Gregory II in 726 advised Boniface, the missionary bishop to Germany, that if a wife were too sick to perform her wifely duty it was best that her husband practice continence, but if this was impossible he might have another wife provided that he took care of the first one. Boniface himself recognized desertion as grounds for divorce, as well as adultery and entrance into a convent or monastery. Other popes of the period, however, protested against what they considered to be unlawful divorces, and the Italian council of Friuli in 791 strictly forbade divorced men to remarry even if their wives had been unfaithful.

One reason why churchmen became involved in marriage and divorce cases, especially after the popes started sending missionaries into northern Europe, was the difference between Roman and Germanic marriage customs. According to Roman tradition marriage was by consent, and after the consent was given by either the spouses or their guardians the marriage was considered legal and binding. In the Frankish and Germanic tradition, however, the giving of consent came at the betrothal, and the marriage was not considered to be completed or consummated until the first act of intercourse had taken place. Moreover, it was customary for parents to consent to the marriage of their children months and even years before they would begin to live together as husband and wife. This was particularly prevalent among the nobility, who often arranged such marriages as a means of securing allies or settling territorial disputes between them. But it sometimes happened that one of the betrothed spouses would undermine the parental arrangement by marrying someone else before the arranged marriage could be consummated. Bishops who were asked to settle these and similar cases could follow either the Roman or the Germanic tradition in coming to their decision. Under Roman law the arranged marriage was the binding one and the subsequent marriage was adultery, but according to Germanic custom the arrangement between the parents was only a nonbinding betrothal and the second marriage was the real one. Even before any marriage was arranged young people might consent to marry and then claim that they were not free to marry the partners their parents picked out for them, whereupon the parents might appeal to the episcopal court for a decision. In still other cases people sought to rid themselves of unwanted spouses by claiming that they had secretly contracted a previous marriage, which would make their present marriage unlawful. The legal question that had to be decided in each case was: Which marriage was the real marriage? And underlying the practical matters was the more theoretical question: Are marriages ratified by consent or by intercourse? For a long time there was no uniform answer to that theoretical question, and both episcopal and royal courts decided the practical matters according to which tradition they were accustomed to follow.

As Charlemagne initiated legal reforms in his European empire, both church and civil governments made an effort to impose stricter standards for marriage. Late in the eighth century the regional council of Verneuil decreed that both nobles and commoners should have public weddings, and a similar council in Bavaria instructed priests to make sure that people who wanted to marry were legally free to do so. In 802 Charlemagne himself passed a law requiring all proposed marriages to be examined for legal restrictions (such as previous marriages or close family relationships) before the wedding could take place. When the false decretals of Isidore were "discovered" in the middle of the ninth century, they contained documents purportedly from the patristic period aimed against the practice of secret marriage. A decree attributed to Pope Evaristus in the second century read, "A legitimate marriage cannot take place unless the woman's legal guardians are asked for their consent, . . . and only if the priest gives her the customary blessing in connection with the prayers and offering of the mass." Another decree represented the third-century pope Calixtus as saying that a marriage was legal only if it was blessed by a priest and the bride-price was paid. The proponents of reform in the Frankish empire used these spurious documents to support their efforts to outlaw secret marriages, and they were partially successful. Laws were passed making marriages legal only if guardians gave their consent and were present at the wedding.

In the meanwhile, however, Rome continued to follow its own tradition. In 866 Pope Nicholas I sent a letter to missionaries in the Balkans who had asked about the Greek church's contention that Christian marriages were not valid unless they were performed and blessed by a priest. In his reply Nicholas described the wedding customs that had become prevalent in Rome: the wedding ceremony took place in the absence of any church authorities and consisted primarily in the exchange of consent between the partners; afterward there was a special mass at which the bride and groom were covered with a veil and given a nuptial blessing. In Nicholas' opinion, however, a marriage was legal and binding even without any public or liturgical ceremony: "If anyone's marriage is in question, all that is needed is that they gave their consent, as the law demands. If this consent is lacking in a marriage then all the other celebrations count for nothing, even if intercourse has occurred" (*Letters* 97). According to Rome, then, it was the couple's consent, not their betrothal by their parents or their blessing by a priest, which legally established the marriage.

Charlemagne had wanted Roman practices to become normative in his empire, and in the years that followed, a Roman-style nuptial liturgy sometimes began to be included in the festivities that followed a wedding, though it was never very prevalent. Moreover, the pope's insistence that only consent constituted a marriage was initially ignored or largely unknown in the rest of Europe. Hincmar, the bishop of Rheims during this period, decided a number of marriage and divorce cases among the Frankish nobility, and he generally followed the opinion of the false decretals that legal marriages had to be publicly contracted. He also followed the Germanic tradition in ruling that marriages had to be consummated by sexual intercourse, and he allowed that people who had been given in marriage but who had not yet lived together could be legally divorced.

For a while, divorce regulations in northern Europe became more stringent under the impetus of ecclesiastical reform. As early as 829 a council of bishops at Paris decreed that divorced persons of both sexes could not remarry even if the divorce had been granted for adultery. By the end of the century a number of other councils in France and Germany passed similar prohibitions, and the penitential books were revised accordingly. But at the same time in Italy, popes and local councils continued to allow divorce and remarriage in certain circumstances, especially adultery and entering the religious life. Then in the next two centuries the trend in northern Europe reversed itself, and councils at Bourges, Worms, and Tours again allowed remarriages in cases of adultery and desertion.

During this same period, moreover, ecclesiastical courts were slowly gaining exclusive jurisdiction over marriage and divorce cases. As Charlemagne's short-lived empire dissolved into a disunited array of local principalities, more and more marriage cases were appealed to church tribunals. Eventually the secular courts came to be bypassed altogether, and by the year 1000 all marriages in Europe effectively came under the jurisdictional power of the church.

Church Control of Marriage Ceremony

There was as yet no obligatory church ceremony connected with marriage, but in the eleventh century this began to change. In order to insure that marriages took place legally and in front of witnesses, bishops invoked the texts of Popes Evaristus and Calixtus in the false decretals to demand that all weddings be solemnly blessed by a priest. It gradually became customary to hold weddings near a church, so that the newly married couple could go inside immediately afterward to obtain the priest's blessing. Eventually this developed into a wedding ceremony that was performed at the church door and was followed by a nuptial mass inside the church during which the marriage was blessed. At the beginning of this development the clergy were present at the ceremony only as official witnesses and to give the required blessing, but as the years progressed priests began to assume some of the functions once relegated to the guardians and the spouses themselves, and many of the once secular customs in the wedding ceremony became part of an ecclesiastical wedding ritual.

By the twelfth century in various parts of Europe there was an established church wedding ceremony that was conducted entirely by the clergy, and although there were numerous local variations it generally conformed to the following pattern. At the entrance to the church the priest asked the bride and groom if they consented to the marriage. The father of the bride then handed his daughter to the groom and gave him her dowry, although in many places the priest performed this function instead. The priest then blessed the ring, which was given to the bride, after which he gave his blessing to the marriage. During the nuptial mass in the church itself the bride was veiled and blessed, after which the priest gave the husband the ritual kiss of peace, who passed it to his wife. In some places the priest also pronounced an additional blessing over the wedding chamber after the day's festivities had concluded.

Along with the church's liturgical and legal involvement in marriage came a growing body of ecclesiastical laws about premarriage kinship, the wedding ceremony itself, and the social consequences of marriage and divorce. The medieval system of government and inheritance emphasized property rights and blood relationships arising from marriage, making it important for ecclesiastical judges to know who was legally free to marry, who was married to whom, and who could have their marriage legally dissolved. In the eleventh century the discovery and circulation of the Code of Justinian led to increasing acceptance of the idea that marriage came about by the consent of the partners, and this idea was reflected in the new rituals for church wedding in which the priest asked the bride and groom, not their parents, for their consent to the marriage. But the growth of the consent theory also led to an increase in the number of secret marriages, which brought legal difficulties about the legitimacy of children and their right to inherit their father's property, as well as pastoral difficulties when women and children were deserted by men who claimed they had never intended to establish a marriage.

In response to these difficulties some church lawyers defended a different theory about when a marriage legally took place, based on the old Germanic notion that intercourse was needed to ratify a marriage. As it was taken up and developed by the law faculty at the University of Bologna, this theory proposed that a real marriage did not exist unless and until the couple had sexual relations. But the opposing theory, that consent alone made a marriage, also had its staunch defenders, mainly at the University of Paris.

Around 1140 Francis Gratian published his collection of canonical regulations known as the *Decree* in which he tried to bring some order into the sometimes conflicting decrees and decisions of popes and councils dating back to the patristic era. He was aware of the two schools of thought about what constituted a marriage, and he tried to harmonize them by suggesting that the consent of the spouses or their parents (in the case of bethrothal) contracted a marriage and that sexual intercourse completed or consummated it. His opinion was that a marriage could be legally dissolved before it was consummated but not afterward, and in this respect he sided with the Bologna school. But he also agreed with the Paris school's contention that a binding marriage could be made in secret, without any public ceremony or priestly blessing. In his opinion such a marriage would be illicit or illegal because it flouted the laws of the church, but it would nonetheless be a real marriage, initiated by consent and consummated by intercourse.

Gratian's work clarified but did not settle the issue. In Italy, for example, church courts continued to dissolve marriages if it could be proven that no sexual relations had taken place, but in France the courts refused to dissolve any marriage once the partners' consent had been given. It was not until later in the twelfth century, when a noted canon lawyer of the period became Pope Alexander III, that a definitive solution was worked out and legislated for the whole Latin church. Because it offered a clearer criterion of an intended marriage between two individuals, Alexander sided with the ancient Roman practice that was defended by the Paris school, and he decreed that the consent given by the two partners themselves was all that was needed for the existence of a real marriage. This consent was viewed as an act of conferring on each other the legal right to marital relations even if they did not occur, and so from the moment of consent there was a true marriage contract between the two partners. In and of itself it was an unbreakable contract, but since the church had jurisdiction over it by the power of the keys, it could also be nullified or annulled by a competent ecclesiastical authority if sexual relations between the spouses had not yet taken place.

The decision of Alexander III became the legal practice of the Catholic church. It was reinforced by further papal decrees in the thirteenth century and has remained in effect in canon law through the twentieth century. With the exception of the "Pauline privilege" by which non-Christian marriages could be dissolved if one of the spouses converted to Catholicism, henceforward the church would grant no divorces whatever. Henceforward the marriage bond would be considered indissoluble not only as a Christian ideal but also as a rule of law. Henceforward if Catholics wanted to be freed from their spouses they would have to prove that their marriage contact could be nullified, declared to be nonexistent, either for lack of intercourse or for some other canonically acceptable reason.

But the pope's decision and the support it received in subsequent centuries did not rest only on the practical needs of ecclesiastical courts. Rather, the indissolubility of Christian marriage in the mind of Alexander and later churchmen rested also on a firm theological ground, the sacramentality of Christian marriage. For it was precisely around this time—the late twelfth century and the early thirteenth century—that marriage came to be viewed as one of the church's seven official sacraments.

Marriage as a Sacrament

What Francis Gratian did for canon law in his *Decree,* Peter Lombard did for theology in his *Sentences*. Lombard's collection of theological texts did not solve many of the theological problems of the Middle Ages but it did go a long way toward defining what they were and how they should be treated. Marriage was treated in the section on the sacraments, for by this time in mid twelfth-century France there was an established Christian ritual for marriage which was not unlike the other rituals that Lombard classified as sacramental.

When the book of *Sentences* was first published, however, many theologians still had difficulty in accepting the idea that marriage was a sacrament in the strict sense which was then being developed, and Lombard himself believed that it was different from the other six sacraments in that it was a sign of something sacred but not a cause of grace. One reason for the diffculty was that marriages involved financial arrangements, and if marriage was counted as a sacrament like the others it looked like grace could be bought and sold. Another reason they hesitated to call marriage a sacrament was that it obviously existed before the coming of Christ, and so it could hardly be said to be a purely Christian institution like the other seven. But the third reason was the most crucial, and it was that marriage involved sexual intercourse.

Throughout the early Middle Ages most churchmen held virginity in higher esteem than marriage. On the one hand Christians could not deny that God had told Adam and Eve to increase and multiply, and so marriage itself had to be good. But on the other hand marriage, as Paul said, distracted one from the things of the Lord, and he seemed to suggest that people should marry only if they could not quench the fire of sexual desire. So marriage in the Middle Ages was often viewed negatively as a remedy against the desires of the flesh rather than positively as a way to become holy, and those desires themselves were viewed as sinful or at best dangerous. Some bishops who blessed newly married couples recommended that they abstain from intercourse for three days out of respect for the blessing; others told them not to come to church for a month after the wedding, or at least not to come to communion with their bodies and souls still unclean from intercourse. Most writers held that sexual activity which was motivated by anything but the desire for children was sinful, but most of them also believed that even here children could not be conceived without the stain of carnal pleasure.

So the western theological tradition through the eleventh century taught that marriage was good even though sexual activity was usually sinful. Three things happened in that century, however, which forced them to reexamine that view. The first was the rise of a religious sect in southern France which, like the Manicheans in the patristic period, taught that matter was evil and so marriage was sinful because it brought new material beings into the world. The Albigensians did not accept the Christian concept of God, they denied the value of church rituals, and their leaders attacked the Catholic clergy as corrupt, so they were first denounced and later burned as heretics. And in combating the Albigensian view of marriage Christian writers began to propose more strongly than before that intercourse for the sake of having children was positively good. The second thing that happened during this century was the development of a Christian wedding ritual which, by the presence of the clergy and the blessing they gave, implied that the church officially sanctioned sexual relations in marriage. And the third thing was the rediscovery of the writings of Augustine on marriage in which he developed the idea that marriage was a *sacramentum*. To the early schoolmen it seemed to suggest that marriage was a sacrament in the same way that baptism and the eucharist were sacraments.

Augustine had taught that marriage was a *sacramentum* in two ways. It was a sign of the union between Christ and his church, and it was also a sacred pledge between Christ and his church, and it was also a sacred pledge between husband and wife, a bond of fidelity between them that could not be dissolved except by death. It was something like a character on the souls of the spouses which permanently united them, and it was this permanence of their union which symbolized the eternal union of Christ and the church. It seemed to the schoolmen, therefore, that the Christian marriage ritual should be open to the same kind of analysis that they gave to the other sacraments, namely that in marriage there was a *sacramentum,* a sacred sign, a *sacramentum et res,* a sacramental reality, and a *res,* a real grace that was conferred in the rite. It took most of the twelfth century for the scholastics to satisfactorily fit marriage into this scheme, but by the time they did it the Catholic concept of sacramental marriage had become the theological basis for the canonical prohibition against divorce.

Defining the Sacrament

But what was the *sacramentum,* the sacramental sign in marriage? At the beginning it seemed to many of the schoolmen that it should be the priest's blessing since in the wedding ritual it corresponded to the part that was played by the priest in the other sacramental rites. Later, others suggested that it should be the physical act of intercourse between the spouses since this physical union could be taken as a sign of the spiritual union between the incarnate Christ and his spouse, the church. Still others felt it should be the spiritual unity of the married couple since this union of wills was closer to the actual way that Christ and the church were united with each other. However, each of these suggestions met with difficulties and had to be abandoned. It was objected that the priest's blessing could not be the sacramental sign because some people were truly married even though they never received the blessing, for example, people who married in secret. The schoolmen who still believed that sexual relations even in marriage were venially sinful objected to intercourse's being considered the sign because this would paradoxically raise a sinful act to the dignity of a sacrament. And it was objected that the union of wills in the married life could not be the sacramental sign because sometimes this spiritual unity was minimal at the beginning of a marriage and altogether lacking later on.

Eventually, because of the growing acceptance of the consent theory of the canon lawyers, the *sacramentum* in a sacramental marriage came to be viewed as the consent that the spouses gave to each other at the beginning of their married life. This mutual consent was something that had to be present in all canonically valid marriages, even those that were unlawfully contracted in secret. Both the canonists in Paris and Pope Alexander in Rome insisted that a real marriage existed from the moment that the consent was given, and theologians such as Hugh of St. Victor argued that a real marriage would have to be possible even without consummation in intercourse since according to tradition Joseph and the Virgin Mary had been truly married even though they had never had sexual relations. In addition, locating the *sacramentum* in the mutual consent kept it within the wedding ritual for most Christian marriages, and it made it possible to look upon the union of wills in a happy married life as a ''fruit'' of the sacrament even if it was not the sacrament itself.

But the greatest theological consequence of seeing the act of consent as the *sacramentum* in marriage was that it made it possible to regard the marriage contract or bond as the *sacramentum et res*. According to canon law the bond of marriage was a legal reality which came into existence when the two spouses consented to bind themselves to each other in a marital union. Now, in theology, the bond of marriage could also be understood as a metaphysical reality which existed in the souls of the spouses from the moment that they spoke the words of the sacramental sign. Following the lead of Augustine, the scholastics argued that this metaphysical bond was unbreakable, since it was a sign of the equally unbreakable union between Christ and the church. It was not, as in the early church, that marriage as a sacred reality *should* not be dissolved; now it was argued that the marriage bond as a sacramental reality *could* not be dissolved. According to the church fathers the dissolution of marriage was possible but not permissible; according to the schoolmen it was not permissible because it was not possible. Thus the absolute Catholic prohibition against divorce arose in the twelfth century both as a canonical regulation supported by sacramental theory, and as a theological doctrine buttressed by ecclesiastical law. The two came hand in hand.

Even through the beginning of the thirteenth century, however, many theologians found it hard to admit that marriage as a sacrament conferred grace like the other sacraments. The traditional view of marriage was that it was more of a hindrance than a help toward holiness, a remedy for the sin of fornication rather than a means of receiving grace. Many theologians accepted Augustine's idea that original sin was transmitted from one generation to the next through the act of intercourse, and so even sexual relations for the sake of having children were often seen as a mixed blessing. Alexander of Hales was the first medieval theologian to reason that since marriage was a sacrament and since all the sacraments bestowed grace, then marriage must do so as well. But William of Auxerre believed that if any grace came from marriage it must be only a grace to avoid sin, not a grace to grow in holiness. William of Auvergne and Bonaventure both agreed that the effect of the sacrament must be some sort of grace, but both of them also held that the grace came through the priest's blessing.

Nevertheless, under the influence of reasoning like Alexander's and the desire to fit all the sacraments into a single conceptual scheme, theologians from Thomas Aquinas onward admitted that the sacrament gave a positive assistance toward holiness in the married state of life. That grace was first of all a grace of fidelity, an ability to be faithful to one's marriage vow, to resist temptations to adultery and desertion despite the hardships of married life. It was also even more positively a grace of spiritual unity between the husband and the wife, enabling him to love and care for her as Christ did the church, and enabling her to honor and obey him as the church did her Lord. It was true, of course, that even non-Christians could be faithful to one another and achieve marital harmony, but for Aquinas Christians were called to an ideal of constant fidelity and perfect love which could not be attained without the supernatural power of God's grace.

Aquinas also realized as did the other scholastics that marriage existed long before the coming of Christ, but for him this was no different from the fact that washing existed before the institution of Baptism or that anointing existed before the sacraments that used oil. It was thus, like the other sacraments, something natural that had been raised in the church to the level of a sacramental sign through which grace might be received. But this also meant for Aquinas that the *sacramentum* in marriage was not just the act of mutual consent in the wedding ritual but the marriage itself, which came into existence through the giving of consent, was sealed by the act of intercourse, and continued for the remainder of one's life. As a sacramental sign it was therefore permanent, as was the sacramental reality of the marriage bond which was created by consent and made permanent through consummation. As a natural institution marriage was ordered to the good of nature, the perpetuation of the human race, and was regulated by natural laws which resulted in the birth of children. As a social institution it was ordered to the good of society, the perpetuation of the family and the state, and was regulated by civil laws which governed the political, social, and economic responsibilities of married persons. And as a sacrament it was ordered to the good of the church, the perpetuation of the community of those who loved, worshipped, and obeyed the one true God, and was regulated by the divine laws which governed the reception of grace and growth in spiritual perfection. The "matter" of the sacrament was therefore the human reality of marriage as a natural and social institution since this was the natural element, like water or oil, out of which it was made. And the "form" of the sacrament consisted of the words of mutual consent spoken by the spouses, since these were what signified the enduring fidelity which would exist between them, just as it existed between Christ and the church.

Most of the other things that Aquinas had to say about marriage—and this was true of the other schoolmen as well—had to do with the ecclesiastical regulation of marriage, with the laws governing who may and may not lawfully marry, with regulations regarding betrothal and inheritance, and so on. For marriage in the Middle Ages was viewed not so much as a personal relationship but as a social reality, an agreement between persons with attendant rights and responsibilities. Thus Aquinas and the other thirteenth-century scholastics occasionally spoke of marriage as a contract, and in the centuries that followed the legal terminology of canon law was further incorporated into the sacramental theology of marriage.

John Duns Scotus, for instance, conceived of marriage as a contract which gave people a right to have sexual relations for the purpose of raising a family, and from this he drew the inference that intercourse in marriage was legitimate not only for begetting children but also for protecting the marriage bond. A woman was bound in justice to give her husband what was his by right, he reasoned, and so she had to grant his requests lest he be tempted to bring discord into the marriage by satisfying his desires with someone else. Other theologians in the fourteenth and fifteenth centuries also came to accept this argument, and by the sixteenth century it was commonly taught that not every act of intercourse had to be performed with the intention of having children. Married people could ask for sex without blame, provided they did it not out of lust but only to relieve their natural needs.

Scotus was also the first theologian to teach that the minister of the sacrament was not the priest but the couple that was getting married. According to canon law people who wed without a priest were validly married even though they went about it illegally, and according to theology people who were validly married received the sacrament. It followed, therefore, that the bride and groom had to be the ones who administered the sacrament to each other when they gave their consent to the marriage. In the fourteenth and fifteenth centuries this view became more widely accepted, but even in the sixteenth century some theologians still maintained that the priest was the minister of the sacrament, for in many places the priest not only handed over the bride to the groom during the wedding ceremony but he also said, "I join you in the name of the Father and of the Son and of the Holy Spirit."

One thing that did not change, however, was the official prohibition against divorce. In the decree that was drawn up for the Armenian Christians during the Council of Florence in 1439, marriage was listed among the seven sacraments of the Roman church and explained as a sign of the union between Christ and the church. It adopted Augustine's summary of the goods of marriage as the procreation and education of children, fidelity between the spouses, and the indissolubility of the sacramental bond. It granted that individuals might receive a legal separation if one of them was unfaithful, but denied that either one of them could marry again "since the bond of marriage lawfully contracted is perpetual."

Nonetheless, Christians in certain cases did separate and remarry. The hierarchy no longer allowed divorces but ecclesiastical courts were now empowered to grant annulments to those who could prove that their present marriage was invalid by canonical standards. If a married person could show, for example, that he had previously consented to marry someone else, the court could decide that the first marriage was valid even though unlawful and that the present marriage was therefore null and void. Marriage within certain degrees of kinship was also regarded as grounds for an annulment even after years of marriage. But the closeness of prohibited relationships varied in different parts of Europe, and so a marriage that might be upheld in one country might be annulled in another. And if the blood relationship or secret marriage was difficult to prove, ecclesiastical courts were sometimes open to being persuaded by financial considerations, generously but discreetly offered.

III: Marriage from the Reformation to the Present

The granting of annulments in dubious cases, for the wealthy and the nobility in particular, was one of the scandals in the Renaissance that led the early reformers to protest against the hierarchy's regulation of marriage. It was not that they denied the sacredness of marriage or disagreed with the practice of church weddings, but they did revolt against the complex canonical legislation that determined who could and could not marry, and they did doubt the justice of the legal system, that allowed dispensations from the law and annulments of established marriages. They could find no justification in the scriptures for most of the ecclesiastical laws about marriage, and when humanist scholars discovered that the early church did not have civil jurisdiction in marriage cases some of the reformers even charged that Rome had turned it into a sacrament just in order to get legal control over it. Most of them called for a more biblical attitude toward marriage, which usually meant a return to allowing divorce in certain cases.

Redefining Marriage As Nonsacrament

According to Martin Luther, since marriage existed since the beginning of the world, "there is no reason why it should be called a sacrament of the new law and the sole property of the church" (*Babylonian Captivity of the Church* 5). Marriage existed even among non-Christian peoples, and there was no record of any such sacrament's being instituted in the New Testament. The Latin text of Ephesians 5:32–33 could not be used to prove the existence of the sacrament, he argued, because *sacramentum* in that text was only a translation of the Greek word *mysterion,* and in that passage Paul was speaking metaphorically about the mysterious union of Christ and the church, not literally about the marriage bond between men and women. So there was no reason to believe that people received any special grace from God just because they were married.

Marriage certainly was instituted by God, said Luther, but not as a sacrament in the Roman sense. Rather it was a natural and social institution which accordingly fell under natural and civil law, not church law. "No one can deny that marriage is an external and secular matter, like food and clothing, houses and land, subject to civil supervision" (*On Matrimonial Matters*). Thus the church should "leave each city and state to its own customs and practices in this regard" (*Short Catechism,* Preface). The role of the clergy should be to advise and counsel Christians about marriage, not to pass laws about it and judge marriage cases. Civil governments, on the other hand, had a right to make marriage laws because all authority ultimately came from God, and so in the secular world they acted in his name. They could not be expected to pass laws that were in strict conformity with the ideals of the gospel, but at the same time they were morally obliged to keep within the bounds set by the laws of nature in enacting legislation for the good of society.

Luther said that he personally detested divorce, but that as a Christian he had to admit that Christ allowed divorce in the case of adultery, and that as a pastor he was inclined to permit it in other serious cases that seemed to have a scriptural justification. Paul, for instance, had allowed the Christian spouses of unbelievers to remarry, and Luther believed this principle could also be followed if a person acted like an unbeliever and deserted his family. Why, he argued, should the innocent suffer for their spouse's wickedness? In his own eyes he saw more harm done to innocent individuals by ecclesiastical annulments, and he saw a greater violation of Jesus' injunction not to sever what God had joined together when the church courts nullified marriages not on moral grounds but on legal technicalities.

John Calvin, like Luther, agreed that marriage was not a sacrament. "It is not enough that marriage should be from God for it to be considered a sacrament, but it is required that there should also be an external ceremony appointed by God for the purpose of confirming a promise," such as the promise of salvation that was confirmed by baptism (*Institutes of the Christian Religion* IV, 19, 34). But in the case of marriage there was no such ceremony and no such promise mentioned in the New Testament. Calvin also believed, therefore, that marriage laws fell within the jurisdiction of civil and not ecclesiastical authorities, but unlike Luther he contended that governments were morally obliged to make marriage and divorce laws in strict conformity with Christian principles. As far as he could see, the only two scriptural grounds for divorce were adultery and acting like a pagan by deserting one's family. Moreover, in these cases only the innocent spouse should be allowed to remarry.

Despite the divorces of Henry VIII which were granted after the English church's break from Rome, Anglican canon law prohibited divorce and remarriage until the middle of the nineteenth century. Nonetheless, English theologians generally agreed with the other reformers that divorce was possible in certain cases, and so when civil courts granted them from the seventeenth century those who were divorced were allowed to remarry in the church. Initially only adultery, deliberate desertion, prolonged absence, and cruelty were recognized as legitimate grounds for divorce, but in later centuries these grounds were expanded. Throughout all this time, however, marriage continued to be commonly regarded as a sacrament, though only as a sacrament of the church which was not instituted by Christ. The English church also continued to uphold the validity of secret marriages, as the Roman church had, until the British parliament passed a law in 1754 requiring weddings to be celebrated in a church, with few legal exceptions.

The Roman Church's Response

The response of the Roman church to the views of the reformers was slow but deliberate. The Council of Trent did not take up the question of marriage until its last session in 1563, but when it did it attempted to vindicate both the sacramentality of marriage and the church's right to regulate it. In a brief statement presenting the Catholic doctrine on marriage, the bishops of the council affirmed that God had made the bond of marriage unbreakable when he made the first man and woman, and that Christ had both reaffirmed this and made marriage a sacrament whose grace raised natural love to perfect love. Thus Christian marriage was superior to other marriages, and for this reason "our holy Fathers, the Councils and the tradition of the universal Church have always taught that marriage should be counted among the sacraments of the New Law." To this doctrinal statement the bishops then added a series of canons condemning anyone who taught the following heresies: that marriage was not a sacrament instituted by Christ which gave grace; that the church does not have the power to regulate who can and cannot legally marry, and to grant dispensations from these regulations; that ecclesiastical courts cannot annul unconsummated marriages or render judgments about other marriage cases; that the church was wrong in teaching that the marriage bond cannot be dissolved for any reason including adultery, or in forbidding remarriage, or in permitting spouses to legally separate without remarrying.

But the bishops also saw that the reaffirmation of traditional doctrine was not all that was needed to make sure that the sacrament was respected and properly administered. The greatest threat to the sacredness of marriage was the continuing practice of secret marriages that enabled people to enter unions which they later renounced, and that allowed them to seek annulments of public marriages on sometimes doubtful grounds. And so the bishops at Trent decided to take a drastic step. In a separate decree they recognized the validity of all previous secret marriages but declared that henceforth no Christian marriage would be valid and sacramental unless it was contracted in the presence of a priest and two witnesses. Those who tried to contract a marriage in any other way would be guilty of a grave sin and treated as adulterers. Furthermore, all marriages now had to be publicly announced three weeks in advance and entered into the parish records afterward in order to be considered canonically legal.

The bishops' decree effectively put an end to secret marriages in the Catholic church, but their institution of a legal requirement for valid marriages—the giving of consent before a priest and two witnesses—raised additional questions for theologians and canon lawyers.

The first was the reappearance of an old question, namely, who was the minister of the sacrament of marriage? In the Middle Ages the possibility that marriages could be contracted without a church ceremony and even without any witnesses led most theologians to conclude that the ministers of the sacramental bond were the couple themselves, although some theologians continued to think otherwise. These argued that the priest was the minister of the sacrament, and so secret marriages were real but not sacramental since no priest joined them in marriage or blessed their union. Then, in the sixteenth century, a Spanish theologian named Melchior Cano developed a theory that in marriage the mutual consent of the partners was the "matter" of the sacrament, while the priest's blessing was the "form." This implied that people could validly marry (as in Protestant countries) even though they did not receive a valid sacrament (since the required form was not followed). It also implied that the priest was the minister of the sacrament since without his presence, which the Catholic church now demanded, the marriage was nonsacramental.

According to Cano's theory, then, the marriage contract was something separate and distinct from the sacrament, and it continued to be defended by Catholic theologians in the seventeenth and eighteenth centuries who wanted to agree that civil governments had the right to make laws governing the secular aspects of marriage even though the church's hierarchy had sole jurisdiction over the sacramental aspects of marriage. These theologians were in the minority, however, for the hierarchy had had complete control over all aspects of marriage for so long that most theologians and canonists continued to assert that only the church had the right to make marriage laws for Christians. Even though Protestant heretics disobeyed those laws they were not really freed from them, the majority argued, for by their baptism they belonged to the church, and the one true church was the Roman Catholic church. This church's laws, therefore, applied to all baptized persons whether or not they acknowledged and followed them.

Civil Marriages

Those who were not under the control of the Catholic hierarchy, however, thought and acted differently. At the beginning of the Protestant era other Christian churches developed their own wedding ceremonies and considered them valid, and as a matter of fact for over two centuries afterward almost all marriages in Europe were church marriages. But late in the eighteenth century this picture began to change. In France the revolution of 1789 brought an end to the ecclesiastical control of marriage, and the Napoleonic Code of 1792 made civil weddings mandatory for all French citizens. During the next century almost all other countries in Europe began to allow people to marry before a civil magistrate rather than a priest or minister, even though most of them did not require it as the French did. And of course governments also continued to regulate the other secular aspects of marriage and divorce (legal registration, inheritance rights, and so forth), as they had done even before the French revolution.

These developments forced the Catholic hierarchy to reexamine the official teaching on marriage and determine more precisely when and how the sacrament was conferred. Technically, according to canon law, all baptized persons who were not married in accordance with Trent's decree were living in sin because their marriages were not canonically valid. But Catholic bishops in Protestant countries began complaining to Rome that this put them in the awkward position of having to regard all Christians who were not married in the Catholic church as adulterers and their children as illegitimate. The popes by this time realized that the Protestant reformation and the civil regulation of marriage were not going to be reversed, and they allowed that the Tridentine decree should be taken as applying only to those who were baptized Catholics and thus still under the legal jurisdiction of the hierarchy.

But what of non-Catholic marriages? Were they, too, sacramental? Even though Rome had reluctantly relinquished control over many of the secular aspects of marriage to other authorities, it did not see how it could surrender its position on the sacramentality of Christian marriage. The church had long recognized that baptisms even by heretics and schismatics were sacramentally valid, and now Rome followed a similar course with regard to non-Catholic marriages. In 1852 Pope Pius IX reacted to the claim of civil governments that all marriages between their citizens were legally dissolvable by declaring that since sacramental marriage was instituted by Christ, "There can be no marriage between Catholics which is not at the same time a sacrament; and consequently any other union between Christian men and women, even a civil marriage, is nothing but shameful and mortally sinful concubinage if it is not a sacrament" (Address, *Acerbissimum vobiscum*). In other words, marriages between Christians were either both valid and sacramental or else they were not marriages at all. The same position was reaffirmed by Leo XIII in 1880: "In Christian marriage the contract cannot be separated from the sacrament, and for this reason the contract cannot be a true and lawful one without being a sacrament as well" (Encyclical, *Arcanum divinae sapientiae*).

One of the reasons why the popes could be confident that the sacrament was identical with the marriage contract was that by this time historical research had shown that through the early Middle Ages the priest's blessing did not have to be given for a marriage to be valid; all that was needed was the mutual consent of the couple. This consent, then, established the contract or bond between the two parties, and so at the same time it had to be the act which established the marriage as sacramental. The theory that the sacrament was conferred by the priest was therefore no longer tenable. Since the contract was established through the giving of consent, the sacrament had to be administered by the bride and groom to each other. And since the sacrament was administered by the bride and groom, even non-Catholic Christians would confer the sacrament on each other whenever they contracted a valid marriage.

But this conclusion raised even further questions for canon lawyers. Were marriages between Christians and non-Christians sacramental as well? If non-Christians became Catholics did they have to be married again, or did their prior marriage automatically become a sacramental one in virtue of their baptism? If non-Catholic Christians divorced and remarried, was their second marriage valid, sacramental, both, or neither? Could a legally divorced non-Catholic validly marry a single Catholic? Could a divorced non-Christian do this? Suppose a

Christian of another denomination became a Catholic and was divorced because of this by the non-Catholic spouse. Could the "Pauline privilege" be applied to this case so the Catholic could remarry? Or suppose that two non-Christians were married and divorced, and later became Catholics. Were they free to remarry or was their previous non-Christian union now sacramental and indissoluble in virtue of their baptism? Questions such as these were actually raised before Catholic marriage courts. They were sent to Rome from countries wherever Catholicism was established, but especially from bishops in Protestant and mission countries. They were cases that had to be decided, and the decisions set precedents for future cases. The ecclesiastical regulation of marriage was becoming more complex than it had ever been.

At the same time, however, the Catholic theology of marriage remained relatively simple. Marriage was a sacrament instituted by Christ in which two legally competent persons became permanently united as husband and wife. The *sacramentum* was the giving of consent, the external rite in which they agreed to the marriage and took each other as their spouse. The *sacramentum et res* was the marriage contract, the sacramental reality which both symbolized the permanent union between Christ and the church and permanently united the couple in the bond of marriage. The *res* was the grace that the couple received to be faithful to each other as Christian spouses, and to fulfill their duties as parents. The primary purpose of marriage was the procreation and education of children; its secondary purpose was the spiritual perfection of the spouses by means of the grace of the sacrament, the mutual support they gave to each other, and the morally permissible satisfaction of their sexual needs.

Marriage in Contemporary Catholicism

Through the beginning of the twentieth century the medieval concept of marriage remained relatively unchallenged in the Catholic church. Whatever challenges there were came from without, and for the most part church leaders reacted by regarding them as dangers to the sacredness of Christian marriage. The Protestant rejection of the sacramentality of marriage was one such challenge; civil governments' regulation of marriage and acceptance of divorce was another. The romantic movement in the eighteenth century exalted erotic love over marital fidelity, and popular writers in the nineteenth century applauded the idea of marrying for no other reason but love. To some extent the church adapted, admitting the rights of governments when it could no longer deny them, and allowing that sexual relations in marriage had some secondary purpose besides conceiving children. But for the most part the official Catholic attitude toward marriage continued to emphasize legal rights and social responsibilities, for that had been the European attitude toward marriage during the centuries when the official doctrines were formed.

During the twentieth century, however, western society began to undergo a change the likes of which it had not seen since the fall of the Roman empire and the rise of medieval society. Its roots were found in the Renaissance humanism of the sixteenth century, the rise of secular nations in the seventeenth century, the industrial revolution in the eighteenth century, and the expansion of the natural and social sciences in the nineteenth century, but it was not until the present century that these developments and the changes they caused began to noticeably alter the basic values and norms of western civilization.

Until the twentieth century Catholics could officially ignore the social transformation that was going on around them. The majority lived mainly in the nonindustrialized countries of Europe and in religious isolation elsewhere. Their intellectuals studied and taught the philosophy and theology of the Middle Ages, and critiqued modern ideas from that perspective. Their church leaders concerned themselves with ecclesiastical affairs and remained aloof from the things of this world. But in the present century this social transformation reached into the everyday lives of Catholics, and it affected the ideas and attitudes of Catholic intellectuals and church leaders alike. Now the challenge to the medieval concept of marriage came from within the church itself, and it could not be ignored.

But during the twentieth century the nature and function of marriage in the west began to change. Before, it was a social duty; now, it was an individual right. Before, it was done in compliance with parents' wishes; now, it was done for personal love. Before, love was expected to begin after the wedding; now, it was expected to precede it. There were parallel changes in the nature and function of the family. Before, it was an extended family with three generations of relatives living in the same house, or at least close to one another; now. it was a nuclear family with parents and children having less and less contact with uncles and aunts, grandparents and cousins. Before, the family was the basic unit of society: families lived and worked together; children were educated, trades were learned, recreation was provided, and most human needs were met, all within and by the family. Now, the family was becoming but one social unit among many: people had jobs that took them away from the people they lived with, children went to schools, most occupations could no longer be learned from one's parents, recreation was brought in by printed or electronic media or sought outside the home, and most human needs except the most basic ones of nurturing and affection were met by people outside the family. In short, marriage was coming to be seen mainly as an expression of love between a man and a woman, and the family was no longer needed to educate children the way it used to be.

After the Second World War Catholic thinkers who were influenced by existentialist and personalist philosophy began to reappraise the traditional teaching that marriage was primarily for the procreation and education of children. Contemporary experience suggested that this was no longer so, and social sciences like psychology and sociology suggested that sex had a much deeper significance in human life than biological reproduction. Catholic personalists primarily in Germany proposed that Christian marriage should therefore be redefined to better fit the contemporary experience and understanding of marriage. As Herbert Doms and others saw it, marriage and sex had meaning in themselves, and so they did not have to get their meaning or justification from the children that resulted from intercourse. The meaning of marriage was the unity of two persons in a common life of sharing and commitment, and the meaning of intercourse was the physical and spiritual self-giving that occurred in the intimate union of two persons in love. Thus the primary purpose of marriage was the personal fulfillment and mutual growth of the spouses, which occurred not only through their sexual relations but through all the interpersonal relations of their married life. Children were thus secondary to the meaning and purpose of marriage, and even though they were to be loved and nurtured for their own sake, neither children nor the absence of children affected the primary meaning of marital and sexual union.

Rome's overall response to this inversion of the traditional Catholic teaching was negative, and various Vatican offices reaffirmed the doctrine that the primary purpose of marriage was the begetting and raising of children. But Pope Pius XII saw some merit in the personalist approach to marriage, and in some of his speeches he granted that interpersonal values like commitment and fulfillment were essential even if they were secondary in Christian marriage. Taking this as an official acceptance of their efforts if not always of their views, other theologians such as Josef Fuchs and Bernard Häring continued to explore the long-neglected personal aspects of marriage in their writings. Some tried to avoid the classic dichotomy between primary and secondary ends in marriage and preferred a more integrated approach. Others tried to translate the traditional scholastic teaching into more contemporary language. All of them tried to move away from a legalistic theology of marriage and sex and toward one that was more scriptural, more personal, and more related to contemporary married life.

Marriage In Vatican II

This change in attitude was reflected in the documents of Vatican II. The council devoted an entire chapter of its pastoral statement on *The Church in the Modern World* to problems of marriage and the family, and although it did not reverse the traditional Catholic teaching on marriage, it did not adopt a more personalistic perspective toward sex. In particular, the council avoided speaking of marriage as a contract or legal bond and instead referred to it in sociological, personal, and biblical terms. It spoke of marriage as a social and divine institution, an agreement between persons, an intimate partnership, a union in love, a community, and a covenant.

Thus on the one hand the council reasserted that "By their very nature, the institution of matrimony and conjugal love are ordained for the procreation and education of children" (*Church in the Modern World* 48). But on the other hand it affirmed that love between spouses is "eminently human" and "involves the good of the whole person." This total love "is uniquely expressed and perfected through the marital act," for sexual relations "signify and promote that mutual self-giving by which spouses enrich each other with a joyful and thankful will" (49). It was because marriage was such a noble and sacred calling that "Christian spouses have a special sacrament by which they are fortified and receive a kind of consecration in the duties and dignity of their state." Through this sacrament "they are penetrated with the spirit of Christ" which "suffuses their whole life with faith, hope and charity" (48). The sacramental nature of marriage, the unity of love, and the welfare of children all imply that marriage is indissoluble. "Thus a man and a woman, who by the marriage covenant of conjugal love 'are no longer two but one flesh,' render mutual help and service to each other through an intimate union of their persons and actions. Through this union they experience the meaning of their oneness and attain to it with growing perfection day by day. As a mutual gift of two persons, this intimate union, as well as the good of the children, imposes a total fidelity on the spouses and argues for an unbreakable oneness between them" (48).

In other documents the council also referred to the traditional analogy from Paul's letter to the Ephesians. Thus, "Christian spouses, in virtue of the sacrament of matrimony, signify and partake of the mystery and fruitful life which exists between Christ and the Church" (*The Church* 11). And in its decree *The Apostolate of the Laity* the council spoke of the sacramentality of marriage in an even broader sense, saying that Christian couples should be signs to each other, their children, and the world of the mystery of Christ and the church by the testimony of their love for each other and their concern for those in need.

As with the other official sacraments, the Catholic marriage rite was revised to allow greater flexibility in different circumstances and more adaptability to personal preferences. Since 1969 couples may choose from a variety of scriptural passages to be read if the wedding is celebrated during a nuptial mass, and they may use a number of different formulas to express their consent to each other in marriage. In some places they are allowed to include poems and other nonscriptural readings which have a special meaning for them, and even to compose their own marriage vows as long as they express the basic Christian understanding of marriage. Bishops in mission countries are encouraged to incorporate native customs into the wedding ceremony as far as possible, and even to draw up new rituals which express the meaning of Christian marriage in the symbols and gestures of their own culture. And church regulations have been revised so that non-Catholics in interfaith marriages may have their own minister as well as a priest preside at their wedding in the Catholic church, a practice that was prohibited in the past.

Since the time of the council in the mid-1960s the theology and the ecclesiastical regulation of Catholic marriage have both tended to become even more liberal, moving away from a uniformly legalistic understanding of marriage and toward a more person-centered theory and practice. In theology there has been a shift away from the nineteenth-century identification of the sacrament with the marriage bond or contract and toward a more liturgical and scriptural identification of the sacrament with the marriage itself. According to Edward Schillebeeckx the wedding ritual should be an occasion for personally encountering the felt reality of divine love in and through the human love that two people have for each other, and for affirming the meaning of Christian marriage as a union in covenant and cooperation. But beyond that, the marriage itself should continue to be a sacramental sign of God's redeeming activity in human life and of the fidelity and devotion between Christ and the church. Karl Rahner takes this a step further and sees Christian marriage as a unique sign of the incarnation, of the mystery that the transcendent reality of God became flesh in the person and life of Christ, just as men and women incarnate the transforming reality of divine grace in their total love for one another. Marriage is therefore a way in which the church, as Christ's continued presence on earth, comes into being; it is an actualization of the nature of the church in and through the everyday incarnate love that married persons have for each other. For Bernard Cooke, too, love in the family is a sacrament of divine love in the same way that Jesus was a sacrament of God to those who knew him, and in the same way that the early Christian community was a sacrament of Christ to the ancient world. On the one hand sexual love is a natural symbol of the life-giving power of divine love in that it is full of vitality and leads to the creation of new life. On the other hand the fidelity and care that two persons have for each other symbolizes the transcendent meaning that God's love has for all persons, and in the measure that that meaning shines forth from the shared life of married persons their marriage is a sacrament to others of the transforming power of grace.

Other theologians and even canonists have been asking about the practical implications of this new sacramental view of marriage. Is it still mecessary to say, for example, that the sacrament is identical with the marriage contract, as Pius IX claimed, and that all the sacramental effects of marriage come from giving one's consent in the wedding ceremony? And is it necessary to hold that the bond of marriage is a metaphysical entity as Augustine and the medieval theologians believed, which cannot be broken except by the death of one of the spouses? If it is not, is the indissolubility of Christian marriage a moral ideal rather than a divine law, as scholars suggest that it was at the time of Jesus and during the first centuries of Christianity? Some have proposed a return to the tradition of the fathers which has been preserved in the Orthodox churches, which regard the permanence of marriage as a norm that allows exceptions in certain cases. Others have proposed even more radically that a marriage is sacramental when it embodies and expresses the kind of love that exists between Christ and the church, and that such a marriage would necessarily be permanent because the love within it would be faithful, forgiving, and self-sacrificing. But if the marriage no longer embodied and expressed that kind of love, it would in fact be no longer sacramental, and by the same token it would be liable to end in divorce.

Divorce among Catholics

Married Catholics, however, have not waited for such theological suggestions to win official church approval, and the divorce rate among American Catholics, for example, is now almost as high as it is for other Americans. Nevertheless, according to present canon law they are not free to remarry and so if they do they are automatically excluded from the sacraments and denied reconciliation until they renounce their adultery. For many of them the matter simply ends there, and they cease being Catholics. But others who have wanted to be reinstated into full membership in the church have tried in increasing numbers to have their first marriage officially annulled.

From the time of the Council of Trent the grounds for annulment were quite well defined and they were interpreted rather strictly. Marriage contracts could be declared null and void only if one of the parties did not fully consent to the marriage (for example, if they were coerced), or if they were not able to fulfill their marital obligations (for example, if they were impotent), or if they had not received a dispensation from one of the canonical impediments to a valid marriage (for example, if they were too closely related). In recent years, however, ecclesiastical courts have been interpreting the grounds for annulment more broadly to mean that if a person was psychologically unable to give a full and mature consent or was emotionally incapable of making a lifetime commitment to another person, then a valid marriage may not have existed from the very beginning. But these broader grounds are quite vague, and they are now in fact sometimes used to nullify marriages which have fallen apart for many of the same reasons that civil courts recognize in granting divorces.

At the present time, then, the only way that Catholics can remarry and remain in the church is to have their first marriage annulled. It is a lengthy and involved process, and despite the willingness of canon lawyers to examine every case for possible grounds, most legally divorced Catholics are either unaware that they might be able to obtain an annulment, or they are unwilling to put themselves through the equivalent of another divorce trial, this time in an eccledisastical court. Over and above this, canonists themselves have noted that if more Catholics petitioned to have their marriages annulled the courts would simply be unable to handle the case load. Some have questioned the way that marriages are now sometimes annulled through what amounts to legal loopholes in canon law, and have been wondering whether the Catholic hierarchy should recognize that church annulments are often being used to legitimate civil divorces. Still others, a minority, have questioned the value and relevance of the whole ecclesiastical judicial system which researches and tries marriage cases, and have recommended that it be dismantled. They would allow marriage and divorce to revert back to civil control, as in the early church, and have the clergy concern themselves only with the pastoral and religious aspects of Christian marriage.

There does not seem to be an easy way out of the dilemma. For the past eight hundred years or so the Catholic church has vigorously maintained that a validity contracted marriage is indissoluble not only by church law but also by divine law. It has become a part of Catholic doctrine, and to change it would call into question the infallibility of the church. During the same period of time the church's judicial system has grown into an elaborate structure of laws and courts intended to safeguard the integrity of marriage. It has become a part of the institutional church, and so to change it would call for a more radical reappraisal than the one that occurred at Vatican II. On the other hand, the historical justification for the present doctrine and judicial system have been called into question by biblical and patristic research. The theological justification for the metaphysical permanence of the marriage bond has been lost in the shift from scholastic to personalist philosophy, and it has been called into question by the view that the sacrament of marriage is not a legally binding contract but a living relationship between two married people. And the practical justification for the impossibility of divorce has been called into question by the fact that the prohibition no longer deters Catholics from obtaining divorces but rather prevents them from remaining Catholics.

But if the dilemma cannot be resolved, steps can be taken to avoid it. The Catholic church has been doing much more than in the past to prepare its members for marriage. In America, for example, Catholic high schools and colleges teach courses not only on the theology of marriage but also on the practical requirements and consequences of being married. Catholic dioceses offer "Pre-Cana Conferences" for engaged couples, and parish priests counsel them about the duties and responsibilities of marriage as well as help them prepare their wedding liturgy. The Christian Family Movement, Marriage Encounter, and similar organizations offer group support for maintaining married and family life. Yet at the same time, these official and unofficial programs reach only a fraction of the Catholics who are about to be married or who are already married. It would seem that much remains to be done if the dilemma is to be averted.

Conclusion

And what of marriage as a "door to the sacred"? Is it that now? Or has it ever been that? It is, and it has been, despite the fact that much of the history of marriage in the Catholic church has been a legal history, and despite the fact that the sacramental theology of marriage was for a long time formulated in juridical terms. Marriage has been a sacrament in the broad sense in two different ways. One of them began in the patristic period, the other in the Middle Ages.

In the very first centuries, Paul's passage on marriage in Ephesians was not used to exemplify the union between Christ and the church, but rather Christ's sacrifice for the church and the church's obedience to Christ were used to exemplify how husbands and wives would behave toward one another. Thus marriage was considered to be sacred but not sacramental since the starting point of the analogy was the divine relationship, which was then used to illustrate the human relationship. Augustine, however, took the analogy both ways: he saw the human relationship as a sign of the divine relationship, and the divine relationship as a sign of the human relationship. Still, his emphasis was on the latter, and so he used the everlasting union between Christ and the church to argue for a permanent union between husband and wife. In the Middle Ages the Pauline analogy continued to be taken mainly in this way, but once marriage was understood to be a sacrament in the strict Catholic sense theologians began to view it also as a sacrament in the broader sense of being a sign of the incarnate spiritual union of Christ and the church. This way of looking at marriage continued in the modern church, and in recent times it has even been emphasized by Catholic theologians who want to affirm the sacramentality of marriage while avoiding the legalism of earlier sacramental theology. Today the analogy is used in books and sermons to bring Catholics to a deeper awareness of what their relationship to Christ and to each other is and ought to be.

The second way in which marriage has been and remains a sacrament in the broad sense is in the sacramental wedding ceremony. It began in the Middle

Ages, evolved through a variety of forms, became stabilized during the Tridentine reforms, and is now evolving again. But wedding ceremonies are always sacramental, at least in the broad sense of being celebrations of the sacred value of marriage, whatever it may be in a given culture, as well as in the sense of being rituals of initiation to a new style of life is honored and meaningful, supported by social custom and religious tradition. In these same ways Christian wedding ceremonies have always been sacramental, for they have celebrated the sacred value of marriage in a Christian culture, and they have initiated men and women into a style of life that was to be modeled on the relationship between Christ and the church. In medieval and modern times that life-style was often understood to be authoritative and faithful on the part of the husband and submissive and dutiful on the part of the wife, for Christ was seen as Lord and master and the church was seen as servant and mistress. In contemporary times that life-style is more often understood to be one of constant self-giving on the part of both husband and wife, for Christ and the church are both seen as incarnations of transcedent love. But however the value and meaning of Christian marriage are understood, the wedding ceremony is always an important and meaningful occasion. Its words and gestures, even the bearing and expressions of its participants, symbolize to the bride and groom and the others who are present the meaning and importance of what is happening and what is about to happen to this couple. They are being transformed, and they are going to be transformed even further. And the wedding is a door through which they enter into that sacred transformation.

II

Contemporary Perspectives on the Theology of Marriage, Commitment, and Fidelity

4.

Christian Marriage: Basic Sacrament

Bernard Cooke

It was only in the late twelfth and early thirteenth centuries that marriage came to be officially regarded as a sacrament. The Council of Trent (1563) and Vatican II (1962–65) reaffirmed this official teaching. However, relatively early in the history of Christianity, marriage was regarded as a sacrament in the broader sense. The marriage ceremony was only one important element within a larger sacramental process. Bernard Cooke retrieves this ancient perspective, which receded with the legalism of the Counter-Reformation.

The starting point for sacramental reflection is the human context of love and friendship. In filial and loving relationships, God's presence unfolds. Marriage is a paradigm of human friendship and personal love. It touches the most basic level of life. It is, as the author suggests, the basic sacrament of God's presence to human life. In their relationship to one another, the couple are a sacrament for each other, for their children, and for all who come to know them. They "give grace" to one another.

The reader will find in this essay the best of personalistic philosophy, the sacraments viewed as process, and marriage contextualized in a redemptive setting of love and friendship.

In the traditional short definition of Christian sacrament, the third element is a brief statement about the effectiveness of sacraments: "Sacraments are sacred signs, instituted by Christ, to give grace." Sacraments are meant to do something. What they do is essentially God's doing; in sacraments God gives grace. Before looking at the sacramentality of human friendship and of marriage in particular, it might help to talk briefly about the kind of transformation that should occur through sacraments.

In trying to explain what sacraments do, we have used various expressions: celebrants of sacraments "administer the sacraments" to people; people "receive sacraments" and "receive grace" through sacraments; sacraments are "channels

Chapter 7 of *Sacraments and Sacramentality* (Mystic, Connecticut: Twenty-Third Publications, 1983), pp. 79–94. Published by Twenty-Third Publications, P.O. Box 180, Mystic, CT 06355 (paper, 256 pp., $9.95).

of grace.'' The official statement of the Council of Trent, which has governed Catholic understandings for the past four centuries, is that ''sacraments contain and confer grace.''

The traditional understanding of grace and sacraments would include at least the following. The grace given was won for us by the death and resurrection of Jesus. Without depending upon misleading images such as ''reservoir'' or a ''bank account,'' it seems that there must be some way that the graces flowing from Jesus' saving action are ''stored up'' so that they can be distributed to people who participate in sacraments. The grace given in sacramental liturgy is, at least for baptized Christians, a needed resource if people are to behave in a way that will lead them to their ultimate destiny in the life to come.

Beneath all such formulations—which we are all familiar with in one form or another—there lurks a basic question: What is this ''grace'' we are speaking about? It is all well and good to say that we receive the grace we need when we come to sacramental liturgy, and that we receive it in proportion to our good will. But what do we have in mind when we use this word ''grace''? We have already begun to see that ''sacrament'' should be understood in a much broader sense, one that extends far beyond the liturgical ceremony that is the focus of a particular sacramental area. Now, with grace also, a deeper examination leads us to the conclusion that grace touches everything in our lives; it pervades everything we are and do.

In trying to get a more accurate notion of grace, it might help to remember a distinction that was sometimes made in technical theological discussions, a distinction that unfortunately received little attention and so was scarcely ever mentioned in catechetical instructions about grace. This is the distinction between ''uncreated grace'' and ''created grace.''

''Uncreated grace'' refers to God himself in his graciousness towards human beings; ''created grace'' refers to that special (''supernatural'') assistance God gives to humans to heal and strengthen them and to raise them to a level of being compatible with their eternal destiny. For the most part, our previous theological and catechetical explanations stressed created grace as a special help that enabled persons to live morally good lives, an assistance to guide and support them when they faced temptations. There was also a frequent reference to ''the state of grace,'' the condition of being in good relationship to God and therefore in position to move from this present earthly life to heaven, rather than to hell. But there was practically no mention of uncreated grace.

During the past few decades, there has been a renewed interest in and study of grace. We have learned to pay much more attention to uncreated grace, that is, to the reality of God who in the act of self-giving and precisely by this self-giving transforms and heals and nurtures our human existence. Along with this new emphasis on God's loving self-gift as *the* great grace, there has been more use of the notion of ''transformation'' to aid our understanding of created grace. Under the impact of God's self-giving, we humans are radically changed; this fundamental and enduring transformation of what we are as persons is created, sanctifying grace.

In various ways, sacraments—in their broader reality as well as in their liturgical elements—are key agencies for achieving this transformation. Though the effectiveness of the different sacraments is quite distinctive, each area of sacramentality touches and changes some of the significances attached to human life. As these significances are transformed, the meaning of what it is to be human is transformed; our human experience is therefore changed, and with it the very reality of our human existing.

This process of transformation is what we now turn our attention to, hoping to discover what sacraments are meant to accomplish in the lives of Christians.

Sacrament of Human Friendship

Explanation of the individual sacraments traditionally starts with baptism. Apparently it is the first sacrament Christians are exposed to, and the one all the others rest upon; it is the one that introduces the person to Christianity, etc. However, as we attempt to place the sacraments in a more human context, there is at least the possibility that we should begin with another starting-point. Perhaps the most basic sacrament of God's saving presence to human life is the sacrament of human love and friendship. After all, even the young infant who is baptized after only a few days of life has already been subjected to the influence of parental love (or its lack), which in the case of Christian parents is really the influence of the sacrament of Christian marriage.

Sacraments are meant to be a special avenue of insight into the reality of God; they are meant to be words of revelation. And the sacramentality of human love and friendship touches the most basic level of this revelation. There is a real problem in our effort to know God. Very simply put, it seems all but impossible for humans to have any correct understanding of the divine as it really is. God is everything we are not. We are finite, God infinite; we are in time, God is eternal; we are created, God is creator. True, we apply to God the ideas we have drawn from our human experience; we even think of God as "person." But is this justified? Is this the way God is?

Some fascinating and important discussion of this problem is going on today among Christian philosophers, but let us confine our approach to those insights from the biblical traditions. As early as the writings of the first chapter of Genesis (which is part of the priestly tradition in Israel that found final form around 500 B.C.), we are given a rich lead. Speaking of the creation of humans by God, Genesis 1:19 says that humans were made "in the image and likeness of God." That is to say, somehow the reality of human persons gives us some genuine insight into the way God exists. But the passage continues—and it is an intrinsic part of the remark about "image and likeness"—"male and female God made them." This means that the imaging of God occurs precisely in the relationship between humans, above all in the interaction of men and women. To put it in our modern terms, some knowledge of the divine can be gained in experiencing the personal relationship of men and women (and one can legitimately broaden that to include all human personal relationships).

The text provides still more understanding, for it points out that from this relationship life is to spread over the earth; humans in their relation to one another (primarily in sexual reproduction, but not limited to that) are to nurture life. And humans are to govern the earth for God; they are to image and implement the divine sovereignty by this nurture of life that is rooted in their relationship to one another. As an instrument of divine providence, human history is meant by the creator to be effected through human community, through humans being persons for one another.

Though the first and immediate aspect of the relationship between Adam and Eve as life-giving is their sexual partnership, the text does not confine it to this. Rather, Genesis goes on to describe the way Adam's own human self-identity is linked with Eve's. As Adam is given the chance to view the other beings in God's creation, he is able to name them, but he is unable to name himself until he sees Eve. The very possibility of existing as a self is dependent upon communion with another.

Implicit in this deceptively simple biblical text is a profound statement about the way human life is to be conducted. If life is to extend to further life, either by creating new humans or by creating new levels of personal life in already-existing humans, it will happen on the basis of people's self-giving to one another. And, if women and men are truly to "rule" the world for God, they will do this by their love and friendship, and not by domination. To the extent that this occurs, the relationship of humans to one another will reveal the fact that God's creative activity, by which he gives life and guides its development (in creation and in history), is essentially one of divine self-gift. Humans have been created and are meant to exist as a word, a revelation, of God's self-giving rule; but they will function in this revealing way in proportion to their free living in open and loving communion with one another.

Whatever small hint we have regarding the way God exists, comes from our own experience of being humanly personal. Our tendency, of course, is to think of the divine in human terms, even carrying to God many of the characteristics of our humanity that obviously could not apply directly to God, for example, changing our minds as to what we intend to do. Excessive anthropomorphism has always been a problem in human religious thinking and imagination; we have always been tempted by idolatry. Even today, when our religious thinking has been purified by modern critical and scientific thought, we still fall into the trap of thinking that God exists in the way we think God does. This does not mean, however, that we must despair of ever knowing God. On the basis of biblical insights (like those in Genesis 1:19 and even more in New Testament texts grounded in Jesus' own religious experience) we can come to some true understanding of God by reflecting on our own experience of being personal.

For us to be personal—aware of ourselves and the world around us, aware that we are so aware, relating to one another as communicating subjects, loving one another, and sharing human experience—is always a limited reality. We are personal within definite constraints of time and place and happenings. Even if our experience as persons is a rich one, through friends and education and cultural opportunities, it is always incomplete. For every bit of knowledge there are immense areas of reality I know nothing about; I can go on learning indefinitely.

Though I may have a wide circle of friends, there are millions of people I can never know; I can go on indefinitely establishing human relationships. There are unlimited interesting human experiences I will never share. In a sense I am an infinity, but an infinity of possibilities, infinite in my completeness. Yet, this very experience of limitation involves some awareness of the unlimited; our experience of finite personhood points toward infinite personhood and gives us some hint of what that might be.

God Reveals Self as Personal

What lets us know the divine is indeed personal in this mysterious unlimited fashion is the fact (which as Christians we believe) that this God has "spoken" to humans; God has revealed, not just some truths about ourselves and our world, but about God's own way of being personal in relation to us. God in the mystery of revelation to humans is revealed as someone. What this means can be grasped by us humans only through our own experience of being human together. In our love and concern for one another, in our friendships and in the human community that results, we can gain some insight into what "God being for us" really means. These human relationships are truly insights into God, but not just in the sense that they are an analogue by which we can gain some metaphorical understanding of the divine. Rather, humans and their relationships to one another are a "word" that is being constantly created by God. In this word God is made present to us, revealing divine selfhood through the sacramentality of our human experience of one another.

One of the most important results of this divine revelation and genuinely open relationships to one another is the ability to trust reality. This might seem a strange thing to say, for reality is a given. Yet, the history of modern times has been one of growing uncertainty and strong distrust of the importance and goodness and even the objective reality of the world that surrounds us, the world of things and especially the world of people. Great world wars, among other things, have made many humans cynical about human existence and have made many others unwilling to admit that things are as they are. There is abundant evidence that our civilization is increasingly fleeing towards phantasy, taking refuge in a world of dreams, so that it does not have to face the real world. It is critically important, perhaps necessary for our sanity, that we find some basis for trusting life and facing reality optimistically and with mature realism.

Most radically, a culture's ability to deal creatively with reality depends on its view of "the ultimate," of God. We must be able to trust this ultimate not only as infinitely powerful but also as infinitely caring, as compassionate and concerned. The only ground, ultimately, for our being able to accept such an incredible thing—and when we stop to reflect, it is incredible—is our experience of loving concern and compassion in our human relationships. If we experience the love and care that others have for us, beginning with an infant's experience of parental love, and experience our own loving concern for others, this can give us some analogue for thinking how the ultimate might personally relate to us. Jesus himself drew from this comparison. "If you who are parents give bread and not a stone to your children when they ask for food, how much more your Father in heaven. . . ."

Experiencing love in our human relationships makes it possible for us to accept the reality of our lives with a positive, even grateful attitude. And this in turn makes it possible for us to see our lives as a gift from a lovingly providential God. If we have friends, life has some basic meaning; we are important to them and they to us. What happens to us and them makes a difference; someone cares. If love exists among people, there is genuine, deep-seated joy, because joy shared by people is the final dimension of love. If this is our experience of being human, then our existence can be seen as a good thing and accepted maturely and responsibly.

All of this means that our experience of being truly personal with and for one another is sacramental; it is a revelation of our humanity at the same time that it is a revelation of God. This experience of human love can make the mystery of divine love for humans credible. On the contrary, if a person does not experience love in his or her life, only with great difficulty can the revelation of divine love be accepted as possible. Learning to trust human love and to trust ourselves to it is the ground for human faith and trust in God.

To say that human love is sacramental, especially if one uses that term strictly (as we are doing), implies that it is a mystery of personal presence. Obviously, in genuine love there is a presence of the beloved in one's consciousness; the deeper and more intimate the love, the more abiding and prominent is the thought of the beloved. To see this as truly sacramental of divine presence means that human love does more than make it possible for us to trust that God loves us. The human friendships we enjoy embody God's love for us; in and through these friendships God is revealing to us the divine self-giving in love. God is working salvifically in all situations of genuine love, for it is our consciousness of being loved both humanly and divinely that most leads us to that full personhood that is our destiny. Such salvation occurs in our lives to the extent that we consciously participate in it, in proportion to our awareness of what is really happening and our free willingness to be part of it.

It is instructive to note that when Jesus, immediately after being baptized by John, was given a special insight into his relationship to God as his Abba, the word used in the gospel to describe his experience of his Father's attitude towards him is the Greek *agapetos,* "my beloved one." This was the awareness of God that Jesus had, an awareness of being unconditionally loved, an awareness that became the key to human salvation. And John's Gospel describes Jesus at the last supper as extending this to his disciples. "I will not now call you servants, but friends."

Marriage, Paradigm of Friendship

Among the various kinds of human friendship and personal love, the one that has always been recognized as a paradigm of human relationship and love, and at the same time a ground of human community, is the relation between husband and wife. There is considerable evidence that humans have never been able to explain or live this relationship satisfactorily, basic and universal though it is. In our own day, there is constant and agitated discussion of the way men and women are meant to deal with one another, and there is widespread talk of a radical shift taking place in the institution of marriage. As never before, the assumptions about respective roles in marriage are being challenged. Marriage is seen much more as a free community of persons rather than as an institution of human society regulated for the general benefit of society; equality of persons rather than respect for patriarchal authority is being stressed. And with considerable anguish in many instances, people are seeking the genuine meaning of the relation between women and men, and more broadly the relationship of persons to one another in any form of friendship.

Questioning the woman-man, and especially the husband-wife, relationship is not, of course, a new phenomenon. As far back as we can trace, literature witnesses to the attempt to shed light on this question. What complicates the issue is the merging of two human realities, sexuality and personal relatedness, in marriage, a merging so profound that people often are unable to distinguish them. We know, however, that in many ancient cultures there was little of what we today consider love between spouses; marriage was a social arrangement for the purpose of continuing the family through procreation. In not a few instances, there was so pronounced a cleavage between love and sexuality that the wife was considered the property of her husband and she was abandoned if she proved unable to bear him children. If men sought human companionship, they sought it outside the home. Apparently the marriages in which something like a true friendship existed between wife and husband were relatively rare.

Sacramentality of Marriage in Israel

In ancient Israel an interesting development began at least eight centuries before Christianity. Surrounded as they were by cultures and religions that worshipped the power of human sexuality, the Israelites assiduously avoided attributing anything like sexuality to their God, Yahweh. At the same time, these neighboring erotic religions were a constant temptation to the Israelites; the great prophets of Israel lashed out repeatedly against participation by Israel's women and men in the ritual prostitution of the Canaanite shrines. In this context it is startling to find the prophet Hosea using the example of a husband's love for his wife as an image of Yahweh's love for his people Israel.

Apparently, Hosea was one of those sensitive humans for whom marriage was more than a family arrangement; he seems to have had a deep affection for his wife, Gomer. The love was not reciprocated; his wife abandoned him for a life of promiscuity with a number of lovers; perhaps she became actively involved in some situation of shrine prostitution. At this point, Hosea was obliged by law to divorce her, which he seems to have done. But then "the word of the Lord came to Hosea," bidding him to seek out and take back his errant wife. And all this as a prophetic gesture that would reveal Yahweh's forgiveness of an adulterous Israel that had gone lusting after false gods.

Once introduced by Hosea, the imagery of husband-wife becomes the basic way in which the prophets depict the relationship between Yahweh and the people Israel. Tragically, the image often has to be used in a negative way. Israel is the unfaithful spouse who abandons Yahweh to run off with "false lovers," the divinities of the surrounding fertility religions. Yet, despite this infidelity on Israel's part, Yahweh is a merciful God who remains faithful to his chosen partner. "Faithful" becomes a key attribute of this God of Israel. Yahweh is a faithful divinity who keeps his promises to Israel. And the husband-wife relation becomes in the prophetic writings an alternative to the king-subject relation that the rulers of Israel and Judah (for their own purposes) preferred as a way of describing the covenant between Yahweh and Israel.

Our particular interest, however, is not the manner in which the use of the husband-wife imagery altered Israel's understanding of the covenant between people and God. Rather, it is the manner in which, conversely, the use of imagery began to alter the understanding of the relation between a married couple. If the comparison husband-wife/Yahweh-Israel is made, the significance of the first couplet passes into understanding the significance of the second couplet, but the significance of the second passes also into understanding the first.

The understanding the people had of their god, Yahweh, and of his relationship to them, the depth and fidelity of his love, the saving power of this relationship, slowly became part of their understanding of what the marriage relationship should be. Thus, a "Yahweh-significance" became part of the meaning of married relatedness. The sacramentality of the love between husband and wife—and indirectly the sacramentality of all human friendship—was being altered. It was, if we can coin a term, being "yahwehized." The meaning of God in his relationship to humans became part of the meaning of marriage, and marriage became capable of explicitly signifying and revealing this God. This meant that human marriage carried much richer significance than before; it meant that the personal aspect of this relationship was to be regarded as paramount; it meant that the woman was neither to be possessed as property nor treated as a thing; it meant that the marital fidelity was expected of both man and woman. Thus the "institution of the sacrament of marriage" begins already in the Old Testament.

Marriage as a Christian Sacrament

With Christianity another dimension of meaning is infused into this relation between wife and husband, the Christ-meaning that comes with Jesus' death and resurrection. Several New Testament passages could be used to indicate this new, deeper meaning, but the key passage probably is the one in Ephesians that traditionally forms part of the marriage liturgy.

> Be subject to one another out of reverence for Christ. Wives, be subject to your husbands as to the Lord; for the man is the head of the woman, just as Christ also is

> the head of the Church. Christ is indeed the savior of the body; but just as the Church is subject to Christ, so must wives be subject to their husbands. Husbands, love your wives, as Christ also loved the Church and gave himself up for it, to consecrate it. . . . In the same way men are bound to love their wives, as they love their own bodies. In loving his wife, a man loves himself. For no one hates his own body: on the contrary, he provides and cares for it; and that is how Christ treats the Church, because it is his body, of which we are living parts. Thus it is that (in the words of scripture) ''a man shall leave his father and mother and shall be joined to his wife, and the two shall become one flesh'' (5:21–32).

In dealing with this text it is important to bear in mind what the author of the epistle is doing. As so often in the Pauline letters, the purpose is neither to challenge nor to vindicate the prevailing structures of human society as they then existed. Just as in other cases the Pauline letters do not argue for or against an institution like slavery, the passage in Ephesians takes for granted the commonly accepted patriarchal arrangements of family authority without defending or attacking them; in a partriarchal culture all authority is vested in the husband-father. However, Ephesians insists that in a Christian family this authority structure must be understood and lived in an entirely new way. The relation between the risen Christ and the Christian community must be the exemplar for a loving relationship between the Christian couple.

This text contains a rich treasure of sacramental and Christological insight that has scarcely been touched by theological reflection. Mutual giving of self to one another in love, not only in marital intercourse but also in the many other sharings that make up an enduring and maturing love relationship, is used in this passage as a way of understanding what Jesus has done in his death and resurrection. He has given himself to those he loves. His death was accepted in love as the means of passing into a new life that could be shared with those who accept him in faith. Jesus' death and consequent resurrection was the continuation of what was done at the supper when Jesus took the bread and said, ''This is my body (myself) given for you.'' Ephesians 5 tells us that we are to understand this self-giving of Jesus in terms of the bodily self-giving in love of a husband and wife, and vice versa, we are to understand what this marital self-gift is meant to be in terms of Jesus' loving gift of self in death and resurrection.

One of the most important things to bear in mind in studying this text is that Jesus' self-giving continues into the new life of resurrection. Actually, his self-giving is intrinsic to this new stage of his human existence. The very purpose and intrinsic finality of his risen life is to share this life with others. The risen Lord shares this resurrection life by sharing what is the source of this life, his own life-giving Spirit. For Jesus to exist as risen is to exist with full openness to and full possession of this Spirit. So, for him to share new life with his friends means giving them his own Spirit. What emerges from this Spirit-sharing is a new human life of togetherness, a life of unexpected fulfillment, but a life that could not have been reached except through Jesus freely accepting his death. So also, a Christian married couple is meant to move into a new and somewhat unexpected common existing, which cannot come to be unless each is willing to die to the more individualistic, less unrelated-to-another, way of life that they had before.

Christ's self-giving to the church is more than the model according to which a man and woman should understand and live out their love for each other. The love, concern, and self-giving that each has for the other is a "word" that expresses Christ's love for each of them. The fidelity of each to their love is a sign that makes concretely credible their Christian hope in Christ's fidelity. In loving and being loved, each person learns that honest self-appreciation which is the psychological grounding for believing the incredible gospel of God's love for humankind. In their relationship to one another, and in proportion as that relationship in a given set of circumstances truly translates Christ's own self-giving, the couple are a sacrament to each other and a sacrament to those who know them.

In this sacramental relationship, a Christian man and woman are truly "grace" to one another; they express and make present that uncreated grace that is God's creative self-giving. Though there certainly is mystery in this loving divine presence, it is revealed in the new meanings discovered in the lived relationship between Christian wife and husband. The trust required by their unqualified intimacy with one another and the hope of genuine acceptance by the other, which accompanies this intimacy, help bring about a new level of personal maturity. But this trust and hope are grounded in the Christian faith insight that open-ended love can lead to new and richer life. Perhaps even more basically, a Christian couple can commit themselves to this relationship, believing that it will not ultimately be negated by death. Instead, Christian hope in risen life supports the almost instinctive feeling of lovers that "love is stronger than death."

Psychological studies have detailed the ways a truly mature married relationship, one that integrates personal and sexual love, fosters the human growth of the two people, and it is not our intent to repeat such reflections here. But these same studies point also to the indispensable role that continuing and deepening communication with each other plays in the evolution of such a relationship. In a Christian marriage the communication is meant to embrace the sharing of faith and hope in that salvation that comes through Jesus. The Christian family is meant to be the most basic instance of Christian community, people bonded together by their shared relationship to the risen Jesus.

All of us can think of marriages where this ideal has been to quite an extent realized, where husband and wife have over the years supported and enriched one another's belief and trust in the reality and importance of Christianity. Various challenges can come to Christian faith, if it is real faith and not just a superficial acceptance of a religious pattern. These challenges can change shape over the years, they can come with suffering or disappointments or disillusionment or boredom, they can come to focus with the need to face the inevitability of death. At such times of crisis, when faith can either deepen or weaken, the witness of a loved one's faith and hope is a powerful and sometimes indispensable preaching of the gospel.

Perhaps the most difficult thing to believe over the course of a lifetime is that one is important enough to be loved by God. Nothing makes this more credible than the discovery of being important to and loved by another human. The fidelity of one's lover, not just in the critically important area of sexual fidelity but also in the broader context of not betraying love by selfishness or exploitation or pettiness or dishonesty or disinterestedness or insensitivity, makes more credible the Christian trust in God's unfailing concern.

One could go on indefinitely describing how a Christian couple ''give grace'' to each other, because the contribution to each other's life of grace (their being human in relation to God) involves the whole of their life together. The sacrament of Christian marriage is much more than the marriage ceremony in the church; that ceremony is only one important element in the sacrament. Christian marriage is the woman and the man in their unfolding relationship to each other as Christians; they are sacrament for each other, sacrament to their children, and sacrament to all those who come to know them. The meaning of what they are for each other should become for them and others a key part of what it means to be a human being.

Summary

If we restrict ''sacrament'' to certain liturgical rituals, it is logical to think of baptism as the initial sacrament. If, however, we realize the fundamental sacramentality of all human experience and the way Jesus transformed this sacramentality, there is good reason for seeing human friendship as the most basic sacrament of God's saving presence to human life. Human friendship reflects and makes credible the reality of God's love for humans; human friendship gives us some insight into the Christian revelation that God is a ''self.''

Within human friendship there is a paradigm role played by the love between a Christian wife and husband. Building on the transformation of a marriage's meaning that began with the Israelitic prophets, Christianity sees the love relationship of a Christian couple as sacramentalizing the relationship between Christ and the church, between God and humankind. God's saving action consists essentially in the divine self-giving. This is expressed by and present in the couple's self-gift to each other; they are sacrament to each other, to their children, and to their fellow Christians. This sacramentality, though specially instanced in Christian marriage, extends to all genuine human friendship.

Questions for Reflection and Discussion

1. What is meant by ''uncreated grace''?
2. Why can we say that human love is the most basic sacrament?
3. What is the revelation expressed in Genesis 1:10?
4. How do we know that God is personal? What does ''personal'' mean when applied to God?
5. When and how was the sacrament of Christian marriage instituted?
6. The Christian couple are the sacrament of marriage. Explain.

5.

Marriage as Worship: A Theological Analogy

German Martinez

Students will need to persevere with this essay. However, uncovering the profound perspective at its core will be the gift and reward that comes from a close reading. Martinez draws a theological analogy between worship and marriage. Both are rooted in historical realities, and yet both transcend them. A coherence exists between marriage and worship from at least three major perspectives: symbolic action, depth of meaning, and self-transcendence. The lifelong process of marital initiation and communion should be seen within this fascinating and mysterious perspective of worship.

For the author the analogy with worship suggests fruitful theological concepts relevant in laying the foundation for a spirituality of marriage. Four vital areas, in particular, are addressed: agape-love, donation (daily dying and rising), communication, and faith sharing in the family. The intrinsic relationship and, in fact, the intersection of marriage and worship become embodied in these vital realities. They provide a fundamental orientation and unity of conjugal spirituality. They become the true hallmarks of an integrated spirituality of marriage.

The essay is challenging and unique in its basic thesis: Marital life and love is an icon, an act of worship, a grammar of godly love and faithfulness.

Marriage and worship are both primary and universal human realities where manifold dimensions converge: body and spirit, earthly elements and symbolic actions, natural phenomena and divine mystery. Though historical by their nature, and always historically related, they essentially transcend the realities they embody. The pledge of the old Anglican wedding ritual, "with my body I thee worship," speaks profoundly of this central meaning. Like worship, which etymologically means "ascribing worth to another being," marriage is the total

From *Worship* 62:4 (1988): 332–353. Reprinted with permission.

validation of the other in the devotion and service, celebration and mystery of a relationship.[1]

Worship, therefore, can be applied to marriage analogously. This analogy suggests theological concepts which are most relevant in articulating the core of a much needed conjugal spirituality. Marriage and worship are, in fact, a typical case of coherence, even at the strictly anthropological level, from at least three major perspectives: symbolic action, depth of meaning, and self-transcendency.[2]

Symbolic actions are the language people use to express the experience of both shared beliefs through worship or the shared love of a marital relationship. The nature, the inner reality, of a marital relationship is to be rooted in love, as the nature of worship is to be grounded in faith. Without belief worship is meaningless; without love marriage is empty. In both cases there is a symbolic and existential relationship. Worship embodies an inner life, a belief through rituals, since the ritual is the cradle of religious belief. It has the proper language and external gesture of inner realities—as does a marital partnership. Its verbal or bodily expressions, its quality and reality of intimacy, its celebrations and struggles, in summary, the whole interaction of self-giving and self-serving relationship of a man and a woman in marriage stems from an inner reality of that love. These are—or are not—embodied in love. Spirituality depends on the truthfulness of that relationship whether in marriage or in worship, and that truthfulness will be apparent in the symbolic and existential language of the relationship.

The depth of meaning is envisioned in the many possible orders of reality to which marriage and worship can refer. That is precisely why symbolic actions and language come into the picture. The ordinary experience of a community of love, or a community of faith, provides the metaphors people use to express their value, and the prophetic revelation describes the covenant of God to his people. Marriage and worship are, in fact, so complex and rich that they are very difficult to define. We all have experienced love—and the primary end of marriage—and we all participate in ritual actions, but we take pains to share with others those kinds of experiences. In this aspect the complexity of love and of religious ritual is very clear. This fact bespeaks their intrinsic link and mutual support in various levels of meaning. "Love," says L. Mitchell, "whether of God or of the girl next door, is all but impossible to express except through outward symbolic action, that is, through ritual acts."[3]

Finally, self-transcendency intrinsically defines worship as it does marriage. Any definition of worship, even from the point of view of many anthropologists, centers around this essential dimension, as expressed in Evelyn Underhill's classical definition, "the response of the creature to the Eternal."[4] Similarly, the total reality of a community of marital life and love cannot be understood without the vertical dimension of that created love, the beyond experienced within the encounter between I and Thou. This intimate encounter by its dynamism is enclosed by and moves toward the ultimate mystery, God himself. God becomes the basis of the encounter, as Martin Buber's interpersonal principle of the external Thou demonstrates.[5] Charles Davis, in his classic study of the nature of religious experience and feeling, reaches the same conclusion from a different perspective. "Human feelings," he says, " by their dynamism, point beyond themselves; they are an expression of self-transcendence."[6] The self-giving marital love, as a symbol of divine love, straining toward God, reveals and celebrates the mystery of God; this is worship. True marriage, like true worship, is rooted primarily, therefore, in self-transcendent love. They both relate to something in the foundations of human consciousness and point beyond their reality. God is thus made present.

This analogical relationship between marriage and worship, which, as stated above, includes symbolic action, depth of meaning and self-transcendency, underlies the biblical understanding of marriage. Conjugal spirituality can only be properly understood against this background. It is against this background of an eminently human love that the biblical perspective of God's covenant reveals its full depth of meaning.

God's nuptial relationship with his people in the Old Testament becomes the paradigm of the most intimate human relationship. The human experience of love parallels that relationship of fidelity and becomes the most telling metaphor of Israel's covenant. "Marriage, then, is the grammar that God uses to express his love and faithfulness."[7] Similarily, in the New Testament perspective, Christ, the lover and husband of humanity,[8] renews the covenant by his sacrifice and becomes the paradigm of Christian marriage. The nuptial reality between Christ and his church is the actual foundation and source of the Christian spirituality of marriage. The Spirit of the risen Lord inspires the daily living of the spouses in a personal and faithful self-donation.[9]

Worship, more accurately spiritual worship exemplifies the ideals of the covenant and consequently of marriage. Israel is called to offer a spiritual worship of obedience to the word and to maintain God's covenant, because this covenant is established after Israel's liberation and passage to faith in Yahweh's love and faithfulness. The "chosen and priestly people" acknowledge this covenant with a life and service of spiritual worship.[10] The prophets and the psalmists use this same language of worship and marital relationship to dramatize that love between Yahweh and Israel.[11]

Christ renews the liturgical conception of the prophets "in spirit and in truth." He dies to establish a new covenant. Very often the liturgical symbolism of the New Testament is rooted in the prophets. The reality of the covenant, definitely concluded in Jesus Christ, is evoked by Paul in Ephesians to establish a parallel between marriage and the union of Christ with his church. The reference here to baptism and the eucharist in the context of the mystery of Christ's cross exemplifies once more the spiritual foundation of marriage as spiritual sacrifice. In fact, Christians are called to be themselves a proclamation of praise of God's love, in the image of Christ, who offered himself up through the Holy Spirit, as the only fitting worship to God.[12]

This application of a biblical-liturgical theology to marriage fills the writings of the early church, which, in turn, inspired the Roman sacramentaries (5th–8th cent.). The later Roman ritual of marriage, however, beginning in the eleventh century makes only sporadic reference to this biblical theology and spirituality of marriage. In its place a contractual mentality arose.[13] In patristic theology, on the contrary, any aspect of marriage as a contract is notoriously absent. The spiritual perspective of marriage centers around the idea of the mystery of marriage seen in the nuptial and sacrificial relationship of Christ and the church. This relationship constitutes the source and foundation of the theology of marriage in the fathers' tradition.[14]

The Eastern Churches are a telling example of a profound and compelling theological spirituality of marriage along the biblical lines of the covenantal relationship of spiritual sacrifice. They have focused on the agape-sacrifice spirituality of marriage, which constitutes the mystical and sacramental reality of marriage. Man and woman's partnership of love is represented poetically as an icon of a mystery of praise. They are ministers in their priestly vocation of an offering of life and love to each other.[15]

In his classic research on marriage in the East, Cardinal Raes has shown the foundation of this biblical and liturgical spirituality. The splendid Eastern rituals present a balance between a highly positive view of conjugal life—an earthly type of spirituality—and a mystical, liturgical theology. Both horizontal and vertical dimensions are rooted in biblical symbolism. It is against the background of the human reality that the church's nuptial offering by Christ at the wedding banquet of the cross is conceived as the foundation of a conjugal spirituality: "Christ-husband, who had married the holy and faithful Church, and had given to her at the cenacle his body and blood . . . raise your right hand and bless husband and wife."[16]

This priestly ministry of Christ has nothing to do with sacred objects. By the same token, marriage as worship goes beyond the boundaries of created sacralization. The "one flesh" is holy from its roots only by the created act of God-agape. In this regard, Catholic theology has long ago acknowledged the impact of desacralization in the New Testament, which sees human values observed from the eschatological reality of Christ. "For his is the peace between us . . . and has broken down the barrier . . . destroying in his own person the hostility caused by the rules and decrees of the law" (Eph 2:14–15).[17]

Worship is, therefore, analogously applied to marriage only under this assumption. In its broad sense it embraces all the secular values, the day-to-day experience and the totality of the challenges of becoming one, if lived as a symbol and reality of the transcendent presence of God-agape. The mystery of life embodied in the mystery of Christ is the background and source of an inclusive spirituality of marriage which demands to be celebrated in the mystery of worship.

This divine dimension of the secular sphere of marriage is neither mythical sacralization nor mystical idealization. E. Schillebeeckx rightly states that "faith in Yahweh in effect 'desacralized,' or secularized, marriage—took it out of a purely religious sphere and set it squarely in the human, secular sphere."[18] Since marriage is foremost secular value, Christian spirituality rejects the false dichotomy between sacred and profane and views the presence of the holy in that secular value. Worship, with its secular character, provides the prism of a larger supernatural context.

Mystical idealization, on the other hand, whether concerned with worship or marriage, devalues the goodness and beauty of the order of creation. True religion is rooted in human experience and feeling, which are very important channels in the encountering of the mystery of the Holy. Davis accurately puts the question in this way: "How is it, then, that religion, mystical religion particularly, has—with reason—been seen as the enemy of the body and the affections?"[19] As people establish communication in worship with the absolute through symbols, so the partners are drawn into the inner life of God through the mutuality of their bodily and spiritual selves, the total being.

These misinterpretations are at the bottom of the ruinous cleavage between faith and life, and are the cause of alienation of many people from the church in the vital sphere of marriage. Furthermore, they run counter to a theology which presents marriage as the natural sign and milieu of God's saving and healing love.

The lifelong process of initiation in an intimate communion of marital life should thus be seen within the perspective of worship: a praise of God-agape. This worship perspective rightly generates a vital spirituality which is essential to a vital marriage, because spirituality is the totality of the personal experiences celebrated as a gift of God and in the intimacy with God which is the very meaning of worship. The spouses, in acknowledging God's intimate presence in each other, accept each other as his permanent gift and perfect each other in their own path of spirituality. This intimacy with God empowers them to a life of holiness, in liturgical terms, to be the "memory" of the Lord.

This theological analogy of worship which sacramentally subsumes all human experiences and from which true spirituality unfolds provides a fundamental orientation and unity of conjugal spirituality. It is not a comprehensive panorama of the complex challenges of marriage and family living, but it integrates four vital realities into a whole: agape-love, donation, communication, and faith sharing.

Agape of Intimacy, Meal and Eucharist

Agape defines the fundamental quality of love and has been applied to the eucharistic meal of intimate fellowship of early Christianity, celebrated at home. This new analogy of worship, based on the spirituality profound, symbolic and transcendent concept of agape, offers fruitful insights to understanding the core of marriage spirituality.

In fact, the sharing of self and the sharing of a meal in married life are not strictly utilitarian bodily functions only because of agape. Body and bread sharing in this respect are gift and communion, that is, agape. It is an intimate sharing which possesses a symbolic quality of donation and devotion, of reverence and care. The Christian eucharist, as agape-meal and agape-offering of the glorious Christ, builds upon the natural significance of that sharing of a fellowship meal. Consequently, mutual intimacy of body and bread are integral to the daily experience of conjugal spirituality, if lived in the spirit of agape as being "in the Lord" (1 Cor 7:39) and celebrated in the source and center of every Christian spirituality, the eucharist.[20]

A cleavage between faith and life in this most important reality of family life on which human growth and fulfillment depend has borne tragic consequences, especially in a world of great change and challenge. This is so because this cleavage cripples conjugal spirituality in its core: the life-giving and life-uniting reality of agape-love of which God himself is the ultimate reality and source, but which has to be lived in the concrete and the human. On one hand, this kind of love is an irreplaceable gift of being, understood and accepted in the deepest self in rejection of the collective depersonalization of modern society. On the other hand, this self-bestowal in the spirit of agape is the common source in which the spouses meet the Holy and their spirituality unfolds in the day-to-day experiences of life. The God-agape is present in both the exceptional and the common experiences of the spouses, whether in the care of a family meal, or the involvement of a psychological and sexual intimacy, or the entering into the self-offering of Christ. There are only three privileged moments from which an all-embracing home spirituality can be enriched.[21]

The importance of this line of thought stems from the clarifying nature of agape in regard to the ambiguities of love in our present culture. Serious psychological analysis provides the background for a solid understanding of the complex reality of love in answering the question: What is agape?

Agape means a way of being; it is a creative process; it never fails because it is unconditional, self-giving and self-sacrificing. This productive activity of loving is not a goddess, says Erich Fromm, because the worshiper of the goddess of love becomes passive and loses his power. In fact, no one can *have* love because it only exists in the *act of loving,* as Fromm rightly describes it.[22] The mode of "having," so pervasive in contemporary human relations, can be deadening and suffocating. The crises of intimacy with all its human counterfeits and psychological barriers depends on such an existential structure in human relations.

The presence of God makes possible the experience of transcendence through love. The New Testament presents this transcendence as a dynamic relationship with which a person is empowered by the mystery of the divine love. The biblical God is a God of intimacy and passion. In fact, God is simply love (1 Jn 4:8, 16).

Marriage, especially in the theology of John (Jn 2, 1–12) and Paul (Eph 5:25, 32), is a total and all-embracing communion after the image of the nuptial sacrifice of Christ, who gives up his nuptial body.[23] The philosophical and psychological analysis often misses this transcendent perspective, but corroborates the teaching of Scriptures on the same basic human realities of agape. Thus they are essential characteristics of a spirituality of marriage: unconditional, sacrificial, caring and self-renewing love. Thus "love never ends" (1 Cor 13) because it tells the other person that he or she will never die.[24]

Stressing the importance of the spiritual nature of agape-love, the reality of eros-love is reaffirmed because that transcendent agape would be meaningless without the sexual desire of eros. The exaggerated fear of "keeping perverse desire within its proper bounds"[25] divorced for centuries, at least in theory, sexual passion, tenderness and intimacy on one hand, and marriage spirituality on the other. J. Ratzinger draws, in this respect, a cogent biblical analogy: "As the covenant without creation would be empty, so agape without eros is inhuman.[26]

Persons are embodied spirits relating symbolically through the language of their body. Profound personal relations demand real personal intimacy in the integrity of a community of love.

A false dichotomy between spirit and flesh has undermined the truth of creation of male and female in the image of God. Furthermore, it has emasculated human sexuality, the capacity to love and procreate, in its physical and spiritual being and doing. The dialogic structure of sexual intimacy, in fact, complements and broadens the being of man and woman,[27] and at the same time, as a saving mystery in faith, symbolizes and causes union with God. Therefore, marital spirituality and sexual intimacy are not enemies but friends, because the author of life and love is present in their sacramental sexual experience. They are called to holiness in intimacy with the divine love through their celebration of the total commitment in a body, soul and spirit intercourse—''nakedness without shame.''[28]

Intimacy expresses the interpersonal capacity of a total life-sharing, not only through the eros of passion, warmth, and mature sexual encounter, but also through the agape of commitment, acceptance and self-disclosure. Intimacy cannot be reduced to tactility or even sexuality. It is a way of being and relating in closeness to the other in the life-process of creating a community. What Erik Erikson and other social scientists state concerning the conditions for interpersonal intimacy has to be taken seriously in the theological underpinning of conjugal spirituality.[29] The spirituality of marriage takes into account what human scientists have to say because of the secular nature of marriage. They present the real and fertile ground in which spouses meet the Holy. Those human conditions, like the attitude of reverence and devotion, the ideal of unselfish love and spiritual nakedness, the spontaneity of freedom and authenticity, all become true hallmarks of conjugal spirituality. By the same token, competition and stagnation, angelism and repression are rejected. To live humanly in the authentic faith of the passionate and committed God of the covenant is the highest form of intimacy.

In fact, no one has better captured this shared intimacy of God as a lover with a passion for truth and life than the prophets, the Song of Songs, the psalmists, or the Christian wisdom of the mystics like Teresa of Avila and John of the Cross.[30] A correlation exists between God's touching us deeply and wholly and the experience of human intimacy. To discover this correlation is to discover the key to marital spirituality.

Existential intimacy is, therefore, absolutely necessary for human life. Paradoxically, many people have lost the capacity to share intimacy because of a highly competitive individualism, and this speaks to us of the need to stress it in reference to family life. It calls for total sexual intimacy in marriage, but touches the larger mutuality of the partners' interaction, especially around the family table.

Meals are especially privileged moments of the intimate presence because of the reality of caring and sharing they entail. Their intimate presence is irreplaceable in a society of high mobility in order for the spouses to remain ''in touch'' and to keep fondness alive. The agape of sharing a meal, as described before, is an integral part of being together. The ''I-Thou'' character of sharing strengthens a love commitment. It is like a "natural sacrament"[31] of love and life which supports family communication and points in faith to the agape of the eucharist. The intimacy of a meal, like the intimacy of the body, rejects idolization as much as it rejects utilitarianism. Body and food are gifts from God which have to be cherished and treated with reverence. The eucharist, which celebrates the Lord's Supper, brings about a healing of human ambiguities and a deeper celebration of the agape of committed love.

Daily Dying and Rising

Living and dying is the basic pattern of all life. Christian faith paradoxically reverses this natural pattern into a mystery of death into life through the transsignification of human life by worship. This mystery of dying to death itself, an act of transcendental worship, thus belongs to the core of spirituality. Ideally, this spirituality means a radically new manner of relating to others, based on inner personal freedom and the passion of love.

This applies specifically to marriage because of its interpersonal reality; a dependability and independence, where the other spouse in his/her interior space is never fully known or validated. That new lifestyle cannot exist without the couple's experience of a free oblatory love. Modern society, with its greater mobility and its greater expectations from personal relationships, its competition and rejection of the idea of self-sacrifice—among other factors—has created a culture where relationships marked by depth of concern and the devotion of faithfulness are the exceptions.

True love is the sign of true freedom, but true freedom in marriage does not exist without a human response to the intimate longing of the other person, and without the coherence of creative fidelity. "There can be no true freedom without our first having emptied ourselves of self, so that we might open ourselves to the only reality capable of fully satisfying our powers of love and knowledge," says Rene Laturelle.[32] The couple encounters that "only reality" in the infinite God through a relationship "in depth" that is characterized by self-control, dependability and commitment. In the Christian perspective it is a matter of faith and love, because through these gifts Christians have been called to freedom in dying and rising each day. The gift of self (love) results in a possession of self (freedom) and establishes the most intimate and most challenging of all human relationships.

This relationship of marriage is influenced throughout life by internal and external factors which affect each partner and, at the same time, by the different stages of development throughout the family's lifetime. Couples experience the ebb and flow, the ups and down, the fervor and dryness of life itself. "Marriage," in the words of Mary G. Durkin, "is not a smooth curve drawn on the chart of life. It is, rather, a series of cycles, of deaths and rebirth, of old endings and new beginnings, of falling in love again."[33] The "happily ever after" mentality of a perpetual ecstasy not only builds up false expectations and leads to disillusionment, it denies the paradoxical reality of the life-giving oblative love and forgiveness, the healing and freeing discipline of self-control, and the redemptive idea of purification necessary for any dynamic and creative love relationship.

In today's society it is more than ever apparent that a marital life span consists of various stages or passages which every couple will experience in the course of their relationship. This development reality must be seen as the necessary starting point of any understanding of the process of marital relationship. Through the years of change and growth the two become many different people with different needs and expectations. Can marital unity and fulfillment still be maintained in this being and becoming? Certain social scientists insist that it is possible through a "creative marriage" which nurtures the basic values of caring, empathy, mutual respect, equality, trust and commitment.[34] These basic values are without a doubt the human thread of a covenantal partnership. The pressure of a career, the challenges of parenthood, the disagreements and tensions of the everchanging modes of modern living, cannot but affect the aliveness of a relationship. A couple hears the call to holiness in this very web of the unending seasons and changing faces of their marital journey.

A spirituality of stability is needed today. It means a spirituality of "staying in love," which embodies, on one hand, the courage to be a partnership[35]—the existentially strengthening effect of self-donations—and, on the other hand, in Christian terms, the wisdom of the cross—the nuptial journey of Jesus through the total donation and service. The former provides a healthy and dynamic tension against the senescence of time; the latter is a rock of stability against external and internal fragmentive dependencies.

Any attempt to grow and endure based solely on endless analysis, or even developmental ideas of a relationship, is bound to fail because the growth of two people is never an automatic process and is always unpredictable and complex. Besides the realistic expectations and lifestyle of each couple, there is a spiritual element which motivates and makes self-knowledge and self-revelation possible. Only authentic faith is transforming and life-giving—no magical effects or external miracles are needed. Nevertheless, since the paradox of the cross, the hallmark of faith, stresses the idea of self-sacrifice and donation, it shows the futility of human self-fulfillment, which actually leads to failure. Religion is not meant to be a guarantee of success, but an unrelenting call to faithfulness. Both the human, which is transient, and the divine, which has an ultimate value, actualize human plenitude.

Dying and rising in marriage reveal the depth of meaning in a spirituality of donation and service. It is the sharing of the paschal mystery. Husband and wife realize that every act of faithfulness is the actualization of the incompleteness of the paschal mystery in them. They marry each other many times in many different ways, always rising to a deeper life. Their living out, in hope, the ideal in the midst of human weakness always brings about growth because of the presence of the cross in their lives, which is their eucharist. The perspective of worship broadens that meaning again, since both eucharistic worship and marriage are covenant signs and actualization of God's love.[36]

From the altar of their everyday experience through which they build an irreplaceable community of love and minister to each other in the domestic church, the spouses offer the concrete unfolding of their spiritual sacrifice to God.[37] The specific applications of this realistic spirituality are numerous. The playful, caring, earthly and meaning-bestowing atmosphere of the house provides the ideal milieu to live authentic worship, but it needs the ritual celebration which can transfigure and heal the fragility of human relations through the force of the celebrated divine life.

Gift-Given Communication

Communication is the lifeblood of the interpersonal communion of the partners. The quality of their relationship has many different degrees of actualization because it rises and falls as the couple's sharing does. It is either an authentic encounter or the failure of encounter. This interpersonal communion is a great challenge because it reveals, on the one hand, the uniqueness of each individual and appears, on the other hand, as a self-gift, which has to be acknowledged as such by either spouse. Created by an inviolable interiority, which constitutes my freedom and my human dignity, I am gifted with words and love to validate the existence of the other. I surrender freely what I freely received—my very self—in the encounter with another person.

Theological reflection is not yet implied, at this anthropological level of one's need for relating to another. A totally new dimension of the spirituality of the spouses in striving for effective communication in giving and sharing is opened up in the light of the mystery of God's self-communication. Analogies which provide a model for a life-generating communication can be drawn from a theology of divine revelation.

Revelation is a happening of God's unconditional love, expressed in mysterious ways, in deeds and words to humankind all throughout history. This happening of God's initiative leads to an encounter with his people, and lets the people know how much he can do for them. In the sharing of the living reality of God himself people participate in the eternal truth resulting in the development of their full potential for being. This participation is actualized in the mystery of worship, since worship is revelation and gift-given communication. And so is marriage.

The forms of communication in the historical revelation of Scripture, as in the marital relationship, are inexhaustible. The analogies of both cases, however, lead to the same conclusive characteristics of essential communication: dialogue, communion, presence and power.

Dialogue through empathy and freedom is vital to human relations. The importance of dialogue as a keystone of effective communication stems from the dialogic structure of human existence. In his existential visions of human reality, Buber not only stressed the intersubjectivity of human communication, but saw this interpersonal reality linked intrinsically to God's reality and the way to him.

God is mysteriously mirrored in the relationship with the other. This interpersonal and transcendent vision has been blurred in a modern consumeristic society.[38]

If the sacredness of personhood is the basis of any meaningful dialogue, the well-being of the other is the goal of marital dialogue. Quality dialogue based on sincere devotion to human dignity and radical acceptance of mutuality leads to effective communication. Couples who develop the right attitudes of creative communication (which is an art, not an information techique), speak in open and honest language, conveying ideas and emotions, listening and sharing, affirming and encouraging.

These are the central aspects of loving-giving communication. The different systems of decision-making, the skills and methods emphasized by modern psychology, and many other helpful suggestions and insights adapted to the variables of a marital relationship are important. On the other hand, the humanistic view will insist on the qualities of a mature and liberated person and on developmental concepts of marital interaction. Nevertheless, these will never work without a conversion of heart and spiritual growth.

A dialogic spirituality in the light of the biblical word makes possible that conversion and growth. Spouses turn to each other and to God. In fact, God takes the initiative in calling husband and wife to dialogue, since both the spouses and God speak the same language of lovers.[39] Like worship, where God and his people engage in dialogue, the spouses "gift" one another in an oblative dialogue.

Depth of communion is the result of meaningful dialogue. The fire of communion in living and loving is maintained through the vehicles of communication: those of mind and body. To establish a relationship marked by the depth of communion of intimacy is a great challenge because it demands total freedom and spiritual nakedness. Furthermore, external pressures and internal ambiguities might compromise that experience of personal freedom. A possibility opens up through God's grace which calls us to true freedom. As Augustine said, "We are not free within ourselves, nor free from ourselves until we have encountered him who causes us to be born ourselves."[40]

Again, the narrative of God's activity, revelation, provides a model for dialogue of communion. Revelation, in fact, is personal communion in which God re-creates the freedom and value of the person and makes possible the encounter with others and with God himself. The barriers of self-centeredness are overcome and communion becomes an ongoing reality. God is the ultimate possibility of communion for the couple who stand as the most telling metaphor of God's self-donation to his people.

Communication as communion occurs, therefore, only when the partners' relationship is characterized by an ongoing, free and intimate relationship. The permanent call to communion breaks the emotional and mental restrictions of mutually alienating attitudes. The partners are free and open, and they give life to one another in the give and take of marriage. Finally, they create an interpersonal intimate living in which their lives are moved by one another. As real and intimate encounter, this type of communication as communion is worship.

Presence is a kind of personal communication in itself. To be together, even without much talk, strengthens a love relationship which, in turn, generates an atmosphere of communication. It can even nurture deep emotions because the body language is the most original and basic human language.

Partners are eminently present to each other through their bond of love and commitment. On one hand, their being present to each other in their daily interaction and human experiences is the essential part of their communication with the complex and rich variety of nonverbal communication. On the other hand, this presence is validated with words which speak their world vision, their desires and frustrations, dreams and joys.

That presence is salvific because the partners through their commitment participate in the concrete, loving intimacy of God himself. This is at the core of marital spirituality.

The analogy to the mysterious presence of God's self-communication further deepens our understanding of this spirituality. The living dialogue with God in the ever-present revelation, and the marital love relationship, are both alive through the awareness of presence. Both realities are foremost a happening, but they are confirmed with words.

Present to each other, husband and wife encounter their lives in a healing and supportive way, and they experience the challenges and demands of their journeying together. Enclosed mysteriously within the presence of God, marriage is actually the fertile ground of the true worship of self-sacrificing love and, therefore, of a true and unique spirituality.

Finally, communication is power because it is always a dynamic force, an ongoing reality which reveals and, at the same time, affects the person. "One of the greatest gifts God has given to spouses," says John G. Quesnell, "is the power to give life to one another. Through the medium of speech, spouses can communicate or destroy life in one another."[41]

Since this power is love generating through an intimate mutual understanding, especially through the quality of listening, it is the closest reality to total communication. But total communication, like complete knowledge of the other, is a lifetime task, never an actual reality. Hence, the challenge of this most powerful human interaction.

The mystery of the inviolable interior space of the other already reveals a divine parenthood.[42] From the beginning God pronounces a word of life to establish a community of love. This word of life is not only creative from the beginning, but it permanently validates human existence as an ineffable power of salvation.

Translated into the art of communication, this theology of revelation creates a most important area of conjugal spirituality: communication as creative power toward perfection. So as the word of God demands faith and trust, communication in marriage is only possible im empathy and love. The spouses empower each other with deep and constant dialogue, as does the God of revelation whom they image in faithfulness. Their commitment to their union and to their fellow humans, if lived out in faith as self-gift of genuine communication, becomes a living sacrifice which worship celebrates.

Faith Sharing in the Domestic Church

Vatican II, referring to marriage and the family as a "domestic church," regained a rich biblical and patristic theology of the priestly dignity of the Christian couple.[43] This priestly dignity stems, not from the exercise of any sacred function, but from the natural sacramental character of their union, lived in the all-embracing nuptial mystery of Christ symbolized by the Christian couple. They meet the Holy through their human core and are empowered by the Holy through their covenant.

Another important vision which derives from that priestly dignity is the ministerial function of the laity as partners in the service of the church and as community builders. This vision, which is being gradually recovered, demonstrates the importance of the theological concept of the domestic church. From this perspective, not only does the couple learn its call to minister to their own concrete existence in the image of the church, but the ecclesial community develops the attitude of respect and mutuality of the family partnership from the model of the domestic church. In fact, the partnership of the family, which "is a living image and historical representation of the mystery of the Church,"[44] stands today as a most compelling metaphor of the church.

This new communal and sacramental perspective will enrich a theology of marriage and the family today, especially in the context of the contemporary crisis of traditional Christian values. The present, in fact, more than ever before, is strong in experiences, but weak in faith. From this perspective of the growth of the family as a living reality of the family of families (that is, the church), two different dimensions of the conjugal spirituality emerge: the inward family living within itself, and the outward faith witnessing beyond itself.

The inward family living in the "sanctuary of the Church in the home"[45] is spiritual if experienced as God's permanent free gift through the different inner components of its reality: a place where love, the law of family living, is experienced in its full and rich meaning; where humanization, identity, and the integration of human sexuality develops; where faith through the word is nurtured and shared, and, consequently, Christ is present; where prayer and contemplation are a reality; and where responsible parenthood and education lead into the integral reality of being human.

The outward faith witnessing depends on the same concept of the family as a living cell of a living organism. As such, the couple manifests the mystery of Christ which is present in its life and love. They make their actual experience visible, and so they become an extension of the church. As a community of hospitality, as a community which witnesses to its faith in the concrete experience, and as a community which takes part in the building up of the church. The fostering of justice and peace, the humanization of culture, and especially the evangelization of society in the sphere of politics, economy and science are essential tasks of the lay ministry.

The life of this small community of worship "receives a kind of consecration" which reaches and transforms all of its conjugal existence because it is called to be a presence and testimony to Christ's ministry which transforms all human reality. The spouses' profoundly humane and historical spirituality becomes a credible sign and instrument of salvation, as they "manifest to all people the Savior's living presence in the world, and the genuine nature of the Church."[46]

The theological analogy between church and family is not simply a useful pious comparison, but is deeply rooted in Christian tradition, both biblical and historical.

The New Testament tradition of the "domestic church"[47] is abundantly testified to in numerous references and centers around the essential meaning of Christian marriage which describes and actualizes symbolically the transcendent mystery of Christ and his church (Eph 5:21–33). This central Pauline affirmation created a conjugal spirituality in the early church which viewed marriage as spiritual worship, donation and service of the partners.

The early literature of the fathers of the church also has abundant references to the same analogy, especially John Chrysostom. His main point of departure is the idea of Christ's privileged presence in "the house of God" (the family), of which the Canaan wedding is a telling example in the light of the theology of John's Gospel. John Chrysostom sees the mystery of Christ and the church in the perspective of marriage as a "substantial image" and "a mysterious icon of the Church."[48] In Paul Evdokimov's words, "The grace of the priestly ministry of the husband and the grace of the priestly maternity of the wife form and shape the conjugal existence in the image of the Church."[49]

The historical tradition goes back to the very dawn of the Christian community and its worship, for at the beginning of Christianity "the Church meets in the house of" a Christian family.[50] Houses were not only literally churches (like Dura-Europos), but households gave birth to communities. The worshiping community, as distinct from the sacred temple gathering, originated in the bosom of the family, around the table, and under the couple's hospitality. They became a prophetic instrument of evangelization. The cases of Lydia and Cornelius are only two telling examples of decisive developments of the nascent Christianity: the former represents the beginning of the expansion of the total church to Europe (Acts 16:14–15) and the latter is the first official acceptance of the Gentile world (Acts 10:1–23).

The theology of the domestic church constitutes, certainly, a sound vision of marriage and family spirituality, especially from the perspective of spiritual worship. As a metaphor of the total church, it struggles to live an analogous reality of caring and hospitality, of celebration and forgiveness, of prayer and peace, and of an unbroken unity in the midst of a broken society. In this family church, "where all the different aspects of the entire Church should be found,"[51] marriage is a concrete experience of Christian life in "faith working through love" (Gal 5:6).

The theological understanding of the intimate faith sharing at home, like the understanding of the other important realities of agape-love, donation and communication, provide the basis for the assertion of the existential and intrinsic relationship of marriage and worship at the core, not only of life, but of the Christian mystery itself. In specific Christian terms, the self-bestowal of the marital embrace and the sharing of one eucharistic cup and bread intersect in the mystery of the cross, the paradigm of Christian worship. Marriage can only be a vital and fulfilling reality of love when lived as a profoundly human and spiritually transcendent experience of spiritual worship in bed and board, children and society.

NOTES

1. James F. White, *Introduction to Christian Worship* (Nashville: Abingdon 1981) 25.

2. For a discussion of the importance of the analogical method and its application to theology and worship, see David Tracy, *The Analogical Imagination* (New York: Crossroad 1981) 405–21.

3. L. L. Mitchell, *The Meaning of Ritual* (New York: Paulist 1970) xii.

4. ''For worship is an acknowledgement of Transcendence,'' she says in her classic study *Worship* (New York: Crossroad 1936) 3.

5. In his existential vision he insists on the dialogical structure of a human being which is the path to the absolute, and which links human beings necessarily to God. Marriage only radicalizes this human interpersonal reality and its transcendency. M. Buber, *I and Thou* (New York: Ch. Scribner's Sons 1970).

6. Ch. Davis, *Body as Spirit: The Nature of Religious Feeling* (New York: Seabury 1976) 16.

7. W. Kasper, *Theology of Christian Marriage* (New York: Seabury 1980) 27. God's plan in Genesis is conceived in the light of Exodus: the mystery of man and woman in love is the symbol of God's covenant of grace, especially dramatized in the prophetic revelation (Hos 1, 3; Jer 2, 3, 31; Ez 16:1–14; Is 54:62).

8. Cf. Ti 2:11; 1 Jn 4:9.

9. Especially in Paul's vision in the light of Exodus and Genesis, of Christ's bridal relationship to the church (Eph 5:21–33).

10. Ex 12:25, 26; 13:5; 19:5; Dt 10:12.

11. Jer 7:22–23; Hos 6:6; Dan 3:39–41; Ps 39:7–9; Ps 49.

12. Cf. Eph 1:4–6; Phil 3:3; Heb 9:14.

13. *Sacramentarium Veronense,* nos. 1105–10, ed. L. C. Mohlberg, RED 1 (Rome 1956) 139–40; *Le sacramentaire gregorien,* ed. J. Deshusses, Spicilegium Friburgense 16 (Fribourg 1971), no. 838ff.

14. Cf. G. Martinez, ''Marriage: Historical Developments and Future Alternatives,'' *The American Benedictine Review* 37 (1986) 376–82.

15. John Chrysostom (4th–5th century) is the best example of this theology, for instance, in Homily 9 on 1 Timothy (PG 62:546ff).

16. R. P. A. Raes, *Le Mariage: Sa célébration et sa spiritualité dans les Eglises d'Orient* (Belgium: Editions de Chevetogne 1958) 12. Cf. E. Schillebeeckx, *Marriage: Human Reality and Saving Mystery* (New York: Sheed and Ward 1965) 344–56.

17. Says Y. Congar in his thorough study of the sacred in the Bible, especially in reference to Ephesians 2:14–15: ''Jesus abolished definitely the separation between the sacred and the profane regarding people, places and times'' (''Situation du 'sacré' en regime chrétien,'' in *La liturgie aprés Vatican II* (Paris 1967) 385–403).

18. *Marriage,* 12.

19. *Body as Spirit,* 34.

20. Cf. *Eucharisticum Mysterium* 3 and 6, *Documents on the Liturgy* (1963–1979): Conciliar, Papal and Curial Texts (Collegeville: The Liturgical Press 1982) 395.

21. Cf. K. Rahner, *Foundations of Christian Faith* (New York: Seabury 1978) 116–33; Jon Nilson, "Theological Bases for a Marital Spirituality," *Studies in Formative Spirituality* 2 (1981) 401–13.

22. Erich Fromm, *To Have or To Be* (New York: Harper & Row 1976) 44–47.

23. For the notion of the nuptial sacrifice of Christ, the author is dependent on P. Evdokimov, *The Sacrament of Love: The Nuptial Mystery in the Light of the Orthodox Tradition* (New York: St. Vladimir's Seminary Press 1985), 122–23.

24. Loving another person means telling him or her: You will not die, according to the French philosopher Gabriel Marcel, quoted by Walter Kasper, *Theology of Christian Marriage* (New York: Seabury 1980) 22.

25. St. Augustine, *De Genesi ad litteram* 9:7, 2, CSEL 28/1 pp. 275–76.

26. "Zur Theologie der Ehe," in *Theologie der Ehe* (1969) 102.

27. We are "sexed being," says Maurice Merleau-Ponty, *The Phenomenology of Perception* (New York: Humanities Press 1962) 154–71.

28. Pope John Paul II, "The Nuptial Meaning of the Body," *Origins* 10 (1980) 303, quoted by Ch. A. Gallagher and others, *Embodied in Love: Sacramental Spirituality and Sexual Intimacy* (New York: Crossroad 1984).

29. Cf. Erik Erikson, *Insight and Responsibility* (New York: Norton 1964) 127–29.

30. Besides the biblical quotations in footnotes 10 and 11, see Ex 10 (Yahweh is a passionate, committed lover), Psalm 139 (intimacy with God) and Song of Songs: 2:5–9; 3:2–5; 4:12–16; 6:3; 8; 8:6–10; Teresa of Avila, *Interior Castle* and John of the Cross, *The Living Flame of Love*.

31. "Centuries of secularism have failed to transform eating into something strictly utilitarian. Food is still treated with reverence. A meal is still a rite—the last 'natural sacrament' of family and friendship" (Alexander Schmemann, *For the Life of the World: Sacraments and Orthodoxy* (New York: St. Vladimir's Seminary Press 1973) 16.

32. *Man and his Problems in the Light of Jesus Christ* (New York: Alba House 1983) 243.

33. A. M. Greeley and M. G. Durkin, *How to Save the Catholic Church* (New York: King Penguin 1984) 126.

34. Such as M. Krantzier who speaks of "marriages within a marriage" when he identifies six natural "passages" corresponding to first years, career, parents, middle and mature years of marriage (*Creative Marriage* [New York: McGraw Hill 1981] 37). These change-patterns have to be taken seriously as "the signs of life" to be read in order to understand a proper spirituality of the family.

35. The author borrows this concept from P. Tillich and applies it to marriage. "Every act of courage is a manifestation of the ground of being" (*The Courage to Be* [New York: Harper and Row 1958] 181).

36. David M. Thomas, in one of the best synthesis of theology and secular values of marriage, warns against overemphasizing developmental concepts and undervaluing the need of a radical acceptance. See *Christian Marriage* (Wilmington: Michael Glazier 1983) 119–20.

37. Cf. Rom 12:1–3 and Pet 2:5.

38. M. Buber, *I and Thou,* 123–68: "Extended, the lines of relationships intersect in the eternal You" (123).

39. See notes 10, 11, and 30.

40. St. Augustine, *Confessions* 10:27, 38. Paul and John's theology stress Christian call to freedom in Christ (Gal 5:1, 13; 4:26, 31; 1 Cor 7:22; 2 Cor 3:17 Jn 8:32, 36).

41. J. Quesnell, *Marriage: A Discovery Together* (Notre Dame: Fides/Claretian 1974) 88.

42. M. Buber, *I and Thou,* 123–68.

43. *Lumen gentium,* no. 11, ed. W. M. Abbott, *The Documents of Vatican II* (1963–1965), 29. Cf also *Gaudium et spes,* nos. 47–52; 248–58; *Apostolicam actuositatem,* nos. 2–10, 489–501, especially *Familiaris Consortio* (*On the Family*) of John Paul II, nos. 13, 55–58 ("Spouses are therefore the permanent reminder to the Church of what happened on the cross." "Of this salvation event marriage, like every sacrament, is a memorial, actuation and prophecy," no 13), *Origins* 11 (1981) 437–66.

44. "In communion and co-responsibility for mission" are the words used to define relationships between ordained and nonordained ministries, which should prevent "continuous wavering between 'clericalism' and 'false democracy'" ("Vocation and Mission of the Laity Working Paper for the 1987 Synod of Bishops," 33–57, *Origins,* no. 17 (1987) 11–14.

45. *Apostolicam actuositatem,* no 11, Abbott, 502–03.

46. *Gaudium et spes* 48, 252.

47. Cf. Rom 16:5; Cor 16:19; Philem 2; Col 4:15, among many other references.

48. *In epistulam ad Colossenses,* PG 62:387; quoted by P. Evdokimov, *The Sacrament of Love* 139.

49. *Evdokimov,* 139.

50. *He kat'oikon ekklēsia* to designate house communities, or faithful gathering in a household, different from "the whole of the Christian movement," or "the household of God." Cf. E. Schillebeeckx, *The Church with a Human Face* (New York: Crossroad 1985) 46–48; N. Provencher, "Vers une théologie de la famille: l'Eglise domestique," *Eglise et Théologie* 12 (1981) 9–34.

51. Paul VI, *Evangelii nuntiandi,* 71, *Origins* 5 (1976) 459.

6.

Varied Meanings of Marriage and Fidelity

James Nelson

James Nelson is a United Church of Christ theologian and ethicist. His writings on sexuality and marriage have been pioneering and profound. This essay falls neatly into three parts. The first part is a survey of some of the perspectives of several contemporary theologians (Thielicke, Bailey, Barth, Pittenger) who have particularly influenced the church's theology of marriage. While the perspective is distinctively Protestant and Anglican, a comparative study of contemporary Roman Catholic theology on marriage will yield some striking commonality.

The second part gives a hearing to voices for change. Four representative critics (Mazur, Francoeur, Roy, Snow) of traditional marital forms and limits argue for an opening and reshaping of the current form.

The third part should provoke lively and imaginative discussion among students. The key issue addressed is fidelity. The author proposes a redefinition that challenges our established sexual categories. Some questions suggested by the essay are:

1. *Are there intrinsic limits to the forms of marriage?*
2. *Are there limits imposed by human nature or by the nature of society?*
3. *What elements are contained in Nelson's understanding of fidelity? Are they convincing to you?*
4. *What marital value does society wish to preserve most? Is it sexual exclusivity or long-term commitment? What do our churches value most?*
5. *Are there any enduring and distinctive qualities a Christian marriage requires irrespective of its future developmental forms?*

In Nelson's book Embodiment: An Approach to Sexuality and Christian Theology *(Minneapolis, MN.: Augsburg, 1978), readers will find an extended treatment of these topics titled, "The Meanings of Marriage and Fidelity," pp. 130–151.*

From *Journal of Current Social Issues* 15, 1 (Spring 1978): 14–22. Reprinted with permission.

There is little doubt that the present confusion about marriage (and sexual morality in general) is intimately linked with the changing functions and perceptions of the family in our society. Typically, the family has had six basic functions: to control sexual access and relations; to provide an orderly context for reproduction; to nurture and socialize children; to provide a context for economic activity; to ascribe social status to its members; and to provide emotional support for its adult members. Without exception each of these functions is now being challenged and changed under the impact of high mobility, new perceptions of sex roles, urbanization, industrialization, and rapid technological developments (importantly including reliable contraception and markedly increased life spans).

Thus, a recent study of family types in the United States produced the following constellation: nuclear families (both parents and their children in one household), 37 percent; single adults without children, 19 percent; single parents (usually divorced or separated) with children, 12 percent; remarried couples with children, 11 percent; childless couples or couples with no children at home, 11 percent; experimental family forms, 6 percent; and three-generation households, 4 percent. In a word, we are moving toward a variety of family types wherein no single form is statistically normative.

If for no other reason than its dramatic increase, this contemporary questioning of marriage deserves to be taken seriously by the church. It would be easy—but mistaken—to write off these signs by labeling them simply as expressions of hedonism, or of wishful and unrealistic thinking, or of the fear of lasting personal commitments. Doubtless, these things are present. But also present is a genuine yearning for more deeply humanizing relationships and sexual bonds.

Regarding the institution of marriage itself, the early church tended to follow the prevailing Roman practice for several centuries; its strong eschatological expectations precluded any vigorous interest in producing new marital forms consonant with its faith. The first traces of distinctively Christian rites of marriage emerge only in the fourth century, and we await the ninth century to find a detailed account of Christian nuptial ceremonies (and even then the order is quite parallel to that used by the ancient Romans). In fact, only at the time of the Council of Trent did the Western church first assert that the use of the Christian ceremony was essential to a valid marriage.

History does not stand still. Institutional change continues and its pace accelerates. But change as such should not be feared. We *are* historical creatures, and the radical monotheism of Christian faith would remind us that no finite, historical form ought ever to be absolutized.

Some Representative Theologies of Marriage

We can begin our quest by examining some of the distinctive perspectives of several contemporary thinkers who have particularly influenced the church's current theology of marriage. While others could certainly be added to the list, I believe that of especial importance are Helmut Thielicke, Derrick Sherwin Bailey, Karl Barth, and Norman Pittenger.

According to Thielicke, marriage is preeminently "the covenant of agape." Yet, it is a "worldly" and not a "sacramental" institution. It "rests upon a primeval order of creation"; a valid marriage does not depend upon the couple's awareness of their union's theological significance. Nor has marriage any redemptive significance—no one is saved through it. It is established for all persons and can be observed independently of faith. It is "an order of preservation" for the whole world—one of those crucial human arrangements which prevent the world from sinking into chaos.

Yet, says Thielicke, *Christian* marriage has particular meanings. Especially, it must be seen as monogamous. The reason for this lies not in any clear biblical laws concerning monogamy (of which there are none, he believes), but rather precisely in the revelation of agape in Jesus Christ. The manner in which agape presses ineluctably toward monogamy is linked with the different sex roles of man and woman. A man's sexual nature is "naturally polygamous." He invests far less of himself in the sex act and "is not nearly so deeply stamped and molded by his sexual experience as is the case with the woman." The woman, in contrast, is "naturally" monogamous, "because she is the one who receives, the one who gives herself and participates with her whole being, is profoundly stamped by the sexual encounter." The woman, then, cannot live polygamously without damage to the substance of her very nature. But if she cannot, then the man cannot either, for his masculinity is a relational term that cannot be defined apart from the woman.

> Once we see that Christian *agape* regards this "existence-for-the-other-person" as the foundation of all fellow humanity, and that it regards man as being determined by his neighbor, it becomes apparent that under the gospel there is a clear trend toward monogamy. Because the wife is a "neighbor," the husband cannot live out his own sex nature without existing for her sex nature and without respecting the unique importance which he himself must have for the physical and personal wholeness of the feminine sex nature.

Karl Barth finds a theology of marriage emerging from the doctrine of the Trinity. God is self-related as three persons in a triunity of being, three persons-in-community. If God is thus communal in being, the human being created in the divine image cannot be complete as a solitary self. Barth says,

> man is directed to woman and woman to man, each being for the other a horizon and focus . . . , man proceeds from woman and woman from man, each being for the other a center and source. . . . It is always in relationship to their opposite that man and woman are what they are in themselves.

What Barth has done is to make a radical interpretation of Genesis 2:18–25. The deficiency of the solitary individual (Adam) is a basic deficiency. What we need, argues Barth, is more than a companion for life. We need a *complement*—one of the opposite sex with whom to share the meaning of history, and in relation to whom we form a marriage which is always greater than the sum of its parts.

Happily, Barth avoids the sharp nature-grace disjunction and the more obvious gender stereotypes which plague Thielicke's interpretation. And the twofold strengths of Barth's approach are indeed of central importance: that we are created not as solitary selves but as beings-in-relationship, destined for communion; and that sexuality is intrinsic to and not accidental to our capacity for such cohumanity.

There is, however, a major problem. At times Barth admits that marriage is a calling which not everyone receives, but at other times he speaks of marriage as a necessity if we are to realize our humanity. In short, he has failed to keep the distinction between a Christian view of sexuality and a Christian view of marriage. The two are closely related, but they are not one and the same. And Barth's failure at this point implicitly casts a long and unwarranted shadow over the single person and over the one who is homosexually oriented. If the sex-role is not as obvious in Barth as in Thielicke, it still seems to underlie this problem.

Either his interpretation of the image of God makes a woman's humanity crucially dependent upon her husband's masculinity and the husband's upon his wife's femininity, or it makes their humanity dependent upon heterosexual genital intercourse, or both. In so doing Barth's interpretation subtly but surely rests upon either sex-role stereotypes or upon the genitalization of sexuality, and probably upon both. And in so doing it unfortunately has squarely linked the doctrine of the image of God with the *alienated* dimensions of our sexuality.

Derrick Sherwin Bailey, who follows Barth in many respects, also makes the same error when he illustrates the meaning of the Genesis account. He appeals to rabbinic pronouncements to the effect that a person who remains unmarried is "no proper person" since this "diminishes the likeness of God." Nevertheless, this is not Bailey's major emphasis. His approach to marriage is essentially focused upon the centrality of the love relationship: "Ethically it is of the utmost importance never to lose sight of the fact that marriage, whatever else it may be, is essentially a personal union of man and woman founded upon love, and sustained and governed by love." Marriage comes into full existence when the psychophysical seal of sexual intercourse has been experienced. Its essential nature is love and its one fundamental purpose is unity—that the man and woman shall become one, creating a common life in which the human sexual duality can find its fullest expression.

Of distinctive importance in Bailey's understanding of marriage is his interpretation of the "one-flesh" union of which the Bible speaks. Ideally, one-flesh occurs in marriage wherein it depends upon the consent of both partners, is based upon responsible love (for each other and for any child that may result), and in which it has the approval of the community. But even when one-flesh is experienced "falsely," as in nonmarital sex or rape, it is always irrevocable.

When intercourse takes place in a loving marriage, it has several qualities and results: it expresses gratitude to God for creation; it is a means of communication more adequate than speech; through it the couple come to understand the meaning of their masculinity and femininity; and it is always an accurate reflection of the whole relationship that the couple has. But when people have intercourse as merely pleasurable, sensual indulgence apart from permanent commitment, "they merely enact a hollow, ephemeral, diabolical parody of marriage which works disintegration in the personality and leaves behind a deeply-seated sense of frustration and dissatisfaction . . ."

I find Bailey's strong emphasis upon love as the central meaning of marriage well taken. Marriage is not centrally a union of two "natures" but of two persons; it is a personal relationship whose meaning is love. And, in spite of his occasional slip, Bailey's treatment gives considerable weight to the egalitarian character of the relationship between woman and man. Furthermore, his interpersonal approach opens him to the possibility of marriage as a context for salvation—it is a significant arena in which God's life-giving, renewing, and healing work goes on, and it is not simply an order of preservation against the chaos of the world.

Yet his exposition of the meaning of one-flesh is not without its difficulties. That intercourse is intended to involve and affect the whole self, I fully agree. That persistent misuse of intercourse in casual and selfish ways damages one's capacities for authentic personal relationships, I have no doubt. But whether each act of intercourse actually does something *irrevocable* to the self and the partner quite apart from the meanings they attach to it is another matter.

Norman Pittenger's interpretation of marriage, like that of Bailey, rests strongly upon the centrality of love—now, however, interpreted from the stance of process theology. Persons are dynamic and not static entities, human becomings more than human beings. The image of God in the self is the capacity for love, the capacity to become more fully shaped into the love that is God's own essence. As embodied selves our sexuality cannot be divorced from our capacity to love, and in the marriage of a man and a woman we have the usual way in which the relationship of embodied love is known and expressed most completely:

> The uniting of man and woman in marriage has made possible an approved way to develop those qualities of love upon which we have insisted, the giving-and-receiving relationship, the mutuality, etc., are all given a setting which makes it much easier to grow in them towards fulfillment. . . .

The conception and birth of another human being, Pittenger claims, discloses the depths of meaning in human sexuality. It is *pro*-creation, creation on behalf of another, God. It is the fecundity of love and a manifestation of divine Charity. Though a great blessing, the procreation of children is not essential to marriage's validity. But sexual intercourse is, because the church has recognized "that a full union, a giving of oneself entirely to another self who gives in return, must of necessity include the body as well as the mind." Thus, marriage is the intimate and faithful communion of two covenanted and embodied persons. What is distinctive about this relationship compared to other types of human relationships is that through its promises and through the radical and intimate self-giving and receiving a sacramental quality emerges—a distinctive participation in the divine Love.

Pittenger's approach to marriage is helpful both in what it affirms and in what is left open. Love is utterly central. Its sexual expression is necessary, and love's fidelity is intrinsic to the union. As an Anglican, he speaks with conviction of the sacramental quality of marriage. It is a decisive human arena in which the healing and humanizing love of God is encountered. Yet, it is not the only sort of human relationship in which this is possible. Nor are gender role and sex identity as such intrinsic to the covenant of grace. It is the person-in-communion, the self-as-lover who is in the image of God, not the male-female relationship as such. This has important implications for an androgynous heterosexual marriage, I believe, and it is clearly essential if single persons and same-sex covenantal unions are to be woven into the same theological fabric that includes a theology of marriage.

In recent years Christians as well as humanists have begun to indict "traditional monogamy" on a variety of counts: its wedding to the nuclear family form; its failure to meet the growth needs of partners through its inherent possessiveness; its propensity toward continuing sexual inequality; its cultivation of a vivid sense of failure when one partner cannot meet all the other's needs or when the relationship itself must be terminated.

A brief look at four representative "voices" (in two instances the writings are by couples) will give a fuller picture of these arguments. In each case the writer identifies with the church.

Ronald Mazur charges that many of the assumptions underlying traditional monogamy are unproven:

> For too long, traditional moralists have been passively allowed to preempt other conscientious lifestyles by propagating the unproven assumptions that we cannot love more than one person (of the opposite sex) concurrently; that co-marital or extramarital sex always destroys a marriage; that "good" marriages are totally self-contained and self-restrictive and sufficient.

Mazur affirms the value of life-long commitment to one other person in a primary marriage relationship but also affirms the possibility and often the desirability of enriching secondary relationships. The latter will have varying degrees of emotional and sexual intimacy and will be engaged in openly and honestly with care that the primary relationship is not jeopardized. This is open-ended marriage.

Anna K. and Robert T. Francoeur advocate essentially the same thing but place their analysis of traditional and new marital forms in the framework of a cultural interpretation—the movement from "hot sex" to "cool sex." Their labels (vivid if somewhat misleading) parallel some of the distinctions that I have made between alienated and reconciled sexuality. Hot sex attitudes (still dominant in our society) are patriarchal, genitally focused, possessive, and performance-obsessed. Cool sex orientation (now making its inroads) affirms the equality of the partners, integration rather than competition and conquest, sexuality experienced as diffused sensuality, an emphasis upon unique personalities and needs, and a tolerance for a pluralism of marital forms. "We are convinced," they write, "that the one-to-one male-to-female marriage is here to stay. We are also firmly convinced of the distinct value of long-term commitments and relationships. . . ." Marriages in the new pattern will create their own intentionally extended "family" of supporting intimate relationships, built upon commitment to open communication, open companionship, and trust.

Rustum and Della Roy argue very similarly: "Our theory demands that we seek to maximize the number of deep relationships and to develop marriages to fit in with a framework of community." Like the others, they seek to expand marriage's intimate and erotic community. They go beyond the others, however, in advocating several institutional changes. Premarital intercourse and cohabitation should be publicly and ecclesiastically accepted for engaged couples, with the provision that no children are to be conceived. An aggressive attempt to incorporate single persons within the total life of a family should be encouraged. The divorce system should be overhauled to make marital termination much less destructive for the two parties. Given the disparity in the sex ratio, polygamy and polyandry should be legalized. And the marriage service should be changed, not to water down the couple's commitments, but to expand them.

John Snow makes a useful distinction between "passive-adaptive" and counterculture" patterns in the changing marital scene. The passive-adaptive pattern is oriented around individual fulfillment and possibly hedonism. Such marriages frequently do not really become part of one's identity. They are active in regard to the concern for self-improvement but passive in relationship to the culture that is regarded as corrupt but essentially unchangeable. Counterculture marriages, in their diverse forms, have in common an ideological revolt against the current economic-political system of the society. If the intentionality or commitment in the passive-adaptive marriage is to personal growth, in the countercultural marriage it is more typically to an ideology.

Snow argues for a third and more Christian alternative: marriages that "are open in the sense of being hospitable, generous, and socially concerned. They do not regard themselves as little islands of intimacy in competition with the world for survival, but as part of a community concerned with the redemption and renewal of that world for whom Christ became incarnate, suffered, died, and rose again."

These, then, are some of the voices of change, in varying ways all advocating expansion of the emotional and intimacy bonds of the traditional marriage and family model. Before engaging in some theological-ethical reflection on such possibilities, it will be helpful to review the conclusions of one of the leading sociologists of marriage. Jessie Bernard, in her thorough and balanced study *The Future of Marriage,* maintains that marriage indeed has a future in our society, for men and women will continue to want and need the lasting bonds of intimacy and mutual support. "Still," she adds, "I do not see the traditional form of marriage retaining its monopolistic sway. I see, rather, a future of marital options."

Are there intrinsic limits to the forms of marriage, limits imposed by human nature or by the nature of society? If these exist, claims Bernard, they are difficult to find. Will future marriages be any happier? Bernard disclaims any sanguine optimism yet maintains that marital behavior, generally, is improving. We have a revolution of rising expectations. "We do not tolerate today forms of marital behavior that were matter-of-fact in the past." She concludes that there is no final solution:

> There will never be an ultimate or last in the sense of final form of marriage. It will go on changing as the times and the people change and as the demands on it change. There is no Ideal Marriage fixed in the nature of things. . . . We are only now getting used to the idea that any form of marriage is always transitional between an old one and a new one.

Now, however, things are changing much more rapidly than before. Sociological estimates of what is or what will be should not be confused with moral judgments concerning what ought to be. While the ethical process, as I have argued, should always give serious attention to all relevant data, there still is a basic difference between facts and values. Thus, no sociologist's prognosis, however carefully researched, equals a Christian (or any other sort of) ethical conclusion. In the face of considerable anxiety about change I believe it needs reiteration that no single form of institutional life on earth, marital or otherwise, ought to be considered final. The sabbath was made for persons and not vice versa. The form or forms which marriage takes ought always to be congruent with the authentic needs of persons—not with their shortsighted or superficial desires, but surely with their needs as persons in the process of becoming that which God intends for humanity.

What Is Fidelity?

This is not the place for analysis of the many and complex issues suggested above. But one particular question claims attention. It threads its way through each of the position statements on marriage, both those of a more traditional nature and those urging change: What is fidelity?

We have already seen how the usual definition of marital fidelity is now being challenged. To summarize, it is argued that possessiveness of an emotional and possibly a genital sort is a major detriment to the marital relationship. Thus, a distinction must be made between infidelity and adultery. Adultery has a straightforward meaning: sexual intercourse with someone other than one's spouse. Infidelity, on the other hand, is the rupture of the bonds of faithfulness, honesty, trust, and commitment between spouses. On the positive side, the argument goes, fidelity is the enduring commitment to the spouse's well-being and growth. It is commitment to the primacy of the marital relationship over any other. Compatible with marital fidelity and supportive of it can be certain secondary relationships of some emotional and sensual depth, possibly including genital intercourse. Perhaps a useful approach toward clarifying the meaning of fidelity is to examine the arguments pro and con regarding sexual exclusiveness in marriage. We begin with its defense.

First, there are biblical and traditional warrants against adultery. It is condemned in the Decalogue (Exodus 20:14), elsewhere in the Old Testament (for example, Deuteronomy 22:22), and Jesus repeats that condemnation (Mark

10:19). While in the moving account of the woman taken in adultery Jesus showed deep compassion and forgiveness, he also instructed her to "sin no more" [John 8:1–11]. To be sure, the Old Testament suggests at numerous points a double standard: the wife is regarded as her husband's property, and her adultery is a violation of property rights. Nevertheless, this cannot be attributed to Jesus' teaching on the issue. Indeed, the manner in which he radicalized the condemnation of adultery to include lust is evidence that he was without ambiguity on this matter. And the weight of Christian tradition sustains this position. The radical nature of sexual intercourse would seem to justify exclusive confinement within marriage. This argument usually rests upon an interpretation of Paul's concept of one-flesh, as in Bailey's understanding of the irrevocable bond established between any sex partners. Others argue similarly. Dwight Hervey Small contends:

> in becoming one-flesh a couple participates in that unity which cannot be reversed ontologically. . . . There is no reversing the knowledge of the shared secret. . . . Two human beings who have shared the sex act can no longer act toward one another as if they had not done so. The unity achieved transcends the biological domain.

Because the sex act is radical it in some way involves the whole person, for good or ill. Psychological opinion can buttress this judgment as Karl Menninger's statement does: "It is an axiom in psychiatry that a plurality of direct sexual outlets indicates the very opposite of what it is popularly assumed to indicate. Dividing the sexual interest into several objectives diminishes the total sexual gratification."

In addition to those arguments for sexual exclusiveness because of something inherent in the nature of the act itself, there are arguments based upon probable consequences. Adultery hurts people. When it is deceptive, it is "bound to corrode all of the honest and fragile personal responses of the married partners," contends Lewis Smedes. Even if it happens in an open marriage where there is no deception, the spouse left out of this relationship is bound to feel rejection. One of the sexual relationships is bound to be neglected, and it will probably be the marriage. And, because of the corrosive effect upon the marital relationship there will be harmful consequences for any children in the family and for society itself which depends upon marital strength for the soundness of the communal life.

The argument of consequences moves from the negative to the positive in Daniel Day Williams:

> [The person] bears the image of God as . . . power to enter an enduring, mutually supportive community which incorporates suffering constructively since self-giving always involves suffering. Love disciplines itself for love's sake.

Let us turn to the other case—the possibility of nonexclusive sexual fidelity. One of the arguments, we have seen, is that unrealistic and impossible demands are placed upon monogamy in today's society. The nuclear family model has carried with it an image of marriage as an encapsulated sphere, hermetically sealed from relationships of emotional depth with those outside it. But it is both unrealistic and unfair to expect that one person can always meet the partner's companionate needs—needs which are legitimate and not merely individualistic, hedonistic, or egocentric. If the gospel truly invites us to the greatest possible realization of human capacity in interdependence with others, then marriage ought to be open to precisely that.

Many have noted that women in particular have suffered from the traditional definition of monogamy. Thus, Rosemary Reuther writes, "Monogamy has especially atrophied the personal development of women who were expected to get their entire emotional feedback through a relationship to a single man while he in turn developed his personality through a multiplicity of friendships, business relationships, and even sex relationships."

Personal growth for either wife or husband may well require intimate friendships besides that with the partner. This does not necessarily mean transmarital sexual intimacy. The important thing is interpersonal emotional intimacy, but sex cannot be arbitrarily excluded.

Given current pressures on marriages, it is argued that the choice for many couples might well be between sexual exclusivity and marital permanence; the latter is the greater value. Jessie Bernard comments, "If we insist on permanence, exclusivity is harder to enforce; if we insist on exclusivity, permanence may be endangered. The trend . . . seems to be in the direction of exclusivity at the expense of permanence in the younger years but permanence at the expense of exclusivity in the later years."

Rosemary Ruether presses this further. Historically, she notes, monogamy has been closely linked with the private property relationship of man over woman in the patriarchal society. And while the church preached a single standard of exclusive sexual fidelity, in many ways it acceded to the double standard. In the modern period, however, the notion of romantic marriage has demanded not only sexual exclusivity but also lifelong companionship and sexual satisfaction. Ruether wonders if we have not lifted up the wrong priority by apparently prizing sexual exclusivity over enduring, intimate companionship and personal fidelity. We might have more of the latter if we were not so insistent as a church upon the former:

> What really is the value that society wishes to preserve most? Is it sexual exclusivity, especially when divorce and remarriage turn this into a serial exclusivity? Or is it not long-term commitment, both for personal support and friendship and for secure child-raising that can provide stable parental figures for the new generation?

Both explicit and implicit in these arguments, then, is an understanding of fidelity that contains several elements: it is a commitment of emotional and physical intimacy with the partner; it means caring for the growth and fulfillment of each as a person; it is commitment to the growth of the marital relationship itself; it requires honesty, openness, and trust; it involves willingness to explore ways of opening the self to the partner at the deepest possible level, risking the pains that may come; it includes openness to secondary relationships of emotional intimacy and potential genital expression, but with commitment to the primacy of the marriage.

In light of these considerations, I believe that the case for the redefinition of marital fidelity cannot be summarily dismissed. Openness to the *possibility* that this might reflect a viable Christian marriage for some is, I think, appropriate. My second concern, however, is one of caution—for several reasons.

Sexuality is indeed an inevitable component of all interpersonal relations. And it is not the case that genital exclusivity in marriage necessarily means that other friendships must, for that reason, be devoid of sensual warmth and emotional depth. Nor must genital exclusivity mean the possessive ownership of the partner. It can be—and frequently is—reflective of the kind of commitment from which emerges a genuine sense of freedom.

Further, the empirical evidence concerning the effects (especially the long-range effects) of sexually nonexclusive fidelity is very limited at present. While a surprising number of persons now write positive testimonies from their own experience, the wider picture has many unknowns.

Again, the possibilities of sexual sanctification are relevant to this issue. Growth in sensuousness, for instance, can mean greater diffusion of sexual feeling throughout the body and throughout the self's relationships. With the diminishing genital focus for one's sexuality, it is conceivable that the desire for genital expression outside of the marriage might be lessened. Growth in androgyny, too, might coincide with sexual exclusiveness, since many instances of extramarital sex seem linked with the desire to reinforce one's sexual self-esteem as either masculine or feminine.

In spite of reconciliation and growth, alienated dimensions of our lives do persist. *Simul justus et peccator*. If law can be a servant of the gospel and rules a servant of wholeness, for most of us the better rule is marital genital exclusivity. But, I would argue, this rule (as is any other) is a presumptive rule that can admit exceptions. If the exception is admitted, the burden of proof must be borne: that this sexual sharing realistically promises to enhance and not damage the capacity for interpersonal fidelity and person wholeness.

Marriage indeed has a future. And its forms will change. What is frequently called traditional Christian marriage is itself a product of historical development, and it is not as old as is often believed. Yet it has served many well and doubtless will continue to do so. But the future will likely be marked by plurality more than uniformity in marital forms.

Nevertheless, I believe that as marital forms take their shape from the reality of God's incarnate love there will be certain characteristic and distinctive qualities about them. They will be covenants and not simply contracts. They will be enduring covenants—pledges of ongoing faithfulness to the well-being and growth of each partner. They will be covenants of intimacy in which eros is undergirded, infused, and transformed by agape. They will be sacramental covenants, whether or not officially sacraments, of the church. They will yet be those unique arenas in which the healing and humanizing love of God is vividly experienced. And they will be covenants that, in one way or another, genuinely enlarge the partners' capacity for communion with others and expand their willingness to be part of God's work of giving new life and renewal to the world.

7.

The Meaning of Commitment

Margaret Farley

During the 1960s and 1970s, it became fashionable in the United States to call commitment into question. Not only were specific forms and concrete situations critiqued, but commitment "in principle" itself came under challenge. We have never been the same as a society since that social and cultural upheaval. Psychologists, sociologists, philosophers, and novelists have attempted to name and shed some light on our area of "the uncommitted." Farley's essay, from a theological perspective, is the most systematic, sophisticated, and sober treatment of the topic available today.

The author sets the question of commitment in the widest social context. The many different forms of commitment, their complex interconnection, and the elements common to them all are laid out with balance and clarity. But what, precisely, is commitment? What purpose does it service? Are there limitations on its binding force? Are there conditions for release from its promise? These are the foundational questions pursued in this essay. The prime case of the purpose of commitment in human love is presented and brings the issue to the center of marital relations.

Some questions to refect on follow:

1. *What is the relation of commitment to the experience of genital sexuality?*
2. *What counts as a legitimate degree of commitment during the engagement period?*
3. *What new dimension of commitment is required between partners if they decide also to become parents?*
4. *What situations or conditions would allow for the release from one's marital commitment-relationship?*

The reason why commitment is such a problem to us is that by it we attempt to influence the future, and by it we bind ourselves to someone or something. Two quite different customs in contemporary society illustrate dramatically how this is

From *Personal Commitments* (San Francisco: Harper & Row, 1986); Chapter 2: pp. 12–22, and Chapter 3, pp. 33–35.

done. If I am arrested by the police and wish not to be in jail while I wait for my trial, I may be able to post bail and so be free until my date in court. In giving money as bail, I am declaring my intention to return for my trial at a future time, and I am binding myself to do so on penalty of forfeiting my bail money. In a wholly different setting, when two persons marry in our society, it is the practice of many to exchange rings. "With this ring, I thee wed . . ." symbolizes the express intention of each to love and to share the life of the other into the future. It symbolizes, moreover, a bonding whereby each gives to the other a claim to the fulfillment of that intention. These examples hold a key to the meaning of commitment. If we look at them closely, we can come to understand what we are doing when we make a commitment.

I find myself hesitant, however, to narrow our focus so quickly to these examples. We need a wider perspective from which to view them. There is, I suspect, something important to be gained from letting our minds roam a bit, trying to see the broadest possible sweep of the forms commitment can take in our lives. This will have the advantage of preventing premature closure on just one meaning for commitment. It will also help to keep any one dilemma of commitment from overwhelming us, or any one celebration of commitment from seducing us into complacency about it.

Indeed, the history of the human race, as well as the story of any one life, might be told in terms of commitments. Civilization's history tends to be written in terms of human discoveries and inventions, wars, artistic creations, laws, forms of government, customs, the cultivation of land, and the conquering of seas. At the heart of this history, however, lies a sometimes hidden narrative of promises, pledges, oaths, compacts, committed beliefs, and projected visions. At the heart of any individual's story, too, lies the tale of her or his commitments—wise or foolish, sustained or broken, fragmented or integrated into one whole.

Surveying the Horizon

Think again, then, of all the ways in which we experience what we with some seriousness call "commitment." Sometimes these are not immediately evident in one person's life at any one given time. When Sheila, for example, thinks of commitment, she tends to focus only on the one area of commitment which right now is difficult for her. Every day she lives with the more and more pressing question of whether or not she should persuade Joshua that they must divorce. Every day she agonizes over her responsibilities, through marriage, to her husband and to their children, to God, and to herself. But she has many more commitments than these, some that intersect with them, some that compete with them, and some that are not in question in relation to them at all. For instance, Sheila is committed to certain truths, to certain principles. From the day she had her first insight into what she now describes as the "equality" of women and men, her conviction has grown regarding this truth. She can no longer act as if she did not believe it. She cannot turn back and live out the roles in her life as if she had never seen the reality of herself and all women in a new way.

Sheila also believes deeply in the obligation of persons not to harm one another unjustly. Her perception of the value, and the need, of persons in relation to one another goes beyond this, to a desire and a sense of responsibility positively to help others. She has often said, "Life is hard enough for anyone to get through. I figure we can't do it alone. We have to help one another." Her compassion is based on conviction, and it does violence to her not to take account of others' needs—especially Joshua's, her children's, the people she meets in her volunteer work and in the political action groups to which she belongs. Genuine caring and compassion serve to motivate her involvement in organizations that oppose racism and violence and that promote economic rights and peace.

There is a sense, also, in which Sheila is committed to herself, though she is almost afraid to think in these terms. Her growing anger at what she perceives to be Joshua's indifference to her and to the children keeps generating in her mind the question of whether this is how she is meant to live. The frightening realization of how destructive her marriage has been to her keeps pushing her imagination to find alternatives, "ways out."

The story of Sheila's life and commitments could go on and on. But even a brief glimpse into it enables us to begin to see the many different forms, and complex interconnections, possible in our commitments. There are, of course, commitments to other persons—some made explicitly, some assumed implicitly. But there are other kinds of commitments, too. For example, there are what can be called "intellectual" commitments—to specific truths, sometimes to "truth" in general (a commitment that can undergird a pursuit of truth wherever it is to be found). There are commitments to values—the value of an institution, or the life of a family, or to so-called "abstract" values like justice, beauty, peace. There are commitments to plans of action, whether specific projects, or life-plans such as "living in accordance with the Gospel," or programs of vengeance, peaceful revolutions, or "being a good mother."

The appearance of commitment in our lives is even more extensive and nuanced than this, however, and more elusive when we try to encompass it in our overall perspective. There is, for example, a kind of unrecognized commitment, one that serves as an important background for almost everything we do. We are not explicitly aware of it (or of them, for there may be many such commitments). We may never bring it to a level of consciousness where we can reflect on it. This kind of commitment serves to *constitute* part of the very horizon against which we interpret everything. It may be a commitment that psychologists could describe as "basic trust," or one that philosophers could name a "presupposition." It may be one that any of us might recognize, if and when it is brought into focus—as, for example, taking for granted that the law is to be respected, or valuing without question the progress of human education, or assuming that things should "make sense" whether we understand them or not. These kinds of commitments, prior to any explicit recognition of them in our conscious awareness, can be called "prereflective" commitments.

But if "commitment" can appear in all these forms, what *is* the common meaning that keeps it from being empty as a term? Some clues emerge from what we have just seen. Commitment seems, in our ordinary language, to include a notion of *willingness to do something* for or about whatever it is we are committed to (at least to protect it or affirm it when it is threatened). Suppose I ask, for example, what truths I am committed to. I soon discover that there are many truths that I hold, affirm, am convinced of, but am not "committed" to. What can this mean? I may say, of some insight of mine or some conviction regarding a state of affairs or a direction to be taken, that "I would not stake my life on it." This could mean that I am not completely certain about it. Or it could mean that, though I am certain, it just is not important enough to me to *do* anything about it. It is not important enough to me to let go of anything else for the sake of it (let alone lay down my life for it); it is not important to me even to use my energy trying to defend it through argument.

A willingness to do something seems, moreover, to follow upon our sense of *being bound* to whomever or whatever is the object of our commitment. Our very selves (in greater or lesser degree) are tied up with this object, so that we do not just appreciate it or desire it, but we are in some way "identified" with it. Our own *integrity* seems to demand that, under certain circumstances, we do something. The object of our commitment has a kind of claim on us, not one that is forced upon us, but that is somehow addressed to our *freedom*. Even when we do not feel very free regarding our commitments, when we feel bound "in spite of ourselves" or against our other desires, there is still a sense in which our own initiative is involved when we act because of that commitment.

We could continue to survey the many forms of commitment and to probe the elements common to them all. We need to ask further about the relation of free choice to commitment and the importance of prereflective commitments for the commitments we are aware of making. Now, however, may be the time when it is more useful to take seriously my lantern metaphor and enter the deeper regions of but one form of commitment. To do that, it helps to identify a kind of "prime case," a central form of commitment—one from which all other forms derive some meaning. *Commitment to persons,* when it is *explicit and expressed,* offers just such a "prime case." And it will bring us back to our two examples: the posting of bail and the exchanging of wedding rings.

"Promises to Keep . . ."

By explicit, expressed, interpersonal commitment I mean promises, contracts, covenants, vows, etc. These commitments provide a prime case for understanding all of the forms of commitment because the elements of commitment appear more clearly in them. We recognize an obligation to act in a certain way within these commitments more frequently than in any others. Moreover, here we most often confront dilemmas and the inescapability of wrenching decisions. It is in these commitments that questions of love, of time and change, of competing obligations, seem more acute. The very explicitness of promises, or covenants and contracts, places the experience of commitment in bold relief and offers the best chance for understanding it.

There are interpersonal commitments, of course, that are not expressed in any explicit way (at least not in the making of them). For example, some roles that we fill or relationships in which we participate entail commitments, but they become ours without an original choice on our part. We are born into roles such as daughter or son, sister or brother. Some friendships grow spontaneously and seem to need no promises. Other roles we assume by explicit choice and usually through some external expression—familial roles such as husband or wife, sometimes mother or father, and professional roles like physician or teacher. Even roles we do not at first choose, however, can be understood in great part through understanding the roles we explicitly choose; for roles of whatever kind usually at some point require free and explicit "ratification" or "acceptance."

The Making of Promises

The first thing we must do in exploring explicit, expressed commitments is to ask: "What takes place when we commit ourselves in this way?" What did Sheila *do* when she married Joshua? What will actually *happen* in the moment when Karen vows to live a celibate and simple life within a community dedicated to God? What does Ruth *effect* when she signs a business contract? What *takes place* when Dan speaks the Hippocratic Oath as he begins his career as a doctor? What *happens* when heads of state sign an international agreement regarding the law of the seas? What *happens* when Jill and Sharon pledge love and friendship for their whole lives long? What do Barbara and Tim *do* when they place their names on the lease whereby they rent their new apartment for a year? What do any of us do whenever we make a commitment to one another, whenever we promise, whenever we enter or ratify a covenant?

We can ask this question of our examples of posting bail and exchanging wedding rings. What is happening in each of these cases? In both, I am "giving my word" to do something in the future. But what can it mean to "give my word"? It is surely not like other things I could do regarding future actions. It is not, for example, like a *prediction.* If it were, I would not be *responsible* for the future's turning out as I said it would (except perhaps in some limited situations like that of the weather forecaster, who is not responsible in the sense of being able to control the weather, but who might be considered irresponsible if she did not show professional competence in forecasting). "Giving my word" is also not like making a *resolution,* where I may indeed feel responsible to do what I resolved, but where my obligation would be only to myself (to be consistent in carrying out my decisions), not to another to whom I had given my word.

When I post bail, I give my word that I will return for trial. I declare to someone that I will do this in the future, and I bind myself to do so by giving my money as a guarantee of my word. When two persons exchange rings in a marriage ceremony, they declare to each other their intention to act and to be in a certain way in the future, and they give a ring as a sign that their word has been given and that they are thereby obligated to it.

To give my word is to "place" a part of myself, or something that belongs to me, into another person's "keeping." It is to give the other person a claim over me, a claim to perform the action that I have committed myself to perform.[1]

When I "give my word," I do not simply give it away. It is given not as a gift (or paid like a fine), but as a pledge. It still belongs to me, but now it is held by the one to whom I have yielded it. It claims my faithfulness, my constancy, not just because I have spoken it to myself, but because it now calls to me from the other person who has received it. My money is still my money when I give it as bail. That is why it binds me to come to trial, lest I lose what is still mine. A wedding ring is not just "given away." It belongs somehow to both partners, for it signifies a word that is "the real" in the speaker, begotten, spoken, first in the heart. Belonging to the speaker, the word now calls from the one who has heard it and who holds it. "What is mine becomes thine," but it is also still mine. It is still mine, or it is still my own self, though I have entrusted it to another. That is why I am bound by it, bound to it, and bound to the other.

What happens, then, when I make a commitment is that I enter a new form of relationship. The root meaning of "commitment" lies in the Latin *mittere*—"to send." I "send" my word into another. Ordinary dictionary uses for "commitment" include "to place" somewhere (as "to commit to the earth," or "to commit to prison," or "to commit to memory"); and "to give in charge," "to entrust," "to consign to a person's care" (as in "to commit all thy cares to God"). When I make a commitment to another person, I dwell in the other by means of my word.

Much of the time "all" that we give is our word—not money, not rings, not special tokens that "stand for" us. We stand in our word. Still, when we give just our word, we search for ways to "incarnate," to "concretize," to make tangible, the word itself. It is as if we need to see the reality of what is happening. For example, we sign our name. Our word within a contract is sealed by placing ourselves—in the form of our name, written by our own hand—on the document. In an ancient Syrian form of blood covenanting, a man was required to write his name in blood on material which was then encased in leather and worn on the arm of his covenant partner.[2] Other rites of blood covenanting went even further, attempting to mingle the blood of one with another. For blood was the sign of life, and it was one's own life that was entrusted to the other in a sacred self-binding ritual.

When words seem too weak to carry the whole meaning of a commitment, sometimes we turn the words into chants, as if by repetition they become more solid, more visible in transfer. There is an old betrothal ceremony among the Berber tribes, where the couple alternates in a song that continues for hours:

> I have asked you, I have asked you, I have asked from God and from you.
> I have given you, I have given you, I have given you if you accept my condition.
> I have accepted, I have accepted, I have accepted and agreed. . . .[3]

Commitment, then, entails a new relation in the *present*—a relation of binding and being-bound, giving and being-claimed. But commitment points to the *future*. The whole reason for the present relation as "obligating" is to try to influence the future, to try to determine ourselves to do the actions we intend and promise. Since we cannot completely do away with our freedom in the future (think of the gambler who must choose again and again to keep his promise or not), we seek by commitment to bind our freedom, though not destroy it. How can commitment do this?

By yielding to someone a claim over my future free actions, I give to that person the power to limit my future freedom. The limitation consists in the fact that I stand to lose what I have given in pledge if I fail to be faithful to my promise. I stand to lose the property I have mortgaged, or the bail money I have posted, or my freedom to travel if I am imprisoned for breach of legal agreement. I stand to lose my reputation, or the trust of others, or my own self-respect, if I am unfaithful to even an ordinary and fairly insignificant promise. I stand to lose another's love, or my home, or strong family support, or my sense of honesty and integrity, or my sense of continuity within a culture and religion, if I betray or finally break a profound commitment that is central to my life. I stand to risk the happiness of someone I love, if my fidelity is needed in a commitment made for the sake of another. Sometimes we know fully what we stand to lose, what binds us to our commitments; sometimes we learn what it is or has become only when our fidelity is seriously in question. It is clear that commitments vary, so that in some commitments we stand to lose little but in others we stand to lose everything. Above all, however, as we take our own word seriously, we always stand to lose a part of ourselves if we betray that word.[4]

If we stop here, accepting this as the full meaning of commitment, we are liable to all the dangers regarding commitment. On the one hand, we can see the glorious possibilities of commitment—of gathering up our future in a great love, of belonging to another in a self-expansive way; and we may move too hastily into commitment for commitment's sake. In so doing, our one great commitment can end in a grand, but empty and finally destructive, gesture. On the other hand, the thought of yielding to another a claim over us—great or small—may intimidate us, make us afraid that any commitment we make will narrow our possibilities, leave us with no "way out," give us claustrophobia in a life walled in by obligations and duties.

The essential elements of interpersonal commitment *are* an intention regarding future action and the undertaking of an obligation to another regarding that intended action. But in order to see a reasonable place for it in our lives, and to be able to discern *how* and *when* commitment obligates in specific circumstances, we need to think about the purposes that commitment can reasonably serve and the limitations that it must necessarily have.

A Remedy and a Wager

The primary purpose of explicit, expressed interpersonal commitments is to provide some reliability of expectation regarding the actions of free persons whose wills are shakable. It is to allow us some grounds for counting on one another. As Hannah Arendt observed, "The remedy for unpredictability, for the chaotic uncertainty of the future, is contained in the faculty to make and keep promises."[5]

Commitment as it appears in the human community implies a state of affairs in which there is doubt about our future actions. It implies the possibility of failure to perform acts in the future that are intended, however intensely and whatever firmness, now. "Without being bound to the fulfillment of promises, we would never be able to keep our identities; we would be condemned to wander helplessly and without direction in the darkness of each man's [*sic*] lonely heart, caught in its contradictions. . . .[6] Ours is not the instinctually specified and determined course of animals insofar as they have no freedom; ours is not the unshakable course of the freedom of God.

Because our wills are indeed shakable, we need a way to *assure others* that we will be consistent. Because we know our own inconsistencies, we need a way to *strengthen ourselves* for fulfilling our present intentions in an otherwise uncertain future. Yielding to someone else a claim over our future actions provides a barrier against our fickle changes of heart, our losses of vision, our weaknesses and our duplicity. By commitment we give ourselves bonds (and give others a power) which will help us to do what we truly want to do, but might otherwise not be able to do, in the future. A remedy for inconsistency and uncertainty, commitment is our wager on the truth of our present insight and the hope of our present love.

Insofar as promise-making provides assurance to others and strength to ourselves, it facilitates important aspects of human living. It is a device upon which personal relationships depend (in one form or another) and which political life (short of tyranny and total domination by force) requires. It undergirds the very possibility of human communication, for it is the implicit guarantor of truth-telling. As Erik Erikson insisted, "A spoken word is a pact. There is an irrevocably committing aspect to an utterance remembered by others. . . ."[7] It is interpersonal commitment (in a social contract of one kind or another) that has been the instrument of structures designed negatively for *mutual protection*—each person from the other, or a group from an outside threat; or designed positively for *mutual gain*—economic or cultural, through shared labor or property, shared knowledge or aesthetic enjoyment. I need not say again that it is commitment that serves to initiate (sometimes) and sustain (sometimes) *companionship and love*. It is commitment, too, that resides at the center of much of the history of religion, whether in the form of primitive bargainings with feared and hidden gods or of a personally offered covenant: God giving God's word, assuring a people of a divine unshakable will, calling a people to their own consistency in freedom and love.

Limitations on Binding

If we are ever to sort out how and when we are obligated by our commitments, we must have some way of determining their limits. Unless Sheila, for example, decides that there is no way, ever, under any circumstances, that she can justify divorcing Joshua, she needs to be clear about the *extent* of her obligation to him and to their marriage. If all of our commitments are absolutely binding, then we shall expect to be overwhelmed by their competing claims, with no way to resolve them or, ironically, to live them faithfully in peace.

Obviously, not every commitment that we make is of equal importance to us or equally comprehensive in its claim on us. We do set limits to the obligations we undertake. Almost all of our commitments, for example, are provisional in some sense; almost all are partial, conditional, relative. It must be so. In fact, we might well ask whether more than one commitment (at least at one time) can ever be, without contradiction, absolute.

Sometimes there are limits within our commitments of which we are not aware. That is, it is possible for us to think mistakenly that we are committed wholly to something or someone when, in fact, we are not; or the depth of our commitment may be much less than we thought it to be. We are in these instances, like the apostle Peter, surprised at our easy betrayal of what we presumed was our one unquestionable commitment. On the other hand, sometimes we are surprised in the opposite way when we discover, like Judas, that we are more committed, more bound, to someone than we realized; what we assumed was relatively superficial or marginal to our lives, shows itself to be profound and unforgettable. In either case, we may weep bitterly at our discovery and be filled with remorse or with gratitude for what it reveals.

There is perhaps no remedy except time and experience for deficiencies in our own self-knowledge. It is, however, possible to be more reflective about the limits we *intend* (legitimately and necessarily) to include in our commitments. To understand limits is not always to diminish a commitment but may, rather, serve to focus it, to allow it to share in the overall power and hope of a committed life.

It is too soon to try to work out all of the ways in which our commitment-obligations relate to one another. But we can see some general ways in this regard, and at the same time gain an understanding of the possible limits of commitments. The terms that are useful for this are terms like "conditional" and "unconditional," "partial" and "total," and "relative" and "absolute." These pairs of terms are not mutually exclusive, so that they tend to blur into one another at times. Nonetheless, they help to articulate how much we yield in claim to another.

If a commitment is *conditional,* it obligates only under certain conditions. Sometimes we make commitments where we very clearly stipulate the conditions under which they will be binding. I promise to do something *only if,* for example, you reciprocate in kind; or *only when* the building code is met; or *only until* another worker can be transferred to this position; or *only if* my insurance policy will cover my expenses. An *unconditonal* commitment, of course, is one in which I commit myself to another "no matter what" conditions prevail. Thus, for example, I may commit myself to "go where you go and stay where you stay," allowing no conditions to justify changing my mind or my sense of being obligated. We may discover that while it is the nature of every commitment to refuse to count some conditions as justifying a change in the commitment, yet most commitments are at least subject to sheer conditions of *possibility* of fulfillment.

A commitment is either *partial* or *total* depending on what is yielded for claim. It may be partial because of a limitation in time: until next week, or until the weather changes, or when I reach retirement age. It may be partial because it simply is "part" of something larger—as a vow of poverty may constitute part of a total commitment to service of one's neighbor. It may be partial because it yields a claim only to my property and not to my person.

We think of commitment as *total* when it somehow involves the whole person of the one who makes it. These are the commitments that constitute fundamental life-options. These may be some of the commitments we make to love other persons. When we try to describe commitments in this way, however, we soon meet difficulties in expressing our complex experiences. For example, how shall we describe the commitment to love another person that arises from the whole of our being, that is affirmed totally with our very lives, and yet does not entail a total availability to the other for the deeds of love? We hesitate to call such a commitment a partial commitment, and the hesitation has its own rich truth.

The notion of "relative" and "absolute" can be extremely helpful for understanding the nature and limits of our commitments. But they, too, hold a variety of possibilities that are not always easy to keep clear. Thus, a *relative* commitment is just that—*related to* another commitment. It is, at least to some extent, dependent for its meaning on the other commitment. It may be derivative from, instrumental to, or a participation in the other commitment. But even these terms, describing modes of relation between commitments, conceal complex possibilities.

There is a vast difference, for example, between purely instrumental commitments (commitments that are solely a means to some other end, some larger commitment) and commitments to love someone who is perceived as an end in herself, though an end (not a means) whose deepest reality is in relation to God. For example, Joshua may be committed to his wife and children purely because they are necessary to him if he is to sustain a certain status with his business associates. Or he may be committed to them because he sees himself as a dependable, responsible husband and father, and he knows that they need his financial and personal support. Or he may be committed to them because he loves them in themselves; but since he believes them to "live and move and have their being" in relation to God, his commitment to them is an intrinsic part of his commitment to God.

The easiest way to think of an "absolute" commitment is to equate it with an "unconditional" commitment. In this sense, however, some relative commitments could be called absolute (if what they are related to is the object of an unconditional commitment, and if the relationship is intrinsic and necessary). We might also equate absolute commitments with "total" commitments; but here we encounter the same sort of uncertainty that we met with the partial/total distinction. To keep the category pure, we might reserve it for commitments that are both unconditional and total. This would be the kind of commitment Gabriel Marcel describes as "entered upon by the whole of myself, or at least by something real in myself which could not be repudiated without repudiating the whole—and which would be addressed to the whole of Being and would be made in the presence of that whole."[8] However Marcel's way of putting it may strike us, it is not difficult to catch the central point of what he is describing.

I could, of course, simply stipulate meanings for these terms. That would be helpful for my use of them hereafter. However, since my real concern is to show the many ways in which we must and do set limits to our commitments, I prefer to leave the terms open to various correlations and to continuing refinement that accords with the many possibilities in our experience.

Distinctions such as I have suggested thus far may seem already overly refined when all we want to do is to live out our commitments faithfully or discern when they no longer bind. Fidelity and betrayal are not simple matters, however, and our lives always prove more complicated than we wish. Not every giving of our word to another is an unlimited yielding of an unlimited claim. Nor is every commitment as circumscribed as our vague promises to "drop in sometime" to visit old friends. Through distinctions we may be surprised by some simple clearings in the forests of complication.

Commitment and Love

Like any other commitment, a commitment to love is not a prediction, not just a resolution. It is the yielding of a claim, the giving of my word, to the one I love—promising what? It can only be promising that I will do all that is possible to keep alive my love and to act faithfully in accordance with it. Like any other commitment, its purpose is to assure the one I love of my ongoing love and to strengthen me in actually loving. Given the challenges we have seen to the wisdom, if not the possibility, of commitments to love, this purpose bears fuller examination.

Purposes of Commitments to Love

Why should I want commitment if love rises spontaneously, and if I can identify with it by my freedom at every moment? Why should I promise to love if there are risks to the love itself in making it a matter of obligation? Only something at the heart of our experience of loving can explain this.

There are some loves whose very power in us moves us to commitment. "Love's reasons" for commitment are at least threefold, and they go something like this. First, like all commitments, a commitment to love seeks to *safeguard* us against our own inconsistencies, what we perceive to be our possibilities of failure. If we are not naively confident that our love can never die, we sense the dangers of our forgetfulness, the contradictions of intervening desires, the brokenness and fragmentation in even our greatest loves. We sense, too, the powerful forces in our milieu—the social and economic pressures that militate against as well as support our love. We need and want a way to be held to the word of our deepest self, a way to prevent ourselves from destroying everything in the inevitable moments when we are less than this. To give to the one we love our word, to yield to her or him a claim over our love, offers a way.

Love seeks more than this, however. We know that freedom cannot once and for all determine its future affirmation of love. No free choice can settle all future free choices for the continuation of love. Yet sometimes we love in a way that makes us yearn to gather up our *whole future* and place it in affirmation of the one we love. Though we know it is impossible because our lives are stretched out in time, we long to seal our love now and forever. By commitment to unconditional love we attempt to make love irrevocable and to communicate it so. This is the one thing we can do: initiate in the present a new form of relationship that will endure in the form of fidelity or betrayal. We do this by giving a new law to our love. Kierkegaard points to this when he says, "When we talk most solemnly we do not say of two friends: 'They love one another'; we say 'They pledged fidelity' or 'They pledged friendship to one another.'"[9] Commitment is love's way of being whole while it still grows into wholeness.

Finally, love sometimes desires commitment because love wants to express itself as clearly as it can. Commitment is destructive if it aims to provide the only remedy for distrust in a loving relationship. But it can be a ground for *trust* if its aim is honesty about intention, communication of how great are the stakes if intention fails. The decision to give my word about my future love can be part of converting my heart, part of going out of myself truly to meet the one I love (not part of hardening my heart because of excessive fear of sanctions if I break the law that I give to my love). My promise, then, not only verbally assures the one I love of my desire for constancy, but it helps to effect what it assures.

NOTES

1. Of course, not all theorists would describe the meaning of commitment, or of promise-making, in the way I do here. What a promise *is* is closely tied to one's view of how it obligates. There are at least three major positions on this that appear in the history of philosophy and in contemporary discussion of promises: (a) the obligation to keep promises is purely conventional—an agreed upon "practice," or "game" in a given community, sometimes a matter of pretense until it is taken for granted and believed (Hume), sometimes a matter of violent discipline until behavioral conditioning gives it lasting status (Nietzsche); (b) the obligation is produced by the promise itself, for the words of promise are "performative," or "commissive," actually *doing* what they say (Austin, Searle, Melden, Sartre); (c) the obligation to keep promises is ultimately grounded in a more general obligation to respect persons, or to sustain moral community, etc. (Aquinas, Kant, Hegel, Hare). Many philosophers hold a combination of these views—for example, asserting that promising produces its own obligation, but only in a context where the conventions are such that this is possible (in other words, the "performative" depends on there being a "practice" of promising). My description of what "happens" when we make a commitment can be understood as a description of commitment as a performative. But it also, as will be clear in chaps. 6 and 7, assumes a fundamental ground of moral obligation in the reality of persons. Key treatments of these questions include historical works such as David Hume, *Treatise of Human Nature,* ed. L. A. Selby-Bigge (Oxford: Clarendon Press, 1968), Book III, Part 2, Sec. 5; Georg Hegel, *Philosophy of Right,* trans. T. M. Knox (New York: Oxford University Press, 1967), 57–63; Friedrich Nietzsche, *On a Genealogy of Morals,* trans. W. Kaufman and R. J. Hollingdale (New York: Vintage Books, 1967), 57–61; linguistic approaches such as J. L. Austin, *How To Do Things With Words,* ed. J. O. Urmson (New York: Oxford University Press, 1962); John R. Searle, *Speech Acts* (Cambridge: University Press, 1970), esp. chaps. 2 and 3; contemporary philosophical discussions such as Pall S. Ardal, "'And That's a Promise'" and "Reply to New on Promises," *The Philosophical Quarterly* 18 and 19 (July 1968 and July 1969); John Rawls, "Two Concepts of Rules,"

Philosophical Review 64 (1955): 3–32; Joseph Raz, "Promises and Obligations," in *Law, Morality and Society: Essays in Honour of H. L. A. Hart,* ed. P. M. S. Hacker and J. Raz (Oxford, 1977); G. J. Warnock, *The Object of Morality* (London: 1971), chap. 7; relevant to contract law, Patrick Atiyah, *The Rise and Fall of Freedom of Contract* (Oxford: Clarendon Press, 1979); Charles Fried, *Contract as Promise: A Theory of Contractual Obligation* (Cambridge: Harvard University Press, 1981). A key treatment important for the whole question of promise-making and promise-keeping is the classic study of Josiah Royce on loyalty: *The Philosophy of Loyalty* (New York: Macmillan, 1924).

2. H. Clay Trumbull, *Blood Covenant: A Primitive Rite and Its Bearings on Scripture* (London: George Redway, 1887), 5 and *passim.*

3. As quoted in Edward Westermarck, *Marriage Ceremonies in Morocco* (London: Macmillan, 1914), 40–41. This same kind of repetition occurs in pre-1965 ceremonies of vows in Roman Catholic religious communities. "Suscipe me, Domine," sang those making their vows, and they repeated this three times.

4. A further discussion of the nature of this obligation, and of what one risks losing, appears in chap. 7.

5. Hannah Arendt, *The Human Condition* (Chicago: University of Chicago Press, 1958), 237.

6. Arendt, ibid.

7. Erik Erikson, *Identity, Youth and Crisis* (New York: Norton, 1968), 162.

8. Gabriel Marcel, *Being and Having* (New York: Harper Torchbooks, 1965), 45–46.

9. Søren Kierkegaard, *Works of Love,* trans. Howard and Edna Hong (New York: Harper Torchbooks, 1962), 45.

III

Marriage: Change, Character, and Context

8.

Marriage Versus Living Together

Jo McGowan

Students will find the following essay by a young woman interesting and controversial. Jo McGowan lays out the two positions—i.e., that marriage is an outdated formality and that it is for several reasons an important step. She makes clear her own position about the importance of marriage as a public and communal act. The questions that follow may be helpful in reflecting on her essay:

1. *Why does the author claim that living together is like taking on all that is difficult in marriage without taking the helps that marriage offers?*
2. *The author says that living together implies that the central relationship of one's life is nobody's business but one's own. Why would you agree or disagree?*
3. *How accurate are the things the author says about the typical wedding?*
4. *How common in North America is the author's conviction that marriage (not just the wedding) is a communal matter?*
5. *What are your reasons for accepting or not accepting the author's conviction "What the community does not bless, it doesn't feel responsible for."*

Some months ago we had a beautiful young woman from India (my husband is also from India) staying with us. At dinner, the conversation turned to the question of marriage versus simply living together. Smita, the Indian woman, maintained that marriage was nothing more than a convenience, a way to avoid the censure of society; that if two people were willing to commit their lives to each other, then marriage was an unnecessary formality, signifying nothing.

To engage in the kind of discussion that followed is to risk sounding foolish. One talks of "marriage" as an institution and yet it is apparent that one is talking out of personal experience that cannot help but be narrow and unimposing compared to the subject itself. Having been married not very long myself, I realized how presumptuous it is to say almost anything (even at the dinner table, let alone

From *Commonweal* (13 March 1981): 142–145.

in print) about marriage in general. But when will it become *not* presumptuous—after five years, ten, twenty, fifty? The more years pass, I also realize, the more changes in social and cultural conditions will separate me from those entering marriage then, and so perhaps my reflections would take on a presumptuousness of a different sort. In any case, the discussion that evening was so enlightening to me that I decided to risk my dignity and write down some of the thoughts that emerged.

Apart from anything else, marriage is simply a very practical institution. It is an institution which recognizes and makes allowances for human failings. Since constancy is a virtue that very few of us possess at all times, it is important that we see marriage as something beyond ourselves. The very nature of marriage insists that we see it so: when we marry, we create new life; we go beyond ourselves. We create responsibilities, the weight of which our marriages must be strong enough to bear. Marriage is one of those peculiar things (like God!) which make immense demands of us while simultaneously giving us the strength to meet those demands. It is precisely because marriage is so difficult that we must see it as permanent. It is precisely because we are so likely to give it up that we must promise—at the outset, when everything is wonderful—that we are in it for life. (This is one reason, then, why the extreme prevalence of divorce is so troubling. It not only destroys the marriages of those individuals who choose to separate, but it erodes the *concept* of the permanence of marriage. It makes it that much easier for the next couple to give up.)

Simply living together, without "benefit of marriage," does not provide the security of knowing that this is forever. But if you need that security, our friend Smita says, then the relationship can't be that strong to begin with. Smita and I are both in our early twenties, still young enough to believe in the power of love to overcome all odds. And I do believe that. What I don't believe is that a wife and husband always love each other enough to stay married. There are times when love fails, and in those times, many people just take a deep breath and stay married because they *are* married. And when they come through to the other side, their marriages are stronger and more firmly rooted in love.

Smita grew up in India where divorce is practically unheard of—I grew up in an America where *marriage* is practically unheard of. She can perhaps afford to take marriage for granted. I can't. I have seen far too many of my friends—and even my parents' friends—divorce. I have taken care of too many children whose parents are separated. I have seen the scars that divorce inevitably leaves—the pain and near-despair in grownups; the bewilderment and insecurity in the children. I'm not saying that these couples didn't have problems; I'm sure they did. But no human relation is without problems. And if one enters into marriage, one should do it knowing full well that this is the case and that *in spite of it,* the marriage is forever. Living together does not carry with it the weight of a centuries-old tradition. The content of the relationship—a woman and a man living together sexually—contains all the elements that are present in a marriage, but without its form. It is like taking on all that is difficult in a marriage without taking the helps that marriage can offer. Simply knowing that one is married, that one has promised—before God and the human community—that this is forever, puts a different light on the inevitable problems that one faces. One is more likely (given, of course, a belief in the permanence of marriage) to slog through, to get past whatever it is in the way, to stay together.

Constancy, of course, is not limited to those couples who are formally married. Many of my friends are living together. They have made serious commitments to each other, they have children and, for all practical purposes, they might as well be married. Indeed, several of them have relationships that I consider to be closer to the ideal of marriage than that of most of the married couples I know. But that, I think, is more a function of the kind of people they are, and not of the form of their living arrangements. They are extraordinary people who would probably make a success of any relationship.

Even so, there is, it seems to me, something missing. I wouldn't presume to judge what goes on between two people who have committed themselves to each other—in whatever form they have chosen. I can only look at their relationship as it is perceived by the rest of the human community. It is here, I think, that the strongest argument for marriage as opposed to living together can be made.

Let us assume two couples: One married, one living together. Both have promised a lifetime commitment, both have children, both are trying to live with each other as lovingly, gently, and non-violently as they can. What is the difference?

The difference, as Ravi and I pieced it together that night with Smita, is this: one is a community-building act from the very beginning and the other is not.

To marry, to celebrate a love and a commitment publicly, in the presence of family and friends, is to say that the meaning of one's life can only be found in the context of a community. It is to acknowledge one's part in the human family, to recognize that one's life is more than one's own, that one's actions affect more than oneself. It is to proclaim that marriage is more than a private affair between one woman and one man.

To live together seems to me to imply that the central relationship of one's life is nobody's business but one's own. To live together is a decision reached privately and put into motion alone. There is no community blessing or celebration of the decision.

And what the community does not bless, it does not feel responsible for. Couples who are living together often find themselves quite alone when problems arise in their relationships. Their community may quite properly feel that such problems are none of its business. It was not asked for advice, or even congratulations, at the outset; why should it feel any responsibility now that things are going badly? On the other hand, a community which is asked to witness and bless the beginning of a marriage is far more likely to feel a sense of responsibility to the couple as their marriage grows and develops. I grant that most couples who do actually marry do not ask this of their community. Indeed, most couples think of the people at their wedding simply as guests who have to be fed, and not as participants in a community celebration. More on that in a bit.

The need for privacy, for individualism, looms extraordinarily large in American culture. We have been brought up to believe that it is a sign of weakness to admit that we need others. We have made a virtue of going it alone. Our ideal family is composed of a mother, a father, and one or two children. Grandparents, aunts, uncles, and cousins are all kept at a safe distance, and even neighbors are required, by zoning laws, to be at least an acre away. That this should be reflected in young people's choosing to live together, an essentially private choice, is not surprising. What *is* perhaps surprising is the extent to which most *marriages* are also quite private affairs, all the while purporting to be community events.

Most weddings say very little about the two individuals marrying—or about the community witnessing the union. Most weddings say something about the amount of money the participants have to throw about. They say something about fashion. They say something about respect for authority, in the form of the State, which issues the license, for a fee.

Most wedding ceremonies take place on an altar—so far from the guests who, theoretically, are there to witness the union of these two, that no one but the priest can hear the vows they exchange.

Most weddings are the occasion for bitter arguments: over relatives one cannot abide but invite anyway, seating arrangements at the reception, who pays for what, how many guests each family is allowed. . . . It goes on and on until many couples wish they *had* just decided to live together and skip all the hassle.

What is most telling, though, is the fact that so many weddings do not welcome children. Indeed, many outright discourage them. The phrase "No children, please" can be found frequently on wedding invitations and hard are the judgments passed on parents who dare to bring them anyway. Children, the hope and the future of any community, are an interruption, a noisy distraction, an additional and unnecessary expense—they take away from what is really important.

And what is really important, apparently, is that two grownups want to live together, but before they can, they have to get married.

This alienation of the community from the wedding ceremony, this lack of identification with the bride and groom—who seem more like actors playing prearranged roles than two people expressing their love for each other—serves to depersonalize the celebration. There is a boring sameness to weddings—one goes because one has to, because it is expected. The community is not asked to take part, and it does not. And the wedding sets the tone for the community's role in the marriage itself. The message is clear: It should limit its involvement to making appearances at the appropriate times, giving gifts on the appropriate occasions. Nothing more.

When Ravi and I married, we wanted a community celebration, one involving as many of our friends and families as possible. We wanted our wedding to reflect our religious (Catholic/Hindu) and cultural backgrounds, as well as our social and political concerns; we wanted our wedding to be a celebration of our love, naturally, but also for the community who had come to share our joy.

And it was. What a diversity of talents went into that day—from the wedding invitations and programs we designed, Ravi's side in Hindi and mine in English, to the wedding clothes made by Ravi's cousin and the wedding cake made by my father and a close friend. Ravi's mother performed the Hindu wedding ceremony; two priest friends witnessed the Catholic ceremony. Ravi and I wrote our own vows and selected the readings (from the Hindu and Christian scriptures) that friends and relatives read at the ceremony. Two nuns who had taught me in high school provided their oceanside convent for the day. The vegetarian banquet (both Indian and American foods) was entirely prepared by friends who arrived a few days early to cook . . . and best of all were the children everywhere, behaving exactly as children should behave, especially at a wedding.

It seemed to us then, and it seems even more so now, that our wedding was a symbol of the way we want to live our lives: surrounded by family and friends; giving, and receiving the gifts of time, laughter, advice, and help; sharing food, work, prayer, and celebration; creating a world where children are free and full of joy.

But marriage is a community event. It expresses, in its ideal form, a belief in the goodness of community, a belief in the beauty of two people who love each other coming together to live in communion, a belief in the wonder of human life, a belief so strong that it expressed itself in the creation of new human life.

If two people who say they want to marry do not believe this, then perhaps they should not marry. If they want to "join America"—to live in the suburbs with themselves and their 1.7 genetically screened children, exactly one acre from the nearest neighbor—then perhaps what they want is not marriage but just to live with each other.

If they want to be part of the human community, to start building the kingdom of God here on earth, then marriage is probably what they are seeking. And if it is, then the wedding itself, which is the beginning of marriage, should be an expression of their belief in community.

9.

A Case for Gay Marriage

Brent Hartinger

What is the meaning of marriage? The question initially *appears simple and easy to answer. Our taken-for-granted assumption, however, is that the frame of reference is heterosexual union. This assumption is challenged and the complexity of the question emerges when homosexual unions insist on inclusion in the meaning of the term. This is the position of Brent Hartinger in this article.*

An estimated 10 percent of the U.S. population—about 25 million people—is exclusively or predominantly homosexual in sexual orientation. Some data indicate that 50 percent of the men and about 70 percent of the women are in long-term, committed relationships. The concept of "domestic partnership" seems an inadequate classification for these enduring relationships. In spite of the benefits associated with the concept, it is seriously flawed. A simpler solution, according to Hartinger, is to allow gay civil marriage and to offer the support and endorsement of religious institutions to the bond.

This, in fact, would be a conservative move, the author claims. It would nurture long-term, committed, monogamous gay relationships and provide them with some measure of respect and social support. It would grant recognition to a significant cross-cultural minority and allow them to participate in an important social institution. The interests of the state would be served by legalizing gay bonding and the meaning of marriage enriched by the public celebration of this loving and committing union.

Some questions for reflection follow:

1. *Would homosexual marriage harmonize with our current underlying cultural values?*
2. *Would legalizing and legitimizing gay unions weaken or strengthen the family?*
3. *In view of the AIDS crisis, would gay marriage foster good public-health policy?*
4. *Is the meaning of marriage (as religious sacrament) consonant with the meaning of homosexuality?*

From *Commonweal* 108, 22 November 1991: 681–683.

5. *Are all sexual forms equal with reference to specific social goods?*
6. *Are procreative heterosexual marriages a special and irreplaceable central symbol of the Jewish and Christian traditions?*

In San Francisco this year, homosexuals won't just be registering for the draft and to vote. In November 1990, voters approved legislation which allows unmarried live-in partners—heterosexual or homosexual—to register themselves as "domestic partners," publicly agreeing to be jointly responsible for basic living expenses. Like a few other cities, including New York and Seattle, San Francisco had already allowed bereavement leave to the domestic partners of municipal employees. But San Francisco lesbians and gays had been trying for eight years to have some form of partnership registration—for symbolic reasons at least—ever since 1982 when then-mayor Diane Feinstein vetoed a similar ordinance. A smattering of other cities provide health benefits to the domestic partners of city employees. In 1989, a New York court ruled that a gay couple is a "family" in that state, at least in regard to their rent-controlled housing (the decision was reaffirmed late in 1990). And in October of 1989, Denmark became the first industrialized country to permit same-sex unions (since then, one-fifth of all marriages performed there have been homosexual ones).

However sporadic, these represent major victories for gay men and lesbians for whom legal marriage is not an option. Other challenges are coming fast and furious. Two women, Sandra Rovira and Majorie Forlini, lived together in a marriage-like relationship for twelve years—and now after her partner's death, Rovira is suing AT&T, Forlini's employer, for refusing to pay the death benefits the company usually provides surviving spouses. Craig Dean and Patrick Gill, a Washington, D.C., couple, have filed a $1 million discrimination suit against that city for denying them a marriage license and allegedly violating its human rights acts which outlaw discrimination on the basis of sexual orientation; the city's marriage laws explicitly prohibit polygamous and incestuous marriages, but not same-sex ones.

Legally and financially, much is at stake. Most employee benefit plans—which include health insurance, parental leave, and bereavement leave—extend only to legal spouses. Marriage also allows partners to file joint income taxes, usually saving them money. Social Security can give extra payment to qualified spouses. And assets left from one legal spouse to the other after death are not subject to estate taxes. If a couple splits up, there is the issue of visitation rights for adopted children or offspring conceived by artificial insemination. And then there are issues of jurisprudence (a legal spouse cannot be compelled to testify against his or her partner) and inheritance, tenancy, and conservatorship: pressing concerns for many gays as a result of AIDS.

In terms of numbers alone, a need exists. An estimated 10 percent of the population—about 25 million Americans—is exclusively or predominantly homosexual in sexual orientation, and upwards of 50 percent of the men and about 70 percent of the women are in long-term, committed relationships. A 1990 survey of 1,266 lesbian and gay couples found that 82 percent of the male couples and 75 percent of the female ones share all or part of their incomes.

As a result, many lesbians and gays have fought for "domestic partnership" legislation to extend some marital and family benefits to unmarried couples—cohabitating partners either unwilling or, in the case of homosexuals, unable to marry. In New York City, for example, unmarried municipal workers who have lived with their partners at least a year may register their relationships with the personnel department, attesting to a "close and committed" relationship "involving shared responsibilities," and are then entitled to bereavement leave.

But such a prescription is inadequate; the protections and benefits are only a fraction of those resulting from marriage—and are available to only a small percentage of gays in a handful of cities (in the above-mentioned survey, considerably less than 10 percent of lesbian and gay couples were eligible for any form of shared job benefits). Even the concept of "domestic partnership" is seriously flawed. What constitutes a "domestic partner"? Could roommates qualify? A woman and her live-in maid? It could take an array of judicial decision making to find out.

Further, because the benefits of "domestic partnership" are allotted to couples without much legal responsibility—and because the advantages of domestic partnership are necessarily allowed for unmarried heterosexual partners as well as homosexual ones—domestic partnership has the unwanted consequence of weakening traditional marriage. Society has a vested interest in stable, committed relationships—especially, as in the case of most heterosexual couples, when children are concerned. But by eliminating the financial and legal advantages to marriage, domestic partnership dilutes that institution.

Society already has a measure of relational union—it's called marriage, and it's not at all difficult to ascertain: you're either married or you're not.

Yet for unmarried heterosexual couples, marriage is at least an option. Gay couples have no such choice—and society also has an interest in committed, long-lasting relationships even between homosexuals. An estimated 3 to 5 million homosexuals have parented children within heterosexual relationships, and at least 1,000 children were born to lesbian or gay couples in the San Francisco area alone in just the last five years. None of the recent thirty-five studies on homosexual parents has shown that parental sexual orientation has any adverse effect on children (and the children of gays are no more likely to be gay themselves). Surely increased stability in the relationships of lesbians and gay men could only help the gays themselves and their many millions of children.

Some suggest that legal mechanisms already exist by which lesbian and gay couples could create some of the desired protections for their relationships: power-of-attorney agreements, proxies, wills, insurance policies, and joint tenancy arrangements. But even these can provide only a fraction of the benefits of marriage. And such an unwieldy checklist guarantees that many lesbian and gay couples will not employ even those available.

There is a simpler solution: Allow gay civil marriage. And throw the weight of our religious institutions behind such unions.

In 1959, Mildred Jeter and Richard Loving, a mixed-race Virginia couple married in Washington, D.C., pleaded guilty to violating Virginia's ban on interracial marriages. Jeter and Loving were given a suspended jail sentence on the condition that they leave the state. In passing the sentence, the judge said, "Almighty God created the races white, black, yellow, Malay, and red, and he placed them on separate continents. And but for the interference with his arrangements, there would be no cause for such marriages. The fact that he separated the races shows that he did not intend for the races to mix." A motion to overturn the decision was denied by two higher Virginia courts until the state's ban on interracial marriage was declared unconstitutional by the United State Supreme Court in 1967. At the time, fifteen other states also had such marital prohibitions.

Clearly, one's sexual orientation is different from one's race. While psychological consensus (and compelling identical and fraternal twin studies) force us to concede that the homosexual *orientation* is not a choice (nor is it subject to change), homosexual behavior definitely is a choice, very unlike race. Critics maintain that gays can marry—just not to members of their same sex.

But with regard to marriage, whether homosexual behavior is a choice or not is irrelevant, since one's marriage partner is *necessarily* a choice. In 1959, Richard Loving, a white man, could have chosen a different partner to marry other than Mildred Jeter, a black woman; the point is that he did not. The question is whether, in the absence of a compelling state interest, the state should be allowed to supersede the individual's choice.

Some maintain that there are compelling state interests to prohibiting same-sex marriages: that tolerance for gay marriages would open the door for any number of unconventional marital arrangements—group marriage, for example. In fact, most lesbian and gay relationships are probably far more conventional than most people think. In the vast majority of respects, gay relationships closely resemble heterosexual ones—or even actually improve upon them (gay relationships tend to be more egalitarian than heterosexual ones). And in a society where most cities have at least one openly gay bar and sizable gay communities—where lesbians and gays appear regularly on television and in the movies—a committed relationship between two people of the same sex is not nearly the break from convention that a polygamous one is. More important, easing the ban on same-sex marriage would make lesbians and gays, the vast majority of whom have not chosen celibacy, even more likely to live within long-term, committed partnerships. The result would be more people living more conventional lifestyles, not more people living less conventional ones. It's actually a conservative move, not a liberal one.

Similarly, there is little danger that giving legitimacy to gay marriages would undermine the legitimacy of heterosexual ones—cause "the breakdown of the family." Since heterosexuality appears to be at least as immutable as homosexuality (and since there's no evidence that the prevalence of homosexuality increases following the decriminalization of it), there's no chance heterosexuals would opt for the "homosexual alternative." Heterosexual marriage would still be the ultimate social union for heterosexuals. Gay marriage would simply recognize a consistent crosscultural, transhistorical minority and allow that significant minority to also participate in an important social institution. And since marriage licenses are not rationed out, homosexual partnerships wouldn't deny anyone else the privilege.

Indeed, the compelling state interest lies in *permitting* gay unions. In the wake of AIDS, encouraging gay monogamy is simply rational public health policy. Just as important, gay marriage would reduce the number of closet gays who marry heterosexual partners, as an estimated 20 percent of all gays do, in an effort to conform to social pressure—but at enormous cost to themselves, their children, and their opposite-sex spouses. It would reduce the atmosphere of ridicule and abuse in which the children of homosexual parents grow up. And it would reduce the number of shameful parents who disown their children or banish their gay teen-agers to lives of crime, prostitution, and drug abuse, or to suicide (psychologists estimate that gay youth comprise up to 30 percent of all teen suicides, and one Seattle study found that a whopping 40 percent of that city's street kids may be lesbian or gay, most having run away or been expelled from intolerant homes). Gay marriage wouldn't weaken the family; it would *strengthen* it.

The unprecedented social legitimacy given gay partnerships—and homosexuality in general—would have other societal benefits as well: it would dramatically reduce the widespread housing and job discrimination, and verbal and physical violence experienced by most lesbians and gays, clear moral and social evils.

Of course, legal and religious gay marriage wouldn't, as some writers claim, "celebrate" or be "an endorsement" of homosexual sexual behavior—any more than heterosexual marriage celebrates heterosexual sex or endorses it; gay marriage would celebrate the loving, committed relationship between two individuals, a relationship in which sexual behavior is one small part. Still, the legalization of gay marriage, while not making homosexual sexual behavior any more prevalent, would remove much of the stigma concerning such behavior, at least that which takes place within the confines of "marriage." And if the church sanctions such unions, a further, moral legitimacy will be granted. In short, regardless of the potential societal gains, should society and the church reserve a centuries-old moral stand that condemns homosexual sexual behavior?

We have no choice; the premises upon which the moral stand are based have changed. Science now acknowledges the existence of a homosexual sexual *orientation,* like heterosexuality, a fundamental affectional predisposition. Unlike specific behaviors of, say, rape or incest, a homosexual's sexual behavior is the logical expression of his or her most basic, unchangeable sexual make-up. And unlike rape and incest, necessarily manifestations of destruction and abuse, sexual behavior resulting from one's sexual orientation can be an expression of love and unity (it is the complete denial of this love—indeed, an unsettling preoccupation with genital activity—that makes the inflammatory comparisons of homosexual sex to rape, incest, and alcoholism so frustrating for lesbians and gays).

Moral condemnation of homosexual sexual behavior is often founded on the belief that sex and marriage are—and should be—inexorably linked with child-rearing; because lesbians and gay men are physiologically incapable of creating children alone, all such sexual behavior is deemed immoral—and gays are considered unsuitable to the institution of marriage. But since moral sanction is not withheld from infertile couples or those who intend to remain childless, this standard is clearly being inconsistently—and unfairly—applied.

Some cite the promiscuity of some male gays as if this is an indication that all homosexuals are incapable or undeserving of marriage. But this standard is also inconsistently applied; it has never been seriously suggested that the existence of promiscuous heterosexuals invalidate all heterosexuals from the privilege of marriage. And if homosexuals are more likely than heterosexuals to be promiscuous—and if continual, harsh condemnation hasn't altered that fact—the sensible solution would seem to be to try to lure gays back to the monogamous fold by providing efforts in that direction with some measure of respect and social support: something gay marriage would definitely provide.

Human beings are sexual creatures. It is simply not logical to say, as the church does, that while one's basic sexual outlook is neither chosen nor sinful, any activity taken as a result of that orientation is. One must then ask exactly where does the sin of "activity" begin anyway? Hugging a person of the same sex? Kissing? Same-sex sexual fantasy? Even apart from the practical impossibilities, what about the ramifications of such an attempt? How does the homosexual adolescent formulate self-esteem while being told that *any* expression of his or her sexuality *ever* is unacceptable—or downright evil? The priest chooses celibacy (asexuality isn't required), but this *is* a choice—one made well after adolescence.

Cultural condemnations and biblical prohibitions of (usually male) homosexual behavior were founded upon an incomplete understanding of human sexuality. To grant the existence of a homosexual orientation requires that there be some acceptable expression of it. Of course, there's no reason why lesbians and gays should be granted moral leniency over heterosexuals—which is why perhaps the most acceptable expression of same-sex sexuality should be within the context of a government sanctioned, religiously blessed marriage. But before we can talk about the proper way to get two brides or two grooms down a single church aisle, we have to first show there's an aisle wide enough to accommodate them.

10.

The Passage into Marriage

Evelyn Eaton Whitehead and James D. Whitehead

While the authors of the following essay know that a wedding happens in a single moment of time, they hold that the movement into a marriage is more complex, involving a three-stage process over a period of time. If their theory is correct and if some couples ignore the tasks of one or other stage, they could have a rocky transition into marriage. Some questions suggested by this essay follow:

1. *In what sense is it true that people who prepare gradually and carefully for a career can move into marriage with little preparation?*
2. *How would it be possible for a couple to be engaged and still not actually go through the stage of engagement?*
3. *In what ways do engaged couples need (or not need) the help of a Christian community in helping them prepare for marriage? Of the help offered by the community, which kinds seem to you most important?*
4. *Why do the authors claim the wedding ceremony neither begins the passage into marriage nor concludes it?*
5. *What do they mean by the "frailty of this ritual celebration" of marriage?*
6. *What are the most important features of the third stage of the passage, the one that starts after the wedding ceremony?*
7. *Does it make sense to you that the birth of the first child ordinarily completes the passage into marriage? Why? Why not?*

To recognize marriage as a passage is to see it as a transition that takes time, one composed of different stages, each with its own tasks and challenges. In this chapter we will examine three stages within the complex passage *into* marriage. These stages, in turn, can teach us how a community might more effectively care for its members who are making this passage.

Chapter 8 of *Marrying Well* (Garden City: Doubleday and Co., 1983), pp. 130–142. Reprinted with permission.

Engagement

The first stage of marriage begins in the decision to marry. With or without a formal engagement ceremony, a couple at some point come to a decision to marry. With this decision they enter the process and begin the passage of marriage. Historically, a couple were "betrothed." An agreement—usually between the families of the couple—initiated this first, ambiguous stage of commitment.

This first stage is ambiguous because the couple are committed *to* marry, but are not yet committed *in* marriage. Exactly how "engaged" are we or should we be? Typical of a passage, this stage begins the process of separation from our former life and introduces us to an "in-betweenness." We enter into a serious but partial commitment. Religious people, sometimes apprehensive that such a partial commitment will involve sexual and genital engagement, have often argued against the seriousness of this stage of the relationship. The question of sexual expression is important here and must be considered in the context of a deepening psychological intimacy and commitment.

The peculiarity and ambiguity of this in-between stage of engagement beg for some comparisons. Similar experiences of serious but partial commitment occur for persons in seminaries and religious novitiates (preparing to enter a religious order or congregation). These men and women, often of comparable age to those entering marriage, begin to live a life-style and vocation to which they have yet to fully commit themselves. They live in an in-between stage—acting much like vowed religious or priests but not yet formally or finally committed to such a vocation. This comparison will be odious for some, since it seems to raise the specter of "trial marriage." Yet it remains intriguing for us to consider that we have young persons move gradually and carefully into certain vocations, but rather abruptly and often without preparation into others. Quite apart from questions of sexual expression between an engaged couple, we can ask how the passage into marriage might be structured with some of the care that goes into the vocations of the priest and vowed religious.

During this ambiguous time of engagement, what are the specific tasks for the couple? An obvious task, though it is not always attended to, is that of mutual sharing and self-disclosure. Who are we? What do we hope for in this relationship? What apprehensions arise as we spend time together? The *duration* of this stage becomes important in regard to these apprehensions: only with time can I become aware of my deepest hopes and fears about a lifetime together. Whether these fears concern sexuality, control, or career, it is most valuable that they can be heard and shared now. Our ability to attend to these parts of ourselves now will be an indication of our ability to confront and resolve other difficulties and fears which will arise later in our marriage.

In this early stage of the passage into marriage, a couple might inquire: From what is each of us separating? What must we let go as we move into this new relationship? What, for instance, are our expectations about closeness to our parents and family? Do we have significantly different notions about our independence from, or continuing relation with, these families? It can be very useful to examine these questions for a first time now rather than two years after we are married.

Expectably, an engaged couple's attention is focused on the future. There are questions here, too, that can help them explore the similarities and differences that exist in their visions of life together. How do we think we should make our decisions after we are married? What size family do we hope for? Whatever the content of the reflection, these questions invite the couple beyond the glow of romantic excitement which convinces them that "we like all the same things." Without suggesting that an engaged couple become calculating and overly methodical, we can hope that they use this time well to deepen their intimacy through a continuing and concrete sharing of hopes.

The discussion of intimacy and self-disclosure returns us to the question of sexual sharing. What is appropriate and what is permitted? The notion of marriage as a passage with stages of deepening intimacy and commitment threatens the conventional Christian understanding that all genital expression is forbidden before marriage. In the past, Christians have been concerned that the growing intimacy between the engaged couple not reach genital expression. A more contemporary concern of Christians might be that this intimacy not be limited to genital expression. Many young adults today find sexual sharing easier than psychological and religious self-disclosure. Christians hope for a growing intimacy which is more than improved sexual compatibility: an increase in intimacy which includes a greater openness not only to each other's bodies but also to each other's dreams and faith.

Rites of Passage: Ministering to Engaged Couples

In this important stage of the passage into marriage a variety of secular rites of passage have evolved to signal and assist the transition. A wedding shower celebrates the new relationship and attempts to equip the couple (with gifts) for this new life. A bachelor party confronts the separation of the man from his former relationship with his men friends. Though such a rite need not be effective (his relationship with "the boys" may not change at all), *as a rite* it does acknowledge separation, loss and the beginning of a new relationship.

The Christian community has always been aware that it is called to assist couples in this period of transition. For Catholics, "pre-Cana" programs have traditionally sought to contribute to this passage. The challenge today is how to make this tradition of care, of ministering to marriage, more than merely exhortative. Most married people and ministers today agree that two or three evening discussions with the pastor do not suffice as an educational rite for this stage of the passage into marriage. Many creative efforts are being undertaken to improve and strengthen the Church's desire to minister to this stage of Christian maturing.

The revitalization of this ministry to engaged couples has begun with the recognition that the community of faith, and especially its married members, has an important role to play. The Church's ministry to the engaged cannot be the sole responsibility of the pastor. The community of faith, from its accumulated experience and its lived hope for marriage, has a contribution to make. We must do more than admonish the couple to abstain from genital sharing before the marriage ceremony. We must develop means of assisting couples to explore their own hopes and expectations, to recognize areas of potential difficulty and conflict, to acknowledge the fears and apprehensions that arise during this time. However life-giving we know the Christian vision of married love to be, this vision risks remaining rhetorical and vaguely unconvincing if not made available in ways that are attractive and effective.

One of the results of current efforts in the Church to devise effective means to prepare for Christian marriage has been to reinforce the expectation that engagement takes time. A number of Catholic dioceses now require a six-month engagement period before marriage. If such a requirement is merely one more "marriage law," if it is not supported by compelling opportunities to use this time well, it will necessarily fail in its intent. But in many areas the effectiveness of Engagement Encounter and other programs of marriage preparation developed at the diocesan or parish level are signs that this time is being well spent.

What might the educational rites of passage into marriage include? One appropriate component is education in the skills of communication that are crucial to a lasting relationship. It is fitting that the Christian community take advantage of information the psychological disciplines provide concerning successful interpersonal communication. The skills of listening, empathy, and self-disclosure are neither mysteries nor gimmicks. They are behaviors that can be learned and that assist the development of open and direct communication. We can learn to identify our own feelings more accurately and to share these with others in ways that are appropriate. We can learn more effective ways to support and to challenge the people we love. Communication skills in themselves are morally neutral; they can be used to pursue a variety of goals—expressing love, selling products, manipulating other people to my way of doing things. But mutual manipulation is not the usual or necessary result of improved communication. And in the context of the Christian vision of married love, these skills can become virtues—part of our habitual way of loving one another well. The skills of communication do not magically abolish conflict or do away with all difficulties between us, but they help us deal gracefully with these problems when they do arise.

To some, these skills still seem to be secular techniques, not quite fitting for the Church to include in its ministry. In such an understanding the faith community is left to exhort and encourage but has few practical ways to equip its marrying members for the challenging journey ahead. When we can envision the skills of listening, empathy and assertion as Christian virtues we will begin to develop ways to include these practical and powerful tools in our marriage preparation programs. Part of the weakness that some sense in Christian life today is that our highest values and hopes have been disengaged from the practical means of pursuing these ideals; we have the goals and ambitions without the virtues to pursue them. Regarding marriage, for example, the Church exhorts fidelity, trust, endurance in love; but *how* are we to achieve these? How—practically—are we to deal with our anger, or shame, or frustration, or inability to share ourselves?

These are all questions of virtue; they point to the skills of effective living and loving. Without these, our values and ambitions for Christian marriage may too easily become unattainable ideals.

Marriage is for the virtuous. Our living of marriage will make us stronger, more virtuous, but some strengths are required even to begin the journey. A contemporary rite of passage at this early stage of engagement will include educational programs which provide the skills and virtues that will allow our intimacy to grow and to flourish.

The Wedding Celebration

The ceremony and celebration of marriage—"marrying" in the usual if limited sense—is both a brief stage in the marriage passage and a ritual transition between stages. If a wedding is seen as including the various events that immediately surround a marriage ceremony, it can be understood as having some duration. Being more than the magical transition from unmarried to married "in the twinkling of an eye," this brief stage merits special attention here. Its position between the other two stages of the passage into marriage is significant. Our attitude toward the marriage ceremony reveals our understanding of the larger process of marrying. The ritual celebration of a marriage neither *begins* the passage into marriage nor *concludes* it. This public ritual event, occurring in a single day, ideally celebrates a relationship well begun; it celebrates the conclusion of the work and exploration of a courtship and engagement. Such a ritual should *testify* to the success of the engaging: this is a relationship with recognized potential. Both the couple and the believing community to which they belong should be able to acknowledge this potential—a potential concretely based in increased awareness of who each of these persons is and what their hopes are, as well as in the evidence that they possess the virtues necessary to sustain this relationship. The wedding ceremony is thus both a recognition (of a process well begun) and a promise. It initiates the next stage of marriage and promises that this commitment will endure.

Perhaps most impressive today is our awareness of the frailty of this ritual celebration. From a secular point of view, we recognize that our deepest good will and intentions cannot effect an enduring marriage commitment. We cannot simply *will* our marriage; no matter how loud or emphatic our "I do!" at the wedding ceremony, other forces at work within us and beyond us profoundly influence our survival as a married couple. From a religious viewpoint, we recognize that our firm consent does not guarantee a lasting marriage; seeing marriage as a complex religious passage, we acknowledge that God cannot miraculously bring about in this ritual the strengths of an enduring commitment if these are lacking in what we bring to the relationship.

Christian theologians in the past have focused considerable attention on this ritual and sacramental celebration. Theological convictions about the force of the ritual found expression in such pieties as "God will provide." Indeed, as Christians, we believe that God does and will provide; but today we demand less of this single ritual moment. The grace of marriage—God's intervention into our life, empowering us to love well and enduringly—is not concentrated in a single ritual. This grace is encountered and responded to in the discussions, sharings and conflicts through which engagement grows; this grace also strengthens us as we build our marriage during its first months and years. If we are oblivious to these graces and to God's active presence in the other phases of this passage, it is religious romanticism to search for this grace and presence in the wedding ceremony.

In religious reflection on the sacrament of Matrimony today, as we have noted, there is renewed interest in the role of the couple. As the Church assumed, over the centuries, more and more control over the civic as well as the sacramental reality of marriage, the priest or minister became increasingly central to the celebration. Today sacramental theology is struggling to recall the more ancient Christian realization that it is the woman and man who marry each other. Theologically speaking, the couple perform the sacrament: they are the celebrants and agents of this religious ceremony. Today many priests and ministers—and couples—are striving to recover the proper agency of the marrying couple. Rather than simply being led through a service, couples today often participate in the design of the ceremony and even the formulation of the vows. In this and other ways we are attempting to overcome the caricature of ministry to marriage in our recent past: ministerial neglect during engagement and the first years of marriage, accompanied by a formal control of the sacrament of Matrimony itself.

The intent of these reflections has not been to deny the importance of the wedding ceremony and the sacramental celebration of marriage. Rather we have intended to locate this celebration more firmly in its context—the several-year-long passage into marriage.

Completing the Passage into Marriage

With our wedding vows and ceremony our journey is not completed; it moves into another stage. There is widespread recognition that the first months of a marriage are especially crucial to the relationship. What occurs in this period will powerfully influence the later shape of the marriage and its viability. We would like to explore this period—the first year or two of marriage—as a stage in the passage into marriage. This approach reminds us that the couple are still in the process of marrying. This analysis will help clarify the agenda that is special to this time and show how this stage of the passage into marriage is itself concluded.

This stage likely begins with the honeymoon—a brief period of privacy and intimacy as the couple are dramatically separated from their former ties. The commitment of exclusive intimacy that was promised is now ritualized in this special time together which, fittingly, we take apart from the distractions of jobs and everyday life. The honeymoon is preeminently a time for the couple alone: a time to celebrate and build their intimacy, between the past demands and distractions of their former families and the coming demands and distractions of their future family.

Many couples today are choosing to lengthen, if not the honeymoon, at least the initial phase of their marriage, in which they give special attention to their relationship itself and its patterns of intimacy. Two factors contribute to this. First, we are more impressed today with the need to understand each other's patterns of work and play, of value and stress. In the first year or so of marriage a couple can expect to learn many new things about themselves and each other. We may find that it is necessary for us to reflect together on the importance of a career: What priority does work have in our marriage? What will we sacrifice for your career, for my job? How will work fit in with the demands of raising a family? This kind of decision making may be impossible to accomplish fully before we are married and share the everyday experiences and stress of life together. We recognize the need to clarify some of these questions before we begin our family.

There is another, related factor. Couples today are impressed with the important differences between being a spouse and being a parent. As obvious as this may be, historical assumptions about marriage and family have clouded the practical distinction between these two roles and vocations. When "wife" and "mother" merged as a single expectation of marriage, the important difference between these two roles was obscured. Likewise with the man: "husband" and "parent" both awaited him, but how, precisely and practically, would these roles and tasks differ? Today many couples judge that they need time to learn what it is to be spouse before taking on the responsibilities of being parent as well.

The delay in beginning a family often comes not from an unwillingness to have children, but from a realization that we had best give careful attention to our intimacy—who we are together and how we can support and challenge each other—before beginning our family. In the first few years of marriage such careful attention to the relationship, assisted by the skills learned during the stage of engagement, can build a strong foundation for our future family. Less optimistically, such attention will, in some instances, help a couple to recognize a deep and unforeseen incompatibility. As tragic as this may be, it is less tragic than discovering this after, and perhaps by means of, the birth of children.

The passage into marriage is completed in this mutual exploration of our priorities about work, our styles of lovemaking and our methods for handling everyday decisions. Such a passage cannot happen on a honeymoon. Yet we can hope that in a year or two of careful attention to our mutual love we can build a relationship that is able to support new life. For an effective ministry to marriage, it is important to highlight this period, its expectable duration (certainly different for different couples) and its specific tasks. Such attention will take us beyond exhortation and toward structuring opportunities in the parish and elsewhere for couples to reflect together, concretely and skillfully, on the tasks of intimacy they are confronting in these early years.

This important stage of the passage into marriage is emerging today with greater clarity. As it does, we also see more clearly how this stage is most often concluded. The several-year-long transition, with its focus on the couple and their growing mutuality, often comes to its fruition in the arrival of the first child. Of course, a married couple's intimacy continues to grow and change throughout their life. And, as we shall discuss below, not every couple can or chooses to have children. Yet the usual, expectable conclusion of the initial passage *into* a life of married intimacy occurs in the birth of the first child.

Most couples testify to the radical change in their marriage that accompanies the arrival of this first child. Some experience an abrupt shift in their sexual relations. Their pattern of sexual sharing, already adjusted during the last months of pregnancy, changes further with the presence of a new person in the family. Attention to this new life is fatiguing; sexual interest changes for one or both of us; or our pattern of frequency or timing has to be altered. Other physical or psychological aspects of our life together enter in here: my increased weight makes me feel less lovely and lovable; our exhaustion with nighttime feeding may arouse anger and blame. Along with the joy and excitement of childbearing, a couple can expect also to feel new and considerable strain in their own relationship.

Our understanding of marriage as a passage suggests that one part of this lifelong journey, the *entry* into mutual love, is concluding as the couple's love is turning in a new, outward direction. As the relationship turns in this unfamiliar (though familial) direction, conflict and strain are expectable. An effective Christian ministry to marriage will help couples anticipate this significant change. It will allow couples to recognize this transition as not only a confusing time but one with great potential. By so doing, Christian ministers and communities can lead new parents beyond embarrassment into a recognition that they are in a new place in their marriage. Such ministry, again, is best understood not in terms of exhortation, but as facilitating the sharing of feelings and the effective communication of needs in this time of stress and transition.

The recognition of the birth of the first child as the conclusion of the passage into marriage and the beginning of the family stage alerts us to a merging of ministries at this time. With the arrival of the baby most Christian couples begin to plan for bringing this child into the believing community in Baptism. The parents' interest in Baptism may arise from any number of sources, ranging from social pressure and guilt to a much more mature ambition for their child's religious formation. Whatever the motivation, Christian ministers are recognizing the time of Baptism as a special opportunity not only for the child but for the parents as well.

As Christian ministry itself moves beyond the image of the clergy ministering to a passive laity, parents are being brought into the ministry to the newborn child. And this does not mean only that the parents take a greater part in the baptismal liturgy. In many communities ministers to Baptism invite the parents of the child to reflect on and share their reasons for bringing this child forward for Baptism. Why are we doing this? What do we hope will happen with this ritual and beyond? What of our own faith do we have to share with this child?

Such an exploration, facilitated in a non-threatening and skillful fashion, is not only a ministry to the child; it is also a direct and effective ministry to the couple. The birth of their child is an extraordinary opportunity for them to inquire into what they hold sacred, what they would have their child share and continue. It is a special opportunity for them to say aloud—to each other and even to their community—what they believe. Such explicit faith sharing, for which there is little opportunity in a busy, workaday life, will include belief and unbelief, deep hopes that surprise us and doubts that frighten us. To share these deepest parts of oneself is itself an exceptional act of intimacy—religious intimacy. We are often either too distracted or too embarrassed to share in this way. The birth of a child may be the best opportunity we ever have. This is a part of the sacredness of this special time. The ministry of Baptism is also a ministry to the family: the child's chances of growing up in a vital Christian family are greatly enhanced by the religious sharing the parents are brought to at this time. The paradox of this opportunity and this transition from *marriage-as-couple* to *marriage-as-family* is that as we minister to our child in Baptism, this child ministers to us, inviting us to a deeper and more explicit experience of our faith.

To return to this event as the conclusion of the passage into marriage: with the birth of the child the focus of our marriage is shifted away from us toward this new life. We now become less exclusively *for ourselves,* concluding an important but limited stage in our marriage. Our attention now begins to turn outward, more emphatically beyond ourselves. This shift in focus is crucial for every marriage. The deepest sense of the fruitfulness of a marriage is related to this impulse to make more life, to create and go beyond ourselves. A marriage which fails to develop this external impulse we call selfish: with or without children, a couple can live a selfish, unfruitful life together. This notion of the fruitfulness of a marriage and the transition from marriage to family raises the question of a childless marriage. The phrase itself may frighten us, with its connotations of frustration and emptiness. We are reminded in these connotations that marriage is meant to go beyond itself; lovemaking is to make more life. But we also recognize that every adult life, whether one is married or not, must necessarily be fruitful. Married and unmarried can remain barren—with no generating of new life, no caring for the next generation and no moving beyond their own immediacy. And we know, too, that the biological generation of children does not guarantee human fruitfulness.

A number of marriages will be childless. Some couples unable to bear their own children will adopt children as a means of moving beyond themselves and becoming fruitful. Other couples will choose not to have children. Some, to be sure, may choose this future out of selfish reasons, just as some will give birth to children from self-centered or irresponsible motives. Still other couples will seek other routes to fruitfulness, to having their mutual love go beyond themselves in creative care for the next generation. The selfishness or maturity of a couple cannot be immediately identified in their decision to have few, none or many children. Their maturity or selfishness becomes clear only in what they do during the decades which follow this decision.

In a discussion of rites of passage, it may be important to inquire how a couple who do not have children can signal this transition from an inner focus in their marriage to an outer focus and fruitfulness. What kind of rites might assist them to celebrate this turning of their attention and other resources beyond themselves? Will it be a shift in career, or perhaps a broader involvement in the civic community, or a more visible commitment to some aspect of social justice that will indicate to them, and their closest friends, that this transition is taking place? It is clear that we have much left to learn about the rituals that serve us effectively in adult life.

With the conclusion of the passage into marriage, attention turns more definitely outward. Roles are multiplied as ''parent'' combines with ''spouse.'' Marriage, well begun and now maturing, turns to new challenges.

11.

The Family as a School for Character

Stanley Hauerwas

Students will find the following essay challenging both for its claims about Christian marriage and for its closely packed ideas. It challenges many current assumptions about the nature of love, of marriage, and of child-rearing. Written by a Methodist theologian, the essay shows the connection between the love of an individual couple for one another and the various communities each person comes from: family and church. In the process of discovering one another, some people prefer to look at the beloved as if they are the only ones who ever really knew or loved him or her. However, each of us has been formed by life in various groups, especially the one we call family.

Part I is an important introduction to Hauerwas's convictions, including a central one: that we cannot assume character to be formed primarily by the family. Some of the questions he raises: Could we make the family into a kind of god? Can we invest the family with a magic significance as the key to all problems? Can we ask the family to do more than it is able?

Part II examines the relation of church and family, but from the perspective of church. Hauerwas looks at early church views about family. Singleness was seen to be as good as marriage. It was just as much a calling from God as marriage. The vocation to either comes from one's life in the saving community, the church.

Part III lays out the distinctiveness of Christian *marriage, as a commitment to lifelong fidelity in imitation of the fidelity of God's own self. In a Christian marriage a couple have children, not out of biological necessity, but out of hope in God's goodness, Parenting is a vocation that comes from one's life in the saving community, the church.*

In Part IV the writer asks what sort of character is necessary to sustain the life of marriage and singleness and how such character is formed. He asks not what kind of love makes marriage possible but what kind of fidelity makes love possible. The problem in marriage is not one of being able to say, "I love you," but of having a character formed in knowing how to be faithful.

In Part V, Hauerwas denies in various ways that love creates marriages, claiming instead that character creates marriages and marriages create love.

From *Religious Education* 80:2 (1985): 272–285. Reprinted by permission of the Religious Education Association, 409 Prospect St., New Haven, CT 06511–2177. Membership information available upon request.

Communities of love are needed to help us sustain our commitments. A very similar idea about communities and fidelity can be found in Jo McGowan's "Marriage Versus Living Together," reading #8.

I

In my past work I have stressed the significance of character and marriage for understanding the nature of our moral existence, but I have not tried to make explicit how they may be related.[1] I have relied on our intuitions that there is probably some very important relation between the development of character and the family, but I think it important now that I try to make candid what I understand that relation to be because I suspect that many of our assumptions about the relation between morality and the family may not only be descriptively mistaken, but theologically suspect as well.

For even though I am sure there is a relation between character and the family, I suspect it is not the one we think. The accepted account about the relation runs something like this: the family is where we learn to be moral or develop character, thus good families produce good children and adults, bad families produce bad children and adults. We know there are exceptions to these generalizations. We all know of people who have overcome extraordinarily bad families; we also know of tragedies where very good families are beset by extremely unpleasant children. But we persist in believing that the generalization on the whole is true.

Moreover the generalization underwrites a certain disquiet about our contemporary situation, for it seems that we live during a time of familial breakdown. If the family is destroyed, so is our morality; if our morality is destroyed, so is our society. So if we want to begin to do something about renewing our society, we should begin with the family. Therefore the concern with many about the family is really a concern about the very foundation of our civilization.

I think, however, that there is quite a number of things wrong with this set of assumptions about the relation of character and the family. First, it may be true that marriage and the family are, like suffering, a test of character, but I do not believe these are the setting where character should be formed. Or to put it more accurately, I do not think that the kind of character developed by the family is sufficient to sustain the moral demands made by being married or by learning to live, as we all must do, as part of a family.

Put simply, no one gets married or begins a family in order to develop character. It may be that being married or learning to raise a family becomes for many the decisive occasion for moral growth; but that is quite different from saying that the family, in and of itself, is the institution that should provide the context for our most decisive moral development. For example, I suspect it is true, as many attest, that they never realized how self-centered they were until they had children. Yet I do not think that our self-centeredness is necessarily cured by our relation with our children. Indeed, too often, our selfishness is only transmuted into more virulent forms as *my* children become a moral legitimation for me to ignore the claims of others in my life.

Though I will suggest that the character necessary for marriage may be enhanced through formation by the family, I do not believe that our character should be formed primarily by the family. Of course, part of the difficulty about such claims is that, as yet, we have little idea about what is meant by character. I have been using character as equivalent with morality; but as I hope to show, to use the language of character challenges many contemporary assumptions about how we should think about morality. But more of that later.

The second reason I think it a mistake to assume a direct correlation between the development of character with the family, and it is related to the first point, is that such an association can too easily turn the family into an idolatrous institution. Too often the church is supported because people care about the family. They assume the church is good because it produces a good family. This attitude is best exemplified by the horrendous claim that the "family that prays together stays together." That makes prayer valuable for some other reason than it being a crucial way of our making ourselves available to God. To give the family such significance is idolatrous as it means God is worshipped as a means to help sustain what we really care about—the family.

Ironically, when we turn the family into that kind of god the result is the destruction of the family. For when the family is asked to carry such moral significance, when the family is thought to be the one place in the moral wilderness in which we live that provides an anchor of moral stability, it is broken by being made to do more than it is able. When the family is invested with such significance, it cannot help but be morally tyrannical; and the tyranny is made all the more perverse by being clothed in the form of love and care.[2] It is no wonder that children, seeking to be free from such care, rebel against everything their parents care about. In doing so, of course, the person they often end up hurting the worst is themselves, but sometimes even injury to self seems better than having no self at all.

Therefore, to spell out how the family should contribute to the development of character I need to say a good deal more about how Christians should understand the nature and status of the family. In the process I hope not only to illumine questions concerning the family, but also the nature and kind of character that should be associated with being a Christian.

II

In order to begin to understand the place of the family for Christians, we cannot begin with the family. Indeed, one of the interesting things about the New Testament is not what it has to say about the family, but that it has so little to say about the family. We have to look hard to find texts to analyze, and even then we often seem to end up with half a picture; or even worse, it seems that at least some of the texts have a positive anti-family ring. We often forget that the disciples, after hearing Jesus on marriage and divorce, suggested that it would be better not to marry; and Jesus' reply did nothing to dissuade them from that notion (Matthew 19:10–12). Moreover we tend, on Mother's Day, not to preach on texts that suggest that our only primary relations are to those who are members of the kingdom of God—"who are my mother and brother but those that do the will of God" (Mark 3:31–35).

It is not my intention to try to develop an account of the family by referring to one text after another. While valuable, I do not think that such a procedure can give us the kind of overview that we need in order to appreciate the kind of revolution the early Christians perpetrated in relation to the status of the family. The account I develop, while trying to be true to the scripture, is an attempt to provide a more general account of the early church's understanding of the family.

Therefore when I say that, to understand the early church's views about the family, you cannot begin with the family, I mean exactly that. For the first way of being Christian for the early church was not in the family, but by being single. Protestants have tended to forget this inasmuch as part of our political power has depended on our renaturalizing the family as the essential institution for social order. We have therefore selectively interpreted the scripture, trying to show that we Christians, like most people, are on the side of the family. But that is just not so.

The most startling fact of the early church is that singleness was regarded as good as marriage, if not preferred to it, for those who joined this new community. I simply think the evidence for that is undeniable. The question is, how we are to understand it, as well as try to appreciate its significance for us? Many attempt to explain the emphasis on singleness by noting it was a pragmatic necessity. The early Christians were, after all, a minority under persecution and there was much to be done. Therefore, some remained single in order to devote all their time and energy to the work of the kingdom. When times got better, there was less need to stress singleness.

While such considerations no doubt played their part, they are clearly not sufficient to explain the early, as well as the later, emphasis in the church on the importance of singleness as a vocation. For pragmatic reasons are really only the result of profounder theological convictions which made, and continue to make, singleness so crucial to the nature of the church. For the church, we believe, is an eschatological community that lives through witness and conversion. That some, and perhaps many, remain single in the church, is therefore not accidental, but a crucial sign of the kind of people we have been called to be. For we believe that the church grows, not through a socialization process rooted in the family, but through God's constant call to the outsider to be part of his kingdom. Therefore singleness stands as a remarkable witness to the world that this is God's community, God's people, and not another human invention.

Indeed, from this perspective you can tell the church is in deep trouble just when it tries to make the family a substitute for its obligation to be a witness. No longer confident we have a message that can attract others, we rely on the family to bring new members into the church by making sure our children never seriously consider they have other alternatives. Of course, in the process we ironically begin to look more and more like our surrounding culture, as the familization of the church increasingly makes Christians end up looking just like every one else.

Though the emphasis on singleness may appear odd in a Protestant context, I assume it is intrinsic to the free church tradition. For it is the free church that refuses legal support by the nation to make clear that the church has a loyalty that cannot be captured by the nation. We have little choice not to be an American, or a Canadian, or a Frenchman. Of course, nations in the liberal tradition try to give us the illusion that we freely choose to be American, but in fact we know that we are determined to be so. Just try not to pay your taxes. But the free church stands as the challenge to all such determinism by maintaining that one can only become a member of the church through being called and by responding through willing commitment. Essential to maintaining that freedom is the refusal to turn the family into an idol in the interest of becoming a cultural force.

It is only with this account of singleness in the background that we can understand the church's transformation of marriage. For no longer is marriage a "natural" institution for the preservation of the species or a paradigm for interpersonal relations. Rather, now marriage is a practice put in service to a wider community—and that community is not the "society" but the church. Marriage, in effect, like singleness, becomes a vocation to which some, perhaps the majority, of the church are called. In that sense, marriage is a reality prior to any couple's decision to be married. For marriage is a set of expectations carried by the community that offers an opportunity for some to be of service in the community.

The reason that singleness makes marriage possible is that, only when we realize that we do not have to marry, does marriage between Christians become a vocation rather than a necessity. Marriage is of service to the community by reminding the church that if we live by hope we do so as a patient people. As much as we long for the kingdom, for justice, we know it is not we who make the kingdom come. We cannot storm the walls of God's will to force the world to be ruled by love. God's kingdom will come when God wills it to come. In the meantime we can rest secure in the conviction that we have everything we need through the life, death, and resurrection of Jesus of Nazareth. As a result we can take the time, even in a world as unjust as ours, to be faithful to one another and to have and raise children.

III

For marriage between Christians is shaped by our commitment of lifelong fidelity to another and by our willingness to be open to new life. Christians believe that such fidelity is the hallmark of marriage.[3] By pledging ourselves to another for a lifetime we believe we learn something of God's fidelity. Marriage thus stands as a fitting vocation in the church to remind us of the faithfulness of a God who has made it possible for us to be open to the call of another without qualification. Children are but result of that openness, as we do not believe the fidelity and love demanded of those who would be married can be true unless it is orientated to and expressed in new life. Christians have children not because they believe that God has willed the indefinite life of the human species, but because they are hopeful and patient people. Ironically, when marriage is justified in the name of the perpetuation of the human species, we lose exactly the kind of hope that makes it possible for the human species to continue to live. For in that process we deny that it is God, but the human species, who is our ultimate loyalty.

Putting this in a somewhat different way, the first family of every Christian is not what we call the "biological" family, but the church. In that sense we must not forget the kind of family appropriate to our Christian convictions cannot help but challenge the conventionally secular assumptions which sustain family habits outside the church.

Let me try to illustrate this by telling you what I do sometimes when I teach a course on marriage and the family. I read to the students a letter I have composed from parents to friends which describes a family tragedy. It says that their son was doing very well. He had been an outstanding student and athlete. After serving in the military he had gone to law school and seemed to have an unlimited future ahead of him. There was even talk of his having high political office. However, he had recently joined a religious sect from the east and he had turned his back on his former life and family. He said that he now wanted nothing to do with the world or with his family. Moreover, he claimed that his whole ambition was to be of service to his religious brothers and sisters and in fact he never intended to marry. The family says they are just heart-sick from all this and do not know what to do.

I ask my students who they think wrote this letter. They usually assume it is from a family whose son has become a Moonie or joined one of the many Hindu sects. I then point out to them it could have quite easily been written by a fourth-century Roman senatorial family about their son's conversion to Christianity. We forget that from the Roman point of view the church struck at the heart of the moral presuppositions that underlay the grand achievement of Rome. From Rome's perspective, the family was an absolute necessity for providing not only the numbers but the kinds of people necessary to run a vast empire. The church decisively challenged those assumptions by making the family a secondary loyalty next to loyalty to the church.

From the church's perspective, Rome no longer had first claim on their children; but then neither do, as is often claimed, the biological parents. Given the church's creation of marriage, parenting is now no longer understood biologically. Rather, just as marriage is a vocation, so is parenting. That means that parenting is an office of the whole community and not just of those who happen to have children. Both the single as well as the married exercise parental responsibility for the community; they just exercise different forms of that responsibility. Therefore those who teach, those who take care of the sick, those who stand as moral examples, all perform a parental role.

That such is the case is exactly why the church can expect those who marry to be open to the possibility and prospect of having children even under times of hardship. For it is assumed that the whole community is ready to support those who have children by relieving them of total and exclusive responsibility. It has been, and often still is, assumed that those who have biological children have primary, and even sole, responsibility for their care; but that assumption is not because of biological identity. Rather Christians assume that they are called to care for those children because that is part and parcel of their vocation by entering the institution of marriage.

It is only against this background that we can understand the church's stand on abortion. Often the general Christian "no" to abortion is interpreted as a legalistic and heartless judgment that ignores the genuinely tragic circumstances associated with some births. But the church's "no" to abortion is in fact a "yes" about the kind of community we want and are intended to be. For we do not assume that the biological mother, unless she wishes, has to care and raise the child. We, as a community, stand ready to provide such care exactly because we do not believe biology makes a parent in the first place.

Indeed, I sometimes think the deepest failure of the contemporary church derives from our willingness to underwrite the pagan assumptions about parenting. We thus train people to think that they have a special relation to biological children. That is why people who adopt are often asked, "now which are your 'real' children?" From the perspective of the church, all our children are our real children and we will abandon none of them to the kind of possessiveness based on biology. Precisely, that same possessiveness is the kind of demonic assumption that makes us assume, as Christians, that we can kill some other's children in the name of our responsibilities to "our children."

IV

But what does all this have to do with the relation of the family and morality; or more exactly, what does it have to do with how the family is or should be a school for character. So far all that I seem to have done is to make some rather bizarre claims about marriage that have almost nothing to do with the working assumption of most Christians in our society. However, I now hope to show that what I have said to this point is important if we are to understand why character is so crucial for sustaining the Christian understanding of marriage and the family.

For, in effect, what I have suggested is that *the character necessary to sustain the life of marriage or singleness is not formed by the family but by the church.* I do not deny the obvious power of the family for making us who we are; but note that the kind of character necessary to sustain the kind of family Christians care about involves more substantive convictions than the family itself can provide. Put simply, if we have not first learned what it means to be faithful to self and other in the church, then we have precious little chance of learning it through marriage and the family. Marriage and family may help reinforce, or even awaken us to, what we have learned in the church; but it cannot be the source of the fidelity necessary for either marriage or family.

Let me try to illustrate this by challenging one of the basic working assumptions that forms a great deal of pastoral practice. I suspect that most pastors, as they counsel people preparing to be married, primarily probe to determine if in fact they are "in love." The assumption is that what Christians primarily care about is whether the marriage will be constituted by a loving and caring relation between the couple, which will then extend to their family. We thus underwrite the general cultural assumption that the warrant necessary to justify two people deciding publicly to commit themselves to each other for a lifetime is the extent of their emotional attachment.

Nothing could be farther from the truth, nor could any practice betray more fundamentally what the church cares about in relation to marriage. What the church cares about is not love, but whether you are a person capable of sustaining the kind of fidelity that makes love, even in marriage, a possibility. Such fidelity, the kind of constancy that promises we will be present through good and evil times, through sickness and health, is not developed in marriage; it is first of all demanded by and developed by being part of a people who have the right to demand it of us. It is a fidelity that comes by being formed by a community whose life is sustained by a God who has proved faithful to us through the call of his people, Israel, and the establishment of the new age in Jesus Christ. Only a people so formed are capable of the kind of promise we make in marriage—that is, of lifelong fidelity.

Moreover, we believe that only when we have that kind of fidelity, that kind of character, are we capable of love. For love, in the first of it, is not some affection for another that contributes to my own sense of well-being. Love is rather the steady gaze on another that does not withdraw regard simply because they fail to please. The paradigm of such love is not learned in marriage and the family; it is first learned through being required to love our brothers and sisters who, like us, are pledged to be disciples of Christ. The love that we have toward our spouses and our children follows, rather than determines, the kind of love that we learn in the church through our being a people pledged to be faithful to God's call.

Without such training we lack the ability to have character in the first place. The problem with the claim that "I love you" is not that I am insincere in my avowal, but that I lack a self, I lack a character, sufficient to the claim. I may well love another, but I simply am not substantive enough to be capable to know what I have said. That is why fidelity necessarily precedes love: it is only through learning to be faithful in relation to God's faithfulness that I have a character capable of the declaration of love.

Character, and our growth in character, places the emphasis in ethics not on decisions about this or that, but rather on our having a history that makes a whole range of what many consider to be decisions no longer possibilities. Thus the Christian must enter marriage with the assumption that divorce is simply not a matter of decision. We do not become married with our fingers crossed or with the assumption that there is always an "out" somewhere down the line. Only by being so formed do we honestly face the reality of the "other" in a way that his/her differences can act to expand the limits of my life so that I might claim that I genuinely love another.[4] And through such an expansion I discover that, rather than being less than who I am, I increasingly have a history that makes me more than I had ever dare hope.

I do not pretend that any of us ever have a character sufficient for marriage when we enter a marriage, but I am contending that at least some beginning has to have been made if we are to have the ability to grow into the kind of person capable of being called to undertake the church's understanding of the vocation of marriage. Indeed, that is why marriage is only possible if it is sustained by a community more significant than the marriage itself. We are sustained not only by convictions about what marriage is about, but by concrete human relations that give us the support we need to face the demands of sharing a history with another human being for a lifetime.

V

I am aware that all this may seem a bit abstract, ideal, or foreign to those who must advise people who are seemingly caught in impossible marriages; so let me call your attention to some examples that may help make the position I am trying to develop more concrete and practical. First, I ask you to consider the fact that for most of the church's life it has legitimated marriage of people who often did not know each other well before the wedding. Indeed, they often did not know each other at all. Though many may think that this is a practice we have well left behind, I think it unwise to overlook the profound moral presuppositions that made it possible for us to bless arranged marriages.

In the first place it is a doubtful empirical claim that, in fact, we have ceased blessing arranged marriage. For example, that is what Notre Dame-St. Mary's, Southern Methodist University, and many other colleges and universities, are about. Students are sent there to meet people of the opposite sex of approximately the same social background, economic potential, and religious affiliation. Such a situation has the advantage of being a form of "arranged marriage" that gives the illusion of choice. There is certainly nothing wrong with such a system except that, like all illusions, it tends to be the breeding ground of self-deception. For as soon as difficulties begin to appear in the marriage, one is invited to think that they simply chose wrongly—forgetting that they were only "choosing" in the first place under very narrow constraints.

More importantly, I think the reality of arranged marriages reminds us that the early church had no illusions about love creating or legitimating marriage. The assumption was that those called to marriage would, drawing on their convictions schooled by the church, have the character to be married faithfully and lovingly. Character preceded marriage; it was not created by marriage. Indeed, one can even push the matter further by noting that, not only has the church blessed arranged marriages, but it even arranged marriages between very young people who had had as yet little chance to develop character one way or the other. While I in no way wish to underwrite this as a good idea, it nonetheless reminds us that such practice rightly assumes that marriage is only possible if it assumes a prior community sufficient for sustaining an ethos for the growth of those who will someday come to maturity and discover they are married.

In this respect I suspect that the family is a crucial institution for the development of character required by the church. In this sense it is a "nature" that makes us prepared for the reception of God's grace. For the one unavoidable fact we all must face is that we do not choose our parents nor do we choose our brothers and sisters. We awaken to discover that we are simply stuck with certain people with no good reason or justification. One way or the other, most of us learn that we simply have to make the best of it. I have no doubt that such a learning is the beginning of growth in moral character for most of us. For it requires that we learn to be faithful to others, to love others, even though *we* did not choose *them* as particular objects worthy of our care of love.

The difficulty occurs when we have no way of explaining to ourselves or even knowing how to describe that formation. We often say that such commitments are simply "natural," but such a description fails to indicate adequately the powerful moral presuppositions that make us turn those "natural" affections into a way of life sponsored by a formative community. Indeed we are becoming increasingly aware, through the tragic evidence of child battering, that such "natural" affections can hardly be relied on to sustain our moral commitments. For what we require is a community that has the convictions as well as the habits that help us to integrate our "nature" into a wider set of expectations that will sustain and enhance our "natural" desires. We need to be taught the "natural affections" we feel are the gracious pull and source necessary for us to be on the road to being people capable of being faithful to ourselves and others—thus people of character.

VI

In a last attempt to suggest that all this may not be quite as crazy as it sounds, let me call your attention to one undeniable fact about marriage and the family that I think helps sustain my account—namely, no one knows what they are doing when they get married and/or begin a family. Do not misunderstand me. I am for marriage preparation. I am for people testing their emotional responses to one another. I am particularly for them working on the planning of the wedding, for as the impeccable authority Miss Manners reminds us, you really only find out about the other person when you have to decide whom you will or will not invite, what color the dresses should be, and so on.

Even more, I am for people being forced to have to consider how they will work out their financial future. For nothing determines our commitment more than our willingness to share our financial destiny. Indeed, I wish the church was more rigorous in suggesting to young people that they may well be in love, but they simply lack the financial resources to enter into the responsibility of marriage and family. Such a challenge would at least help remind us of the kind of commitment which you are making.

Yet I think what finally must be admitted is that no matter how much we may have been prepared to be married, no matter how much we may have considered every possible question, in fact no one can be prepared to be married or for being a family. For how can you prepare for lifetime commitment? How can you prepare for countless small annoyances that you could not know without the experience of marriage itself? How could you prepare yourself for the goods of marriage and the family without the actual experience itself to know? In fact, we must admit that we simply cannot know what we are doing, what we are promising, when we accept the call to be married.

And that is why we so desperately need character, and a community of character sufficient to sustain our growth in character, if we are to enter the extraordinary adventure we call marriage. But what is interesting about this aspect of our commitment is that it reminds us that our acceptance of a call whose implications we do not fully understand is not all that odd. That, in fact, is what almost any important moral commitment involves. It is a promissory note that we can be relied on, even when we do not fully understand the implications of the life we were taking up. Being courageous may well make our world more dangerous than we had anticipated, but that is no reason not to be courageous.

For, finally, marriage and character gain their significance from our recognition of the inherently temporal character of our existence. Morally we are pulled into the future by the commitments we have made, whose implications we hardly understood or anticipated. But that is the way it should be for a people who believe that we serve a God who forever calls us from a life of sin to a new life, free from our fears and our obsessions with safety and control. To serve such a God certainly does require that we be people of constancy, of steadfastness, in the face of the continual temptations to deny that we have a future. But exactly because we are Christians who have experienced God's future in the church, we can continue to accept the call to marriage and family knowing that God has given us all we need to face the challenges of marriage and family with joy, and perhaps even confidence.

NOTES

1. For example, see my *A Community of Character: Toward a Constructive Christian Social Ethic* (Notre Dame: University of Notre Dame, 1981), pp. 155–174.

2. One of the greatest failures of Christians in respect to marriage has been our acceptance of love as the primary defining feature. As a result, the fact that marriage involves power is unacknowledged, so that the language of love often becomes the way coercive relations are legitimized. Phyllis Rose has seen this with great clarity as she says "I believe marriage to be the primary political experience in which most of us engage as adults, and so I am interested in the management of power between men and women in that microcosmic relationship. Whatever the balance, every marriage is based upon some understanding, articulated or not, about the relative importance, the priority of desires, between its two partners. Marriages go bad not when love fades—love can modulate into affection without driving two people apart—but when the understanding about the balance of power breaks down, when the weaker member feels exploited or the stronger feels unrewarded for his or her strength." *Parallel Lives: Five Victorian Marriages* (New York: Alfred Knopf, 1983, p. 7). Power, however, must be understood not simply as physical force, but the ability to impose on the other one's understanding of the relation. Again as Rose suggests, "easy stories drive out hard ones. Simple paradigms prevail over complicated ones. If, within marriages, power is the ability to impose one's imaginative vision and make it prevail, then power is more easily obtained if one has a simple and widely accepted paradigm at hand. The patriarchal paradigm has long enforced men's power within mar-

riage: a man works hard to make himself worthy of a woman; they marry; he heads the family; she serves him, working to please him and care for him, getting protection in return. This plot regularly generates its opposite, the plot of female power through weakness: the woman, somehow wounded by family life, needs to be cared for and requires an offering of guilt. The suffering female demanding care has often proved stronger than the conquering male deserving care, but neither side of the patriarchal paradigm seems to bring out the best in humanity. In regard to marriage, we need more and more complex plots'' (p. 9). The most important contribution the church makes to marriage and family life is to provide the institutional context in which such plots can be developed and told. Unfortunately the realism originally associated with Christian views about marriage has largely been lost in the interest of underwriting modern fantasies about ''happy marriages.''

3. Such a commitment makes possible the ideal of living in a nonviolent way for the resolution of conflicts in marriage. For without such a commitment every conflict invites the temptation to manipulate the other or to engage in fantasies of escape. No marriage can or should be without conflict. The only interesting question is whether those conflicts can be honestly confronted and acknowledged so that they do not become the source of destruction of one another and/or the marriage itself. This point I owe to John Howard Yoder.

4. As Phyllis Rose suggests, ''I am tempted to say that divorce makes marriage meaningless—which doesn't mean I would wish there to be less divorce, just less marriage. When divorce is possible, people no longer need to conform themselves to the discipline of the marital relationship. Instead the law is pressured to authorize more personalized and meaningful forms of relationship. This is the wrong way round. People should be able to hide within the thickets of the law, in Thomas More's phrase. The attempt to make laws supple enough to accommodate the wrinkles of their personalities and desires may be quixotic. Since it was in the nineteenth century that the attempt was first made to humanize the marriage laws, it was in the nineteenth century that marriage as an institution began to lose meaning. Bad enough to choose once in a lifetime whom to live with; to go choosing, to reaffirm one's choice day after day, as one must when it is culturally possible to divorce, is really asking a lot of people. Perhaps better the old way, indissoluble unions with a great deal of civilized behavior—in other words, secrecy, even lying—for the sake of harmony.'' *Parallel Lives,* p. 18.

12.

The Church as the Context for the Family

Robert N. Bellah

Robert Bellah is a sociologist-professor who teaches and writes in California, probably best known for the study he organized of the way adults think and speak, Habits of the Heart. *Here he echoes some of the convictions of Stanley Hauerwas in the previous essay.*

Questions that might help students get at Bellah's main ideas include these:

1. *What institutions are necessary for the support of the family?*
2. *In addition to the ones Bellah names, what systems can you list that influence us all?*
3. *Which of the "solutions" or reactions to the powerful pressures on the family do you find the most appealing? Why?*
4. *What problems does Bellah find in the culture and ideology of romantic love?*
5. *Why does the author say a traditioning community like the church is ultimately indispensable to a committed marriage and family life?*
6. *Why does Bellah pay so much attention to family meals? Are they really so important?*
7. *If a particular family were, as Bellah recommends, a community of interpreters, what would it do different from other families?*
8. *Which of Bellah's ideas have the most or the least appeal to you?*

I think I can take it for granted that we believe in the family. We see it as a basic unit in both social life and spiritual life. We can say this without any nostalgia for an allegedly perfect ''traditional family,'' knowing that the family, like all other human institutions, is subject to sin and corruption and always in need of thoughtful reform. But just the simple idea of a man and a woman vowing to stay with each other ''till death do us part'' and trying to raise children together seems

From *New Oxford Review* (December 1987): 6–13.

daunting. Marriage is difficult. A recent *New Yorker* cartoon showed a clergyman celebrating the marriage service and saying to the couple, "till death do you part or the going gets hairy." And how is one to bring children into a world such as ours? How can we transmit to them a sense of moral responsibility and a religious understanding of life? How can we create within the family a moral and religious atmosphere that can withstand the pressures of the larger world in which we live?

It is bad enough that television programs like "Dallas" and "Dynasty" come into our homes with their sordid tales of ambition, corruption, and cruelty. But these days the newspapers and television news bring into our homes real stories that make "Dallas" and "Dynasty" seem like fairy tales. We read about Ivan Boesky and other inside traders in the top investment banking firms on Wall Street. And we read about the National Security Advisor and the Director of the CIA undertaking or condoning actions Congress had not authorized and which are at variance with our expressed policy. And we read about officials all the way up to the President trying to deny what has plainly happened. The family seems like a very fragile institution if it is expected to adhere to a higher morality than that of our central economic and political institutions.

Let me say at once that I believe the family, or to give it a physical location, the home, is too small and too vulnerable to sustain the moral life of its members unassisted. If it is to succeed over time in providing meaning and coherence it will have to be included in and supported by larger social structures. I want to emphasize the church in particular as a context for family life, but it is not the only institution that is necessary for the immediate support of the family. We must include the school, the neighborhood, and the larger public realm as well. What all these institutions have in common, if I may borrow some terminology from Jurgen Habermas, is that they are part of what he calls the life-world. What is distinctive about the life-world is that its medium of communication, what he sometimes calls its steering mechanism, is language.

In contrast to the life-world, Habermas speaks of the systems, in particular the market economy and administrative bureaucracy. It is characteristic of the systems that their medium of communication, their steering mechanism, is non-linguistic. In the case of the economy it is money; in the case of bureaucracy it is power. Ideally money and power are means that should be used to attain ends established through linguistic communication, through discussion, in the life-world. But as we know, these particular means easily become ends in themselves. Those devoted to them become obsessed with them. In our recent scandals we have seen how certain investment bankers sought to accumulate money beyond any capacity to spend it and how certain administrators sought power without any restraint by law or Congress. These are possibilities deeply rooted in the human soul, in what Christians call sin. St. Augustine, in his masterly analysis of the disorders of the soul, spoke of concupiscence, that is, wishes that are out of control, and *libido dominandi,* the desire to dominate. These are endemic possibilities for human beings, but the powerful systems of the modern world have the capacity to objectify these disordered motives, and to tempt individuals, sometimes almost irresistibly, in the direction of inordinate desires for wealth and/or power. Finally, Habermas argues, the systems become so powerful in our modern society that they begin to, as he puts it, colonize the life-world—that is, subordinate the life-world to the nonlinguistic steering mechanisms of money and power.

A number of incidents reported in *Habits of the Heart* can be interpreted in these terms. Brian Palmer (we do not use real names), whose story is recounted in Chapter 1, was so preoccupied with rising in status and income in his corporation that he badly neglected his wife and children to the point where his marriage broke up and he had to rethink everything about his life. Jim Reichert, whom we find in Chapter 8, had for several years raised money for the local YMCA in order to build recreation facilities for Chicano youngsters with nowhere else to play, an activity that gave him a great sense of fulfillment in contributing to the life of his community. But now he finds the bank for which he works requires him to leave the community of which he has come to feel so much a part or sacrifice monetary advancement and a sense of achievement in his occupation.

One need not turn to *Habits of the Heart* to think of examples of how pervasively the systems colonize the life-world. I am thinking of the constant anxiety about achievement and competition our society fosters almost from nursery school. Will my grades be good enough to get me into a good college? Will I get a good job? Will I get promoted? Will I become CEO? Or, in my own profession, will I get tenure? Will I be recognized in the profession? Will I get a named chair? And I am sure there are comparable pressures on the clergy. Under these conditions it is not surprising that our society offers so many palliatives for anxiety. We turn to drugs when we need relief, to alcohol, tobacco, or cocaine, all of which are enormously profitable businesses. Or we turn to television or music, whose only redeeming features are that they distract us from our worries. Or we think that one more purchase of one more commodity will make us feel better. Both the anxiety and the common palliatives we use to fight it undermine our capacity to function adequately in the family and other parts of our life-world.

Much in the way our society is organized makes us think the pressures we feel are beyond our capacity to resist. Competition is a powerful force and survival a powerful motive. Dare we turn down a monetary advancement, given the high cost of college education for our children and the uncertainty of retirement plans, just because our husband or wife and children would be better off in the place where we presently live? And how can we control what our children are doing? Do we have the authority? Can we set guidelines on the television they watch, or, perish forbid, the music they listen to? And how do we know what they are up to anyway? Powerful outside pressures fragment and atomize the family so that each member goes his or her own way and the home is only a temporary stopping point between lives that are basically unconnected.

There are three ways of dealing with this situation, in ascending order of difficulty, and I think you can already guess that I will be advocating the third. The first is essentially to surrender, and allow the family to be buffeted by whatever waves from the larger society sweep over it. This indeed does not take much effort, but we can hardly be surprised if it produces broken marriages and neglected or abused children.

The second solution is harder and requires considerable institutional support beyond the family. This is the fundamentalist response. It attempts to create a separate culture in the family with as little penetration from the larger society as possible. Certainly the church will be necessary for this solution and probably "Christian schools" as well. The trouble with this solution is that it is undiscriminating with respect to what it keeps out and ultimately with respect to what it lets in as well. It settles for easy and superficial absolutes, believing, for example, that the Bible speaks for itself and needs no interpretation. And it accepts stereotypical views of "traditional values" and the "traditional family," which turn out to date back only to the 19th century, or even to the 1950s, and represent only an earlier version of American values—not necessarily more validly Christian than our current values.

The third response to the pressures on the family coming from the larger society is more difficult, for it neither surrenders to those pressures nor imagines that one can simply erect a wall to keep them out. This third alternative involves a much more active process of discussion and discrimination, one which involves certain basic understandings and then a great deal of negotiation about particular cases. Because this alternative is so difficult, the family alone is not likely to be able to sustain it. Here, as in the second response the church will be of decisive importance, that is, the church as context for discussion, decision, and commitment.

I can only suggest some of the things this third alternative means to me. To begin with, we have to realize that the idea of Christian marriage is not very well understood in our society. Love is a central word in the Christian vocabulary and it is a central word in the vocabulary of romantic love in our society, but though the meaning of the term overlaps in the two contexts it is certainly not identical. In the culture of romantic love, love is primarily a feeling—indeed, an overwhelming feeling that sweeps us away. But unfortunately, it is a rather fragile feeling that often disappears rapidly: according to some social scientists, in its acute form it lasts 90 days and in its more attenuated form about a year.

Jesus tells us to love God and our neighbor. Indeed, he tells us to love our enemies. It is not that Christian love is not a matter of feeling, but it is certainly not only a matter of feeling. You cannot command a feeling, so that cannot be all that Jesus meant by love. And certainly we are not overwhelmed by an emotion of love toward our enemies. Without beginning to exhaust the idea of Christian love, we can see that it involves an element of intention and will that are largely absent in the idea of romantic love. Love intends the good of the loved one. It is more focused on the other than on the self and its transient feelings. Love is prepared to persist in the face of adversity. (Think of the classic phrases in the marriage service in the Book of Common Prayer.)

If we think of love as the basis of marriage, and I think we should, then these two different conceptions of love will imply two very different conceptions of marriage. If marriage is based only on the idea of romantic love, then when the intensity of the feeling begins to diminish, as it inevitably does, the basis of the marriage is threatened. Popular psychology tells us that marriage partners can be friends and that that is an important basis of a good marriage. That is fine and may help things survive for a while, but what if a third party comes along who rouses the intense feelings of romantic love? In the ideology of romantic love, not only is it all right to leave the original partner and go off with the new love, it is actually immoral not to. I believe that this ideology has much to do with the high divorce rate in our society. If romantic love is the chief basis of marriage and intense romantic love is inevitably transient, then marriage will inevitably be transient. This is what is called serial monogamy, but perhaps it should better be called serial polygamy.

The idea of Christian marriage is quite different. For Christians marriage is a contract to a noncontractual relationship. Most Christians believe this so strongly that they consider marriage a sacrament. In any case, its intention is indissolubility: not for as long as the excitement lasts, but till death do us part. Yet, Christian marriages can fail, not because we see marriage as inherently transient but because we know that we are sinners, that we do those things we ought not to do and do not do those things we ought to do. In a society saturated with the idea of romantic love, of course Christian marriage cannot ignore romance. We can be grateful for all the therapeutic advice as to how to bring a little romance back into a longstanding marriage. And we must also face the fact that romantic feelings toward third parties will occur in most marriages, and sometimes more than feelings. And not only sexual feelings, but also feelings of anger and hostility, often justified by the behavior of the other partner, are normal in any long-lasting close relationship. But in a genuine marriage, feelings of rejection, normal enough, will be moderated by understanding and forgiveness. Forgiveness is a central component of the Christian idea of love, and certainly forgiveness is essential in any marriage that will endure.

Forgiveness, acceptance, and understanding are essential components of Christian marriage, but it is not a one-sided forgiveness, acceptance, and understanding. "Traditional marriage" often expected these virtues from women but not from men. There is nothing Christian about that. If they are wonderful qualities in women they are wonderful enough to be shared by men.

But a marriage based not only on subjective feeling, but also on concern for the good of the other and on the intention of indissolubility, requires more than the commitment of two individuals, particularly in a culture that is not particularly sympathetic to those ideas and doesn't even understand them very well. Family and friends are important, but only if they understand and support this conception of marriage. A tradition and a community that embody these ideals, such as the church, are ultimately indispensable. This is all the more the case when the marriage fulfills itself by becoming a family, by bringing children into the world.

Although I will define the archetypal family as consisting of a married couple and their children, and I believe a kind of primary dignity belongs to that form of the family, I want to say strongly that that is not the only form of family that exists, or that needs to be affirmed and supported by the church. Indeed these archetypal families are probably in the minority in sheer numbers. Single-parent families are common and need our special care and support. Pairs of adults of the same sex who have committed themselves to live together as a family, and groups of adults who have agreed to do so for whatever reason, are frequent in our larger cities. Wherever we find relationships of faithful mutual care and support it will be important to include them in the larger pattern of family life. It does not detract from the worth of such relationships, however, to give a certain primacy to the relation of a married couple and their children and to see the health of the family in that sense as vital for all other forms of family life as well.

All I have said about relations between the spouses applies to relations between parents and children. Often we must care for our children when we don't feel like it, or when we don't feel like it at a particular time. If we only cared for children when we feel like it they probably would not survive. We may love our children deeply and yet when they are angry, unreasonable, or inordinately greedy, as at times they inevitably are, we may thoroughly dislike them. And however much we may intend not to, from time to time we let them down. So here too understanding, acceptance, and forgiveness are essential. Especially forgiveness, including the capacity to forgive ourselves.

Things really become difficult when children start school and become to a degree independent. Then it is all too easy for each member of the family to go his or her own way. At this time peer group, television, and school begin to have dominant influence on the life of children, and parents may give up the struggle to have any effect on their characters. The only way for the family to have a formative influence is for the family to create a common culture with common symbols and practices. Though the parents must lead in this process, from an early age children can be included in the creation of the common culture, can contribute their own symbols and meanings, and help shape common practices.

The core of the common culture is ritual, what we might call "family sacraments." Perhaps the most central family sacrament is the common meal. Yet today when everyone is so busy and schedules do not mesh, it is very easy to abandon the common meal altogether, which happens in many families. Each individual drifts into the kitchen when he or she needs a bite to eat and then hastens off to the next event. Such eating cannot even be called a meal but has recently become known as grazing. Resisting this tendency is the beginning of the effort to maintain a family culture.

One should probably try to have a common meal at least once a day and at least several common dinners a week. Family meals are not always pleasant. They may be the occasion of conflict between siblings or between parents and children. But they are one of the few places where family members can find out what each other is doing and thinking. Meals should not be a time only for children to show and tell. Parents should express their own concerns and tell of their activities outside the home. But meals should not be an occasion where only the parents talk to each other and the children listen. All should be included. It is on such occasions that children learn to listen respectfully to others and to express what is important to them. The dinner table is a place to learn the rules of civil discourse, and I can assure you that if they are not learned there they are not likely to be learned in school.

But meals are for eating and not just talking. Grabbing a Big Mac on the run is not a meal. Essential to the ritual is the preparation and enjoyment of wholesome food. The preparation of the food, serving it, and cleaning up after the meal are all part of a co-operative effort which makes the meal festive and sacramental.

These are not activities that should devolve only on the wife and mother. Father and children should play an essential part, especially when, as is usually the case today, the mother is employed outside the home.

In a Christian family it will be normal for the meal to begin with a prayer. This need not be tedious but should be more than perfunctory. It is appropriate to mention in the prayer some major event in the life of the family (such as a birthday), an event in the larger world, or some aspect of the liturgical year, and then to include some discussion of this in the following conversation.

It may be possible to have common prayer, perhaps including a Bible reading, at some time other than meals, perhaps at the children's bedtime. Biblical literacy is threatened everywhere in America today, and not only among children. Thus parents should consciously try to bring in Bible stories and verses, perhaps some lines from a psalm or a parable or saying of Jesus', not only on solemn occasions but as part of the daily discourse of family life. What is learned in Sunday School or religious education classes will not be retained if it is not reflected in family life. And without biblical literacy a genuine Christian community cannot survive these days.

Other family rituals include recreation and holidays. Trying to make time for family outings, picnics, or shared sports strengthens the family culture. If these are a natural part of family life, then the annual vacation will be more enriching and less of a trial. It is particularly important for the family to redeem the great Christian holidays from their distorted commercial form. This is especially difficult at Christmas time, but where Advent is used as a time of preparation, Christmas may take on quite a different meaning. At Easter time it is important to remember Good Friday, something Protestants have not always done. The secular culture does not want to think of Good Friday, for it wants resurrection without crucifixion. Children love holidays, and the great Christian feast days can be the occasion of much joyful learning.

On these occasions, as well as on Sundays, family observances ought to be combined with church observances. The family should understand itself as a cell in the body of Christ, mirroring the local congregation as the local congregation mirrors the whole of God's people. But the kind of integration of family and congregation I am advocating may imply a kind of parish life that is no commoner than the Christian family for which I am hoping. Richard Osmer vividly paints the picture that concerns me:

> Noticing what is absent is more difficult than we commonly suppose. If a person grows up in a suburb in which all the trees were leveled when the houses were built, it is highly unlikely that he or she will miss the massive beauty of an old oak or the splendor of a large maple in fall. What is missing remains hidden, for it has never been a part of the person's experience.
>
> The contemporary Protestant church faces the difficult task of noticing the absence of an authentic teaching office in its life. The fact that this essential ministry of the church has been missing for so long makes it even more difficult to recognize what is not present. Its absence, however, is felt in an increasingly widespread sense that something is missing; something of great importance is not present in the church's life.

What Osmer means by the "teaching office" includes sermons and more informal instruction by the pastor and all the forms of Christian education, so it is not literally missing in any local church. What Osmer means by the teaching office, however, involves the whole congregation in the whole of its life searching for God's will in the light of Scripture and the experience of the Christian people, and trying to discern what to do. For most of us most of the time, when we leave church on Sunday morning we are no longer concerned or only subliminally concerned with the teachings of Christ, which leaves not only our work life but even our family life highly vulnerable to the prevailing winds of the surrounding secular culture. What would a more vigorous teaching office look like?

I was recently talking to a leader in the Mennonite Church about the practice in his congregation. It is a bit rigorous, but it is instructive. Sunday service lasts for three hours. First there is an hour of hymn, prayer, and thanksgiving; then there is an hour devoted to the word of God, including the sermon; then there is an hour of discernment, which perhaps has the most to teach us. This Mennonite group believes the worshiping community should be no larger than 100 people, so when the congregation grows beyond that it splits into two separate worshiping groups. This is precisely because of the importance of discernment, for everyone is included in that part of the service. During the period of discernment what has been sung in the hymns and read in the Scripture and spoken in the sermon is thrown open for discussion as to what it means for those assembled. There is the assumption that what has been taught has authority, but that it is a difficult task to discern what that means and that it requires the effort of all to find the way. This particular group breaks down into still smaller segments that meet one night a week in the homes of members. In this smaller prayer and discussion group the teen-age children are encouraged to attend, to learn from the efforts of their seniors, and to add their own comments when appropriate. This idea of teaching by inclusion rather than through one-way instruction is worth pondering.

The point I want to make is that the church as a context for family life as well as the church as a vital point for the transformation of the world cannot operate without what the Mennonites call discernment. We need to know what is in Scripture, tradition, and history and we also need to think together about what that means to us here and now in this confusing and rapidly changing world. In our individualistic culture it is easy to assume that discernment and application can be left to the individual. All he or she needs is an inspiring or "motivating" church service and the rest can take care of itself. But very often either nothing gets carried away at all or only very halting or even distorted ideas get taken away and the individual has no real idea of what to do about it. This is why we need the church not just as a group of passive listeners but as an active community of interpreters able to carry the message into the whole of our lives. And the family too should be, as a microcosm of the church, a community of interpreters.

It is this notion of the church as a community of interpreters that makes the third response to the challenge of culture that I mentioned earlier different from the second or fundamentalist type. The fundamentalist response to all the problems of life is that "Christ is the answer." In practice this means relying on a very superficial reading of the Bible or on the words of an authoritarian pastor. But if

Christ is the answer, he is also the question. If the answer to our problems is to follow him, and I think it is, then we find, if we are serious, that Christ is challenging us very hard to think about what following him actually means under these particular circumstances of our lives. He does not expect us to sink into passivity, but to take up our cross, to take up the way of the servant, and to discern in the midst of all our suffering and confusion what to do. And mercifully he has given us his church to help us in that task. We need to uphold and sustain each other and we need all the wisdom we can gather from our brothers and sisters in Christ. But the answer is almost never obvious or easy. And we will find that we make many false starts and must begin again or correct our direction.

But let us try to be a bit more specific in thinking about the process of discernment, which, by the way, can also be called practical theology or the theology of the laity, or, as far as I am concerned, just theology. Richard Mouw has called attention to a useful set of criteria developed by Gerald Vandezande in a book entitled *Christians in the Crisis*. These are guidelines for what Vandezande calls "biblically faithful stewardship" and were developed initially with an eye to investment policies, but, as Mouw suggests, have much broader implications. Christians, Vandezande says, must attempt: first, to be *gentle* in the way we treat the environment; second, *just* in the way we treat our fellow workers; third, *wise* in the use of the creation's resources; fourth, *sensitive* to the needs of our neighbors as they pursue their vocations and tasks; fifth, *careful* in the way we use technology, so that we do not idolize technical know-how, but use it as a way of serving legitimate human goals; sixth, *frugal* in our patterns of energy consumption, so that we do not waste the building blocks of the good life; seventh, *vigilant* in the prevention of waste; eighth, *fair* in the determination of prices; ninth, *honest* in the way we promote the sales of our products; and tenth, *equitable* in the earning of profit.

Actually that would be an excellent list to begin a discussion group on Christian vocation, on the way we can think about being Christians in our occupations. But it could just as well be the beginning of quite a few good discussions in the family. How do we in the home, as well as in the church and the larger world, show forth the gentleness, the justice, the wisdom, the sensitivity, the care, the frugality, the vigilance, the fairness, the honesty, and the equity God requires of us?

Another way to look at Vandezande's list is to see that he is advocating what Albert Borgmann calls "careful power." Borgmann points out that in much of the world today we are devoted to the exploitation of what he calls "regardless power." When we cut down rain forests for short-term profits even though there may be long-term deleterious effects on the world's climate, that is regardless power. When we support military insurrection without exhausting the possibility of negotiation, that is regardless power. When a corporate headquarters in New York closes down a plant in Oklahoma where 200 women make slacks at $6 an hour in order to open one in Mexico where the women will make 60 cents an hour, that is regardless power. Or when we load our homes with expensive appliances that are used only a few times a week, if then, that is regardless power. Jesus reminds us again and again, in the Beatitudes, in the whole of the Sermon on the Mount, in many of the sayings and parables, that his way is the way of careful power, not of passivity, but of the gentle persistence of love. But in a culture obsessed with what is called ''regaining the competitive edge,'' with winning and dominating and being number one, it is easy for Christians to forget about Jesus or, even worse, to imagine him as a ''member of a winning team'' devoted to regardless power, and so turn him into our image rather than conform ourselves to his. We need the family and the church working together to help us discern a more truly Christian way.

Yet I do not want to give the impression that the family and the church together can provide a little oasis in the larger society where we can maintain a decent Christian life whatever happens. There are societies where that strategy is the only one possible and then it must be carried out as best one can. But in a society where free institutions still survive, institutions to which Christians have made enormous contributions even though we have never come near creating a ''Christian society,'' I believe we have an obligation of citizenship as well as discipleship. Or, I would say, following Fr. John Coleman, that citizenship is part of our discipleship. It will take more than the church to nurture the family, even though I have tried to argue for the church as the primary context for the family. The family is terribly vulnerable to pressures from the larger society that the church itself cannot control. Tax laws, the welfare system, the cost of housing, work schedules, and patterns of remuneration in business and industry all have enormous consequences for family life. The family will be colonized and strip-mined by the systems controlled by money and power unless we can reinvigorate a political life dedicated to the common good.

Not only must the church defend and strengthen the family, it must also lead us into a vigorous discussion of public life so that we can discern those measures that will limit the destructive consequences of regardless power and will shape our lives in accord with the dictates of careful power. It will not always be easy. We must expect to be rejected and persecuted as Jesus was. But if we follow his way we also know that we are the salt of the earth, the light of the world, that we have lost our lives in order to find them in the most joyous fellowship of all.

IV

Attitudes Toward Sexuality

13.

Passionate Attachments in the West in Historical Perspective

Lawrence Stone

Lawrence Stone, a historian at Princeton University, introduces the rich set of readings in this section with a historical examination of passionate attachments in the West. The two most common passionate attachments explored are: 1. romantic love between two adolescents or adults of different sex, and 2. the caring love-bond between mothers and children. We have undergone a revolution in both sets of attachments in the past century and find ourselves in a unique position today. Stone traces the historical trend in the spread of the cultural concept of romantic love in the West and the evolution in the mother-child relationship. This fascinating material gives us an educational reminder: We cannot assume people in the past thought about romance and parenting the way we do.

Study questions to assist in processing this article follow:

1. *What is the nature of romantic love?*
2. *Is it an adequate rationale and basis for marriage?*
3. *Is there a relationship between romantic love and permanence in marriage?*
4. *What is the distinction between romance, lust, and committed love?*
5. *Is there a gender gap in attitudes toward these diverse forms of passionate attachments in our culture?*
6. *What are some of the possible implications of changing sex roles in marriage on parent-child relationships?*

The subject of romantic love is treated with skill and depth analysis by Robert Johnson in his book We: The Psychology of Romantic Love *(San Francisco: Harper & Row, 1983).*

From Willard Gaylin and Ethel Person, eds., *Passionate Attachments: Thinking About Love* (New York: Free Press, 1988): 15–26.

Central to the argument of this chapter is a proposition put forward by my colleague Robert Darnton:

> One thing seems clear to everyone who returns from field work: other people are other. They do not think the way we do. And if we want to understand their way of thinking, we should set out with the idea of capturing otherness.[1]

What this means is that we cannot assume that people in the past—even in our own Western Judeo-Christian world—thought about and felt passionate attachments the way we do.

My remarks will be confined to the two most common of passionate attachments—between two adolescents or adults of different sexes, and between mothers and children. I know there are other attachments—between homosexuals, siblings, fathers and children—but they are not of such central importance as the first two. Before we can begin to examine the very complex issue of passionate attachments in the past, we therefore have to make a fundamental distinction between attachment between two sexually mature persons, usually of the opposite gender, and attachment to the child of one's body.

In the former case, the problem is how to distinguish what is generally known as falling in love from two other human conditions. The first of those conditions is an urgent desire for sexual intercourse with a particular individual, a passion for sexual access to the body of the person desired. In this particular instance the libido is for some reason closely focussed upon a specific body, rather than there being a general state of sexual excitement capable of satisfaction by any promiscuous coupling. The second condition is one of settled and well-tried ties which develop between two people who have known each other for a long time and have come to trust each other's judgment and have confidence in each other's loyalty and affection. This condition of caring may or may not be accompanied by exciting sexual bonding, and may or may not have begun with falling in love, a phase of violent and irrational psychological passion, which does not last very long.

Historians and anthropologists are in general agreement that romantic love—this usually brief but very intensely felt and all-consuming attraction toward another person—is culturally conditioned, and therefore common only in certain societies at certain times, or even in certain social groups within those societies—usually the elite, with the leisure to cultivate such feelings. They are, however, less certain whether or not romantic love is merely a culture-induced sublimated psychological overlay on top of the biological drive for sex, or whether it has biochemical roots which operate quite independently from the libido. Would anyone in fact "fall in love" if they had not read about it or heard it talked about? Did poetry invent love, or love poetry?

Some things can be said with certainty about the history of the phenomenon. The first is that cases of romantic love can be found at all times and places and have often been the subject of powerful poetic expression, from the Song of Solomon to Shakespeare. On the other hand, neither social approbation nor the actual experience of romantic love is at all common to all societies, as anthropologists have discovered. Second, historical evidence for romantic love before the age of printing is largely confined to elite groups, which of course does not mean that it may not have occurred lower down the social scale among illiterates. As a socially approved cultural artifact it began in Europe in the southern French aristocratic courts in the twelfth century, made fashionable by a group of poets, the troubadours. In this case the culture dictated that it should occur between an unmarried male and a married woman, and that it should either go sexually unconsummated or should be adulterous. This cultural ideal certainly spread into wider circles in the middle ages—witness the love story of Aucassin and Nicolette—but it should be noted that none of these models ends happily.

By the sixteenth and seventeenth centuries, our evidence for the first time becomes quite extensive, thanks to the spread of literacy and the printing press. We now have love poems, like Shakespeare's sonnets, love letters, and autobiographies by women primarily concerned with their love life. All the courts of Europe were evidently hotbeds of passionate intrigues and liaisons, some romantic, some sexual. The printing press began to spread pornography to a wider public, thus stimulating the libido, while the plays of Shakespeare indicate that romantic love was a familiar concept to society at large, who composed his audience.

Whether this romantic love was approved of, however, is another qustion. We simply do not know how Shakespearean audiences reacted to Romeo and Juliet. Did they, like us, and as Shakespeare clearly intended, fully identify with the young lovers? Or, when they left the theatre, did they continue to act like the Montague and Capulet parents, who were trying to stop these irresponsible adolescents from allowing an ephemeral and irrational passion to interfere with the serious business of politics and patronage? What is certain is that every advice book, every medical treatise, every sermon and religious homily of the sixteenth and seventeenth centuries firmly rejected both romantic passion and lust as suitable bases for marriage.[2] In the sixteenth century marriage was thought to be best arranged by parents, who could be relied upon to choose socially and economically suitable partners who would enhance the prestige and importance of the kin group as a whole. It was believed that the sexual bond would automatically create the necessary harmony between the two strangers in order to maintain the stability of the new family unit. This, it seems, is not an unreasonable assumption, since recent investigations in Japan have shown that there is no difference in the rate of divorce between couples whose marriages were arranged by their parents and couples whose marriages were made by individual choice based on romantic love. The arranged and the romantic marriage each has an equal chance of turning out well, or breaking up.[3]

Public admiration for marriage-for-love is thus a fairly recent occurrence in Western society, arising out of the romantic movement of the late eighteenth century, and only winning general acceptance in the twentieth. In the eighteenth century orthodox opinion about marriage shifted away from subordinating the individual will to the interests of the group and away from economic or political considerations towards those of well-tried personal affection. The ideal marriage of the eighteenth century was one preceded by three to six months of intensive courting, between a couple from families roughly equal in social status and economic wealth, a courtship which only took place with the prior consent of parents on both sides. A sudden falling head over heels in love, although a familiar enough psychological phenomenon, was thought of as a mild form of insanity, in which judgment and prudence are cast aside, all the inevitable imperfections of the loved one become invisible, and wholly unrealistic dreams of everlasting happiness possess the mind of the afflicted victim. Fortunately, in most cases the disease is of short duration, and the patient normally makes a full recovery. To the eighteenth century, the main object of society—church, law, government, and parents—was to prevent the victim from taking some irrevocable step, particularly from getting married. This is why most European countries made marriage under the age of 21 or even later illegal and invalid unless carried out with the consent of parents or guardians. In England this became law in 1753. Runaway marriages based on passionate attachments still took place, but they were made as difficult as possible to carry out, and in most countries were virtually impossible.

It was not, therefore, until the romantic movement and the rise of the novel, especially the pulp novel, in the nineteenth century, that society at large accepted a new idea—that it was normal and indeed praiseworthy for young men and women to fall passionately in love, and that there must be something wrong with those who have failed to have such an overwhelming experience some time in late adolescence or early manhood. Once this new idea was publicly accepted, the dictation of marriage by parents came to be regarded as intolerable and immoral.

Today, the role of passionate attachments between adults in our society is obscured by a new development, the saturation of the whole culture—through every medium of communication—with sexuality as the predominant and overriding human drive, a doctrine whose theoretical foundations were provided by Freud. In no past society known to me has sex been given so prominent a role in the culture at large, nor has sexual fulfillment been elevated to such preeminence in the list of human aspirations—in a vain attempt to relieve civilization of its discontents. If Thomas Jefferson today was asked to rewrite the Declaration of Independence he would certainly have to add total sexual fulfillment to "Life, Liberty and Human Happiness" as one of the basic natural rights of every member of society. The traditional restraints upon sexual freedom—religious and social taboos, and the fear of pregnancy and venereal disease—have now been almost entirely removed. We find it scarcely credible today that in most of Western Europe in the seventeenth century, in a society whose marriage age was postponed into the late twenties, a degree of chastity was practiced that kept the illegitimacy rate—without contraceptives—as low as 2 or 3 percent. Only in Southern Ireland does such a situation still exist—according to one hypothesis, due to a lowering of the libido caused by large-scale consumption of Guinness Stout. Under these conditions, it seems to me almost impossible today to distinguish passionate attachment in the psychological sense—meaning love—from passionate attachment in the physical sense—meaning lust. But the enormous success today of pulp fiction concerned almost exclusively with romantic rather than physical love shows that women at least still hanker after the experience of falling in love. Whether the same applies to men is more doubtful, so that there may be a real gender gap on this subject today, which justifies this distinction I am making between love and lust.

To sum up, the historian can see a clear historical trend in the spread of the cultural concept of romantic love in the West, beginning in court circles in the twelfth century, and expanding outward from the sixteenth century on. It received an enormous boost with the rise of the romantic novel, and another boost with the achievement of near-total literacy by the end of the nineteenth century. Today, however, it is so intertwined with sexuality, that is is almost impossible to distinguish between the two. Both, however, remain clearly distinct from caring, that is, well-tried and settled affection based on long-term commitment and familiarity.

It is also possible to say something about the changing relationship of passionate love to marriage. For all classes who possessed property—that is, the top two-thirds economically—marriage before the seventeenth century was arranged by the parents, and the motives were the economic and political benefit of the kin group, not the emotional satisfaction of the individuals. As the concept of individualism grew in the seventeenth and eighteenth centuries, it slowly became accepted that the prime object was "holy matrimony," a sanctified state of monogamous married contentment. This was best achieved by allowing the couple to make their own choice, provided that both sets of parents agreed that the social and economic gap was not too wide, and that marriage was preceded by a long period of courtship. By the eighteenth and nineteenth centuries, individualism had so far taken precedence over the group interests of the kin that the couple were left more or less free to make their own decision, except in the highest aristocratic and royal circles. Today individualism is given such absolute priority in most Western societies, that the couple are virtually free to act as they please, to sleep with whom they please, and to marry and divorce when and whom they please to suit their own pleasure. The psychic cost of such behavior, and its self-defeating consequences, are becoming clear, however, and how long this situation will last is anybody's guess.

Here I should point out that the present-day family—I exclude the poor black family in America from this generalization—is not, as is generally supposed, disintegrating because of a very high divorce rate of up to 50 percent. It has to be remembered that the median duration of marriage today is almost exactly the same as it was 100 years ago. Divorce, in short, now acts as a functional substitute for death: both are means of terminating marriage at a premature stage. It may well be that the psychological effects on the survivor may be very different, although in most cases the catastrophic economic consequences for the woman remain the same. But the point to be emphasized is that broken marriages, stepchildren, and single-parent households were as common in the past as they are today, the only difference being the mechanism which has brought about this situation.

The most difficult historical problem concerns the role of romantic love among the propertyless poor, who comprised about one-third of the population. Since they were propertyless, their loves and marriages were of little concern to their kin, and they were therefore more or less free to choose their own mates. By the eighteenth century, and probably before, court records make it clear that these groups often married for love, combined with a confused set of motives including lust and the economic necessity to have a strong and healthy assistant to run the farm or the shop. It was generally expected that they would behave "lovingly" towards each other, but this often did not happen. In many a peasant marriage, the husband seems to have valued his cow more than his wife. Passionate attachments among the poor certainly occurred, but how often they took priority over material interests we may never know for certain.[4]

All that we do know is that courting among the poor normally lasted six months or more, and that it often involved all-night sessions alone together in the dark in a room with a bed, usually with the knowledge and consent of the parents or masters. Only relatively rarely, and only at a late stage after engagement, did full sexual intercourse commonly take place during these nights, but it is certain that affectionate conversation, and discussion of the possibilities of marriage, were accompanied by embracing and kissing, and probably also by what today is euphemistically called "heavy petting." This practice of "bundling," as it was called, occurred in what was by our standards an extremely prudish, and indeed sexually innocent, society. When men and women went to bed together they almost invariably kept on a piece of clothing, a smock or a shirt, to conceal their nakedness. Moreover the sexual act itself was almost always carried out in the "missionary" position. The evidence offered in the courts in cases of divorce in the pre-modern period provide little evidence of that polymorphous perversity advocated in the sex manuals available in every bookstore today.

What is certain is that even after this process of intimate physical and verbal courtship had taken place, economic factors still loomed large in the final decision by both parties about whether or not to marry. Thus passion and material interest were in the end inextricably involved, but it is important to stress that, among the poor, material interest only became central at the *end* of the process of courtship instead of at the beginning, as was the case with the rich.

If an early modern peasant said "I love a woman with ten acres of land," just what did he mean? Did he lust after the body of the woman? Did he admire her good health, administrative and intellectual talents and strength of character as a potential housekeeper, income producer, and mother of his children? Was he romantically head over heels in love with her? Or did he above all prize her for her ten acres? Deconstruct the text as we wish, there is no way of getting a clear answer to that question; and in any case, if we could put that peasant on the couch today and interrogate him, it would probably turn out that he merely felt that he liked the woman more because of her ten acres.

Finally, we know that in the eighteenth century at least half of all brides in England and America were pregnant on their wedding day. But this tells us more about sexual customs than about passionate attachments: sex began at the moment of engagement, and marriage in church came later, often triggered by the pregnancy. We also know that if a poor servant girl was impregnated by her master, which often happened, the latter usually had no trouble finding a poor man who would marry her, in return for payment of ten pounds or so. Not much passionate attachment there, among any of the three persons involved.

The second type of passionate attachment is that which develops between the parent, especially the mother, and the child. Here again as historians we are faced with the intractable problem of nature versus nurture, of the respective roles of biology and culture. The survival of the species demands that the female adult should take optimum care of the child over a long period, to ensure its survival. This is particularly necessary among humans since the child is born prematurely compared with all other primates, because of its exaggerated cranial size, and so is peculiarly helpless for an exceptionally long period of time. Moreover, experiments with primates have shown that it is close body contact in the first weeks of life which creates the strong bond between mother and child. A passionate attachment of the mother for its child therefore seems to be both a biological necessity for survival and an emotional reality.

On the other hand recorded human behavior indicates that cultural traditions and economic necessity often override this biological drive. For over 90 percent of human history man has been a hunter-gatherer, and it is impossible for a woman to carry two babies and perform her daily task of gathering. Barring sexual abstention, which seems unlikely, some form of infanticide must therefore have been a necessity, dictated by economic conditions.

Other factors came into play in more recent times. From at least classical antiquity to the eighteenth century it was normal in northwest Europe to swaddle all babies at birth—that is, to tie them up head to foot in bandages, taken off only to remove the urine and feces. This automatically reduced body contact with the mother, and therefore presumably the bonding effect between mother and child. Secondly, all women who could afford to do so put their infants out to wet-nurse from birth to about the age of two. The prime reason for this among the more well-to-do was undoubtedly the accepted belief that sexual excitement spoils the milk. Few husbands were willing to do without the sexual services of their wives for that length of time; hence the reliance on a wet nurse. But this meant that for all except the tiny minority who could afford to take the nurse into the house, the child was removed within a few days of birth and put in the care of a village woman some distance from the home. Under these conditions affection between parents and children could not begin to grow until the child returned to the home at about the age of eighteen months or two years, and the child might well have a more passionate relationship with its nurse than with its mother—as was the case with Shakespeare's Juliet.

In any case, the child's return to its mother would only take place if it did not die while with the wet nurse. There is overwhelming evidence that the mortality rate of children being wet-nursed was very much higher than that of children being breast-fed by their mothers, and contemporaries were well aware of this. It is difficult to avoid the suspicion that one incentive for the practice, particularly for its enormous expansion in France in the nineteenth century, was as an indirect method of infanticide, out of sight and out of mind. This suspicion is reinforced by the huge numbers of children in the eighteenth and nineteenth centuries who were abandoned and deposited in workhouses or foundling hospitals, only a small fraction of whom survived the experience. Whatever the intention, in practice the foundling hospitals of London or Paris acted as a socially acceptable means of family limitation after birth. Few women other than those who gave birth to bastard children practiced infanticide themselves, if only because the risks were too great. But overlaying and stifling by accident while in the same bed during the nights, putting out to wet-nurse, abandoning to public authorities, or depositing in foundling hospitals served the same purpose. Unwanted children of the poor and not so poor were somehow or other got rid of in all these socially acceptable ways.[5]

These common eighteenth and even nineteenth century practices, especially prevalent in France, raise questions about the degree of maternal love in that society. This is not an easy question to answer, and historians are deeply divided on this issue. Some point to evidence of mothers who were devoted to their children and seriously disturbed by their premature deaths. Others point to the bleak statistics of infant mortality: about 25 percent dead before the age of two, a percentage deliberately increased by wet-nursing, abandonment, and infanticide by neglect—practices which have been described as "post-natal family planning." A mid–nineteenth century Bavarian woman summed up the emotional causes and consequences:

> The parents are glad to see the first and second child, especially if there is a boy amongst them. But all that come after aren't so heartily welcome. Anyway not many of these children live. Four out of a dozen at most, I suppose. The others very soon get to heaven. When little children die, it's not often that you have a lot of grief. They're little angels in heaven.[6]

Another question is how kindly children were treated if they did survive. I have suggested that sixteenth and early seventeenth century societies were cold and harsh, relatively indifferent to children, and resorting to frequent and brutal whippings from an early age as the only reliable method of discipline. Calvinism, with its grim insistence on original sin, encouraged parents and schoolmasters to whip children, in order quite literally to beat the Hell out of them. I have argued that only in the eighteenth century did there develop a more optimistic view of the infant as a plain sheet of paper upon which good or evil could be written by the process of cultural socialization. The more extreme view of Rousseau, that the child is born good, in a state of innocence, was widely read, but not very widely accepted, so far as can be seen—for the rather obvious reason that it is contradicted by the direct experience of all observant parents.

To sum up, first there is ample evidence for the widespread practice of infanticide in societies ignorant of contraception, a practice which, disguised in socially acceptable forms, lasted well into the nineteenth century. Second, children, even of the rich, were often treated with calculated brutality in the sixteenth and seventeenth centuries, and again in the nineteenth, in order to eradicate original sin; the eighteenth and twentieth centuries are two rare periods of educational permissiveness. As for the poor, they have always regarded children very largely as potential economic assets and treated them accordingly. Their prime functions have been to help in the house, the workshop, and the field, to add to the family income, and to support their parents in old age. How much room was left over from these economic considerations for passionate attachment, even with the mother, remains an open question.

Passionate attachments between young people can and do happen in any society as a by-product of biological sexual attraction, but the social acceptability of the emotion has varied enormously over time and class and space, determined primarily by cultural norms and property arrangements. Furthermore, though there is a strong biological component in the passionate attachment of mothers to children, it too is often overlaid by economic necessities, by religious views about the nature of the child, and by accepted cultural practices such as wet-nursing. We are in a unique position today in that society, through social security and other devices, has taken over the economic responsibilities of children for their aged parents; contraception is normal and efficient; our culture is dominated by romantic notions of passionate love as the only socially admissible reason for marriage; and sexual fulfillment is accepted as the dominant human drive and a natural right for both sexes. Behind all this there lies a frenetic individualism, a restless search for the sexual and emotional ideal in human relationships, and a demand for instant ego gratification which is inevitably self-defeating and ultimately destructive.

Most of this is new and unique to our culture. It is, therefore, quite impossible to extrapolate from present values and behavior to those in the past. Historical others—even our own forefathers and mothers—were indeed other.

NOTES

1. Darnton, R. *The Great Cat Massacre and Other Episodes in French Cultural History* (New York: Basic Books, 1984), 4.
2. For further discussion of these issues, and references, see my book *The Family, Sex and Marriage in England 1500–1800* (New York: Harper & Row, 1977).
3. *Journal of Family History,* 8, 1983, p. 100.
4. Flandrin, J.-L. *Les Amours Paysannes* (XVI–XIX Siècles) (Paris: Gallimard, 1975).
5. The literature on infanticide (rare), infant abandonment, and early death by deliberate neglect or wet-nursing in Western Europe up to the nineteenth century is now enormous. See, for example:

de Mause, L. *The History of Childhood* (New York: Psychohistory Press, 1974).

Delasselle, C. 'Les enfants abandonés à Paris au XVIII siécle,' *Annales E.C.S.*, 30, Jan.–Feb. 1975.

Flandrin, J.-L. 'L'attitude devant le petit enfant . . . dans la Civilisation Occidentale,' in *Annales de Demographie Historique,* 1973.

Sussman, G. D. *Selling Mother's Milk: the Wet-nursing Business in France 1715–1914* (Champaign: Univ. of Illinois Press, 1982).

6. Medick, H. and D. W. Sabean, eds. *Interest and Emotion* (new York: Cambridge University Press, 1984), 91.

14.

A Revolution's Broken Promises

Peter Marin

Sex has acquired a prominence in our society unrivaled in past time or in other cultures. Sexual fulfillment, likewise, has been elevated to the highest level in the list of human aspirations. We have witnessed, in the last quarter century, a sea change in sexual attitudes and behaviors. Peter Marin, in this essay, takes a critical look at what came ashore with this sexual revolution in the 1960s.

Marin is one of our favorite writers. He is an insightful and passionate cultural critic. His essay is a severe indictment of what passes for sexual liberation today. With the collapse of most traditional (social and religious) restraints upon sexual freedom, our human lives have become marred by pretense, desperation, and an immense amount of "bad faith." The author chronicles our culture's restless search for the ideal and the demand for instant gratification. The path is inevitably self-defeating and the results ultimately destructive. The essay will spark lively discussion, and its counterculture stance will challenge the student reader.

The following questions will help to focus the discussion:

1. *Where do you see people acting out the sexual images and ideas provided for them, projected upon them, by others today?*
2. *Are these images/ideas liberating toward authentic personhood or a new form of imprisonment to selfishness?*
3. *Where are the casualties of the sexual revolution displayed today?*
4. *Are constraints, generosity, and kindness the new taboos in sexual relations?*
5. *Is sex a private affair, or does it have social and public dimensions?*
6. *What images/values can our religious traditions offer to the humanizing of sexual activity?*

A similar analysis and critique of our sexual lives is made by Rollo May in his classic work Love and Will *(New York: Norton, 1969), Chapter I.*

From *Psychology Today* (July 1983): 50–57. Reprinted by permission of the author.

Mention the sexual revolution to a dozen people, or to 100, and you get a dozen or 100 different analyses, conclusions, and complaints. And mixed in with these responses, there usually is a shrug or a grimace or a bitter smile: "What revolution?" people ask. The response does not mean that changes have not occurred. Obviously, they have. The rueful question means rather that the sexual freedom established during the past couple of decades has not been accompanied by the increase in happiness that many people assumed would follow from a freeing of sexual mores.

There have been obvious and important gains, of course. But there have been losses as well, many of which are suggested by the story of a friend of mine, Colin, who decided in the late '70s to have a sex change. He was then in his middle 30s, recently divorced from his wife, and separated from his son and daughter. One afternoon he showed up at our house and announced that he was going to have a sex change.

Years before, it turned out, he had idly picked up a book about sex and sex changes, and realized that he, too, like the subjects of the book, had felt since childhood as if he were a woman trapped in a man's body. Some time after that, he said, he made an agreement with his wife: When she went out with other men, as she often did, he would dress up in her clothes at home, pretending to be a woman. Then, after he and his wife separated, he met a woman who ran a clothing store for transvestites, and she taught him how to walk and talk and smile like a woman and to relearn, as a woman, all of the things that he did as a man. And now, he explained, he was going to have an operation to change his sex physically: He was going to become a woman.

A year later, one night in a bar in Los Angeles, a tanned, long-haired, muscular young man came up to me and said: "There's someone with me who knows you." At his table I saw a pretty, middle-aged woman, in a cashmere sweater, a string of pearls around her neck. "Hello, Peter," said the woman in a high, rather artificial voice, and I realized that it was my old friend Colin, now become Claire.

I saw him, or her, from time to time after that, and she seemed neither happy nor unhappy, only much the same as before. I remember once being taken aback when, in a discussion of how her sex change affected her children, she said: "Oh, but I just want to be a mother to them."

I did not see her for about a year, when she came again to the house one afternoon with a woman friend who might also once have been—I was not sure—a man. This time Claire was not happy. She was tired, and the feminine surface that she had so carefully cultivated seemed to be slipping. One could almost see through it, as if she was unable to muster the energy required to keep her femininity intact. Her voice kept sliding down into the lower registers; her hair kept coming undone; even her gestures had become again, at least for the moment, a man's. Her operation, it turned out, had not gone well. The doctor had botched it, though I did not get the details; and when I asked her how she was feeling in general, she said: "I had hoped I'd be happier. To tell the truth, I seem to be trapped in any body."

I think of my friend now, and it seems to me that there is a sense in which he or she was trapped in a body—but not one of flesh and bone. It was, instead, an idea of a body that had been sold to him as surely as his car or house had been sold. He was acting out before the operation, and she was acting out afterward, not only a social role defined by others, but also a set of images imposed upon him. The mechanical devices that were now Claire's—the pumped-up silicone breasts, the carved vagina lined with the skin of what had been a penis—were no different, really, from the gadgets hawked in the marketplace: the various objects and accoutrements that we accept without question as a necessary part of our modern lives.

In essence, most of us are no different from Colin-Claire. Everywhere we act out the sexual images and ideas provided for us, projected upon us, by others. Whether it is men with their Marlboro mustaches, lesbians in their bull-dyke janitor's outfits, male homosexuals with their clone look, or adolescent girls in tight jeans, we move somnambulistically through roles and rituals, responding to every whim and wind in the cultural air. We have been liberated from the taboos of the past only to find ourselves imprisoned in a "freedom" that brings us no closer to our real nature or needs.

It is this that explains the grimaces and shrugs when one mentions liberation. For many people, the idea of liberation—whether it is sexual, political, or social—is synonymous with happiness or satisfaction. In the instance of sexual freedom, whether it is the work of Freud or that of myriad insistent sex-rebels exemplified by men and women as varied as Margaret Sanger, Havelock Ellis, John Cowper Powys, and Wilhelm Reich, everything said in support of sexual freedom implied that it would transform and restore all aspects of emotional and relational life. Since the absence of a successful sexual life was taken to be a cause of disease and pain, it followed that its presence would inevitably bring joy in its wake and, ultimately, social happiness.

There is something peculiarly bourgeois and hygienic about this line of thought. Sex, which in the culture's past had been associated with evil, was moved lock, stock, and barrel into the camp of goodness. It became an all-purpose healing instrument, a kind of glorifed patent medicine for everything that might ail us. Eventually, by a continuation of this logic, for Wilhelm Reich and the succeeding generation, the ideal orgasm itself became the wellspring of kindness and human decency.

What most of us currently seem to believe is that once restraints are removed from human behavior, "nature" simply asserts itself, like water filling an empty space. We forget that we bring with us, into any kind of freedom, the baggage of the past, our internalized cultural limits and weaknesses. Thus freedom—in this case sexual freedom—increases choice, but it guarantees nothing, delivers nothing. To the extent that it diversifies and expands experience, it also diversifies and multiplies the pain that accompanies experience, the kinds of errors that we can make, the kinds of harm that we can do to one another.

The simple fact is that many of the obstacles to sexual life are not merely the function of repressive attitudes or mores. They are grounded in the complexities of human nature and in the everyday difficulties of living together. And all these natural—one is almost inclined to say "eternal"—difficulties are intensified by the disappearance of traditional sexual roles, the proliferation of sexual choices and styles, the permission to introduce, in public life, the full range of sexual fantasies and yearnings to which we are prey and heir.

I cannot here enumerate the various casualties of the sexual revolution, from the young men and women whom I once saw as a therapist and teacher, who, barely out of adolescence, had slept with so many people that they found themselves frigid or unresponsive beside those whom they genuinely loved, to the middle-aged couples who, spurred on by glowing reports of open marriage, pushed one another too far, into the jealousy and fury that they believed they could leave behind. But I think all of us must acknowledge, however reluctantly, that there was something to those "reactionaries"—starting with Freud's colleagues—who argued that deliberate, broad changes in our systems of sexual remissions and taboos would let loose among us as many troubles as they solved.

Sexual life, which ought to begin with, and deepen, a pervasive and genuine sympathy between men and women, seems instead to produce among us a set of altogether different emotions: rage, disappointment, suspicion, antagonism, a sense of betrayal, and sometimes contempt. It is not so much that one cannot find good feelings in many persons or between many lovers; it is, rather, that the sexual realm as a whole seems somehow corrupted. The general feel to it is one of perplexity, even anger; betrayal rather than gratitude pervades it; and though sex no longer seems to us a curse visited upon us by the devil, few of us seem to experience it continuously, or even often, as a gift. It remains for most men and women a world through which they move warily, cautiously, self-protectively—not a home but an alien land.

Ironically, much of this is the result of the shifting of sex from the private to the public world, which is the hallmark of the sexual revolution. Back in the '50s and early '60s, sex could be an alternative to the dominant culture. It constituted a world in which the mores and fashions of the public realm did not hold quite the power that they did elsewhere, and to enter that world on one's own was, in various small rebellious ways, to leave home, to mark out a territory where one could define oneself. Though that had its own costs (making a rebellion out of behavior that should be natural), it also had its advantages, not least of which was that it often made comrades and friends out of lovers; they were, after all, engaged together in creating a private world.

But the popularization of sex has changed much of that. Sex has become almost entirely socialized, invaded by manufactured images and experts; it is no longer a way of retreating from the public world but a way of entering it. The sexual realm has been corrupted by any number of absurd or destructive ideas, almost all of them put forward by people whose main interest is not sex but making money or names for themselves. The nonsense bruited about is unbelievable; the ignorance passing itself off as wisdom is endless; sexual ideas and techniques are hawked incessantly in the marketplace. Creative masturbation, the ideal orgasm, the clitoral orgasm, the G spot, the joys of sex, the virtues of homosexuality, the virtues of bisexuality, the virtues of sex with children, porpoises and disembodied spirits, the good old missionary virtues of heterosexuality—all of these now have their norms, their measures, their proprieties. We do sex filling in the squares laid out for us by others.

Whereas in the more puritanical past, the darkness and mysteries of sex remained outside the order of things, it has now become a sort of vast Club Med, a vacation paradise into which supposedly anyone can venture successfully and without cost. It is crowded with visitors, each of them seeking an identity and an experience that bears no more relation to things as they are than did the old idea of sex as the devil's playground.

Beneath all this there is one crucial point that we often ignore: that many people are far less driven or drawn by sex than we like to think. The cant and fashion of the age imply that sex is fundamentally important to everyone and a powerful, primary source of pleasure for everyone. But if you listen carefully to what moves beneath people's words, it does not really seem so.

The loneliness and dissatisfaction that most people express, the yearnings they articulate, have much less to do with sex than with an unfulfilled desire for good company or good conversation or the intimacy of shared perceptions and interests. I would say that friendship and community seem more important to most people than genuine sexual passion, and what they accept as a decent sexual life has little to do with the turbulence and confusion and adventurous risk required to live out, deeply and fully, the tendings of one's sexual nature. What seems to dominate their concerns about sex when they do surface is a sort of idealized and sugary notion, brought up to date with erotic trimmings—a child's drawing of security extended to include sex: a house with smoke curling from the chimney, a couple hand-in-hand at the door, and behind them, upstairs, the circular mirrored bed into which, after the day's work has been done and the front and back doors locked, they tumble for a riotous good time.

One does not usually find attached to sex these days the curiosity, adventurousness, and the tolerance for disappointment or capacity for camaraderie that once seemed to mark it and that must always accompany any genuine attempt to keep faith with one's nature. Where excitement does exist, it seems as often as not to come not from the pleasures of sex, but from the situation, the cinematic trappings of "affairs." One is tempted to say that we are a nation of romantics, using sex to create idealized scenarios for ourselves, but it is probably more accurate to say that we are sentimentalists, pining—as James Joyce puts it—for emotions for which we are not willing to pay the price of experience.

Do I exaggerate here? Perhaps a bit. There are moments, of course, for most men and women—both those who are genuinely concerned with sex and those who are not—in which the raw truth of some kind of love breaks through the preconceptions that have ringed it round, and desire sweeps all before it, even our notions of romance. Such moments have nothing to do with fantasy or even images. When they occur, they occur, as the wise Greeks understood, in forms and with consequences we have not anticipated or even wanted: A world is revealed—and with it a sweetness and a self we had not imagined.

But how often does it happen? Once, twice, half a dozen times in a lifetime. Sometimes, for some people, it does not happen at all. At my daughter's school, for instance, they teach the children about sex with the help of a child's picture book that describes orgasm as something akin to a sneeze. The book tells its readers that the children can get an idea of sexual pleasure by imagining first a terrible itch and then the relief of scratching it. What kind of adults could have written such a book? Certainly not those for whom sex has *some* importance. Perhaps the authors simply lack a talent for language, but one suspects that there is more to it than that. The sneeze represents the head and the itch stands for surface sensation, and these seem to be the ways in which many men and women experience sex.

We are, after all, a puritanical people still, whose talent or capacity for sexual feeling falls far short of the attention we pay to it. As a result—despite all our rhetoric, all our manuals, all our universal make-believe—the sexual realm is marred by pretense, desperation, and an immense amount of "bad faith," which constitutes a simultaneous betrayal of both the other person and oneself.

It was not always this bad; it was not this bad even recently. I remember coming to California from the East in the very early '60s, a couple of years before the sexual revolution burst into full bloom and "the greening of America" made adolescence into the model for all adult behavior. I was surprised by the quality of sexual life I found there: men and women who seemed to feel at ease with sexuality and with one another. This was true of men and women in their 40s and 50s—something that I had not seen before.

I had grown up in Brooklyn in the '50s, when almost no one had much of a sexual life—not, at least, in terms of real lovemaking. Our adolescence—which was more openly sexual than the life of our elders—was not as terrible as it has since been made out to be. We were romantic, mildly driven, somewhat frustrated, skewed in various trivial sexual ways, but nonetheless we did not have a bad time of it. Sex for us was straightforward; desire was almost always focused on a particular person. There were crushes, attractions, awkwardnesses, small fiascos, and though we never got to make love (that came later, in college, for us), at least there was a genuine yearning, accumulating and mixing with frustration, forming a preliminary sense of what desire means, of how it might feel.

It was not until I got to California that I came upon large numbers of grown men and women who had about them a casual sexual grace. Remember, these were the early '60s. This was not the California of cranks and encounter groups, idealized sexual abandon, and foolish or apocalyptic zeal. It was an easier and warmer world in which people drank rather than took drugs, and somehow this gave a different tone to things than the one later provided by drugs. There was not the driven sexuality or pornographic desperation back then that would later fill the air.

As a corollary, those who found themselves drawn to sex or one another were more often than not comrades. That seems to me the most significant difference from what now surrounds us. Of course, sex did not often have attached to it, even back then, the deepest intensities, higher kinds of awareness, or the transformative significance that Reich or D. H. Lawrence claimed for it. But it did have kindness and good humor attached, and to meet with someone in the flesh was to enter a shared community of flesh, as if one had met someone far from home with whom one could make at least a temporary home.

That happened to people—by their own accounts—even in casual encounters, not all of the time, but at least part of the time. Men and women seemed capable of tolerating their disappointments and mistakes without holding their failures—as we tend to do now—against those with whom they were involved. If things went wrong in or out of bed, there was little recrimination and much less rage. Expectations were lower, needs not as great; people did not yet think of sex as a panacea, did not expect it to make them whole or pure or healthy in any magical way.

Most important of all was that the only people who bothered with sex were, for the most part, those who liked it. Everyone else left it alone. The mild taboos still intact in those days were not strong enough to discourage those who were genuinely drawn to a sexual life. But they were strong enough to allow those not so drawn to stick to pleasures closer to their own natures. This left the sexual world to those who felt at home in it, whereas today, it is much like the ski slopes on a crowded weekend: mobbed with people who are there for a dozen reasons other than a genuine love of skiing or the slopes.

No doubt it was all too casual, and perhaps it was not all that it seemed. There must have been—there always is—cruelty, exploitation, and pretending, and at the heart of each privacy, the kinds of sorrow, estrangement, and pain familiar to us all. Perhaps women complained less because they did not then have the courage to speak out. But I think there was more to it than that. There was a kind of restraint, as if men and women still understood that what they owed one another, and the way to protect the sexual realm, was not to visit upon one another all of their sorrows and pain.

It is precisely that constraint, a minimal kindness connected to a naturalness of behavior, that is in large part what is missing from the sexual world today. What we have seen on a grand scale during the sexual revolution has been called in another context "the return of the repressed." But what has been repressed for so long is not only animal need. It is also, we have learned as it comes flooding upward to the surface, a raw mix of anger, frustration, bitter disappointment, sullen resentment—the whole underlying plane of feeling that forms itself in those whose world (despite all our talk of liberation) seems to have made no room for their deepest nature. How many, these days, turning to take another person in their arms, have not found themselves confronted by a range of accumulated disappointments, betrayals, and unfulfilled yearning—the living residue left behind not only by mothers, fathers, and lovers, but also the despair engendered by an unlived or falsely lived life?

Caught in the midst of this, people seem to have no one to blame but one another.

What I hear, everywhere around me, are complaints, descriptions of unmet demands, disappointments—that someone has failed them, let them down, is not what they ought to be. This is the strain that runs through much that I have heard as a therapist, teacher, or friend when men and women talk about one another (though men are less articulate, feel less justified than women in their public complaints). Many of these complaints are accurate, of course—we do fail one another. But their accuracy cannot hide the fact that the expectations have less to do with the world as it is or people as they are than with mistaken, preconceived, borrowed or inherited notions about what men and women ought to be or can be.

The tone of all this is not merely one of sadness or unanswered yearning; more often than not it is a tone of judgment, impatience, even contempt. It is as if every lover is also an enemy, as if every companion is less an invited guest than an unwanted intruder.

We have come a long way in the sexual revolution. We have left behind us a great many old illusions and delusions; we know more than we did about the kinds of betrayal, guilt, and confusion that we can survive, and the kinds that we cannot. But what we have not learned—and this is the heart of the problem—is how to be kind to one another, how, in the midst of the confusions we ourselves have created, among the congeries of styles and pretenses of sex that surround us—how we can sustain those we find at our sides or in our arms.

The problem is not that sex has been separated from love, as many people have suggested (though there is some truth to this). The more general problem is that sex—along with countless other activities—has been emptied of generosity. There is nothing specifically sexual about such generosity, nor is there anything unique in the place it ought to play in sex. Yet the hardest thing of all in sexual life, more difficult by far than having the world's finest orgasm, is to leave images and dreams behind, and to learn that the person in one's arms is a poor forked creature, subject to the same confusions and alarums as oneself. Beyond all will, beyond all imagining, beyond all sensation—whether a sneeze or an itch—there remains a human reality that yields itself to a kindness of touch but which remains closed to us, despite all our yearnings, until we can somehow learn to bring to sex, through generosity, precisely what it is that we seek there from others, and without which the sexual world remains a kind of limbo.

Unfortunately for all of us, a capacity for generosity may be no more easily learned than a love of sexual life. Here too is an area where manuals or good advice are not likely to save us. It is one thing to be able to explain where the generosity in a particular culture comes from or what tends to destroy it (and I am not sure, really, that we can do even that). But nobody knows how to interject generosity into a culture whose members no longer seem to feel it on their own.

Of course, in spite of all this, the graces of flesh have not vanished completely and will not vanish. Like any other power rooted in nature, they seem capable of reasserting themselves in spite of anything we do or say. There will always be experiences that sweep away our notions derived from therapy or ideology and liberate us even from our notion of liberation.

It is the imperviousness of sex to ultimate understanding, the way it dissolves understanding, that gives rise to both its curses and gifts, its devils and angels. It remains, in the midst of that "revolution" which has provided neither much equality nor liberty nor fraternity, a troublesome but fecund darkness in which, like lost children, we call out to one another in both fear and delight.

15.

Four Mischievous Theories of Sex: Demonic, Divine, Casual and Nuisance

William F. May

Wiliam May's article easily lends itself to constructive classroom discussion and clarification on the topic of sex. The author lays out four conflicting attitudes on sex. While the attitudes are loosely associated with the behavior of different cultural groups, the divergent views can be found in each one of us.

The author's typology is suggestive and stimulating: sex as demonic, divine, casual, and a nuisance. Each category contains an element of truth, but each is also ultimately fallacious. A theological interpretation and analysis is offered of the viewpoints on the basis of the biblical tradition and Christian heritage. The article is a valuable pedagogical tool to facilitate a self-examination on this important topic.

Some questions for consideration follow:

1. *How do you account for the popularity of the playboy philosophy of sex in the United States? What is the root problem in this attitude? Is there a credible alternative?*
2. *Is there a place for discipline in sex? How would you justify it? Where does discipline end and repression begin?*
3. *What is the current dominant attitude toward sex in contemporary music, movies, church?*

Several conflicting attitudes toward sex beset us today. We loosely associate these attitudes with the behavior of different cultural groups. Whether the groups actually behaved in these ways poses a descriptive question that will not preoccupy me for the moment. I am interested more in the attitudes than in the

From Willard Gaylin and Ethel Person, ed., *Passionate Attachments: Thinking about Love* (New York: Free Press, 1988): 27–39. Reprinted with permission.

historical accuracy of the symbols. The Victorian prude feared sex as demonic; romantics, such as D. H. Lawrence, elevated sex to the divine; liberals tend to reduce sex to the casual; and the British, as the satirists relentlessly portray them, pass it off as a nuisance. I will argue that all these views of sex contain an element of truth; all are ultimately mischievous; and most can be found conflicting and concurrent in ourselves.

Sex as Demonic

Those who fear sex as the demon in the groin reckon with sex as a power which, once let loose, tends to grip and destroy its host; it is self-destructive and destructive of others, a loose cannon, as it were, in human affairs. Our movies and drugstore paperbacks relentlessly mock this view, which we tend to assign remotely to our Victorian forebears and proximately to our parents. While parents, in fact, may fear the explosive power of sex in their adolescent children, it is doubtful whether most parents are quite the Victorians their children assume them to be. Children impute this view to their elders because at some level of their being they partly hold to this attitude themselves.

In any event, this pessimism that emphasizes the runaway destructiveness of sex hardly originated with the Victorians. Religiously, it dates back to the Manichaean dualists of the Third Century of the Common Era. Manichaeans divided all reality and power into two rival kingdoms: the Kingdom of God pitted against the Kingdom of Satan, Good versus Evil, Light versus Darkness. They associated the Absolute Good with Spirit and Absolute Evil with Matter. Originally Spirit and Matter existed in an uneasy separation from one another; but through the aggressive strategies of Satan, the present world and humankind came into existence, a sad commingling of them both—Spirit and Flesh. The world is a kind of battleground between these two rival kingdoms. Man's only hope rests in disengaging himself from the pain and confusion and muck of life in the flesh, and allying himself with the Kingdom of Spirit. I say "man" deliberately because the Manichaeans tended to associate women with the intentions of the Devil; that is, with his strategy to perpetuate this present age of confusion and commingling through the device of sex and offspring. Quite literally, marriage in their view is an invention of the Devil, a scheme for perpetuating the human race and the messy world that we know. Man should achieve a final state of metaphysical *Apartheid,* a clean separation from the toils of the flesh, women, and all their issue.

Manichaean sex counselors thus urged on their followers a rigorous ethic of sexual denial—with, however, an antinomian escape clause since not everyone could lead the wholly ascetic life. If one couldn't totally abstain—here is the twist—the Manichaeans believed it was better to engage in "unnatural sex" so as to avoid the risk of progeny. In the Manichaean vision of things, sex is bad, but children are worse. Reproduction should be avoided at all costs, since it only perpetuates the grim, woe-beset world that we know. (The mythology sounds strange to the modern ear, but the Manichaeans have served as a symbol of pessimism in later Western theology, and rightly so. A reluctance to have children usually blurts out the pessimism—whatever its causes—of those who think little of the world's present and future prospects.)

Christianity rejected this Manichaean pessimism, and thereby confirmed the religious vision it derived largely from the Scriptures of Israel and from the New Testament. Its monotheism differs from a dualism that takes evil too seriously and that identifies evil too readily with the flesh. Its scriptures highly esteem sexual love (the erotic Song of Solomon would jar in a Manichaean scripture); it grants a sacramental status to marriage; and it describes the body as the temple of the Lord. The lowly, needy, hungering, flatulent body is nothing less than the real estate where the resurrection will occur.

But dualism kept reappearing in the Western tradition, often nesting in Christianity itself or appearing in an alluring alternative, the cult of romantic love. On the surface, the ideal of romantic love, Denis de Rougemont once shrewdly argued, appears to be sexually vigorous; it celebrates God's good green gift of sex. But, in fact, it secretly despairs of sex; it always directs itself to the faraway princess—not to the partner you've got, but to the dream person, the remote figure not yet yours. Sex slips its focus on actual contacts between people and transposes to the realm of the imagination. To possess her is to lose one's appetite for her. Love, therefore, feeds best on obstacles. "We love each other, but you're a Capulet and I'm a Montague." And so it goes from Romeo and Juliet, backward to the Tristan and Iseult myth, and forward to Noel Coward's "Brief Encounter" and the mawkish *Love Story*. The poignancy of passion depends upon separation, ultimately upon death. The cult of romantic love locates passion in the teased imagination. The flesh kills; the spirit alone endures; thus Manichaean pessimism hides in its alluring garb.

The post-Renaissance world offered a somewhat drabber version of this dualist suspicion of sex. Social diseases assaulted the Western countries and associated sex with forces that abuse the mind and body. Further, a concept of marriage emerged with middle-class careerism that encourages a Manichaean wariness toward sex. The bourgeois family depended for its stability and life on the career and the property of the male provider. Premarital sex, which distracts a man from his career and leads him prematurely into marriage, severely limits his prospects. Extramarital sex spoils the marriage itself and public reputation. And marital sex leads to too many children with a cramping effect on the careers of those already arrived. Thus, all told, sex severely inconveniences a careerist-oriented society that depends throughout on deferred gratification.

But not surprisingly, bourgeois culture produced not only repression, but also a pornographic fascination with sex. Sex became, at one and the same time, unmentionable in polite society but also an unshakable obsession in fantasy. Geoffrey Gorer, the English social anthropologist, in his often plagiarized article, "The Pornography of Death," nicely defined all such pornographic preoccupation with sex as an obsession with the sex act abstracted from its natural human emotion, which is *affection*. This definition helps explain the inevitable structure of pornographic novels and films. Invariably, they must proliferate and escalate the varieties of sexaul performance. When the sex act separates from its natural human emotion of affection, it loses its tie with the concrete lives of the two persons performing the act; it becomes *boring*. Inevitably, one must reinvest one's interest in the variety of ways and techniques with which the act is performed—one on one, then two on one, then in all possible permutations and combinations, culminating in the orgy. When affection isn't there, it won't do to have bodies perform the act in the age-old ways. Sad variety alone compensates.

(The oft-cited pornographic preoccupation with death and violence today follows the same pattern of escalation. A pornography of death entails an obsession with death and violence abstracted from its natural human emotion, which is grief. Once again such violence, abstracted from persons, inevitably bores, and therefore one must reinvest interest in the technology with which the act is performed. It won't do for James Bond to drive an ordinary General Motors car (as though it weren't death-dealing an instrument enough); he must have a specially equipped vehicle that jets flames out its exhaust. Spies must be killed in all sorts of combinations and permutations. Violence inevitably excalates.)

This ambivalent attitude toward sex that generates both repression and obsession is basically religious—not Jewish or Christian, to be sure, but religious, specifically Manichaean—in its root. It religiously preoccupies itself with sex as a major evil in human affairs.

Sex as Divine

The second of the four attitudes toward sex also qualifies as religious; in this case, however, one elevates sex from the demonic to the divine. D. H. Lawrence offers the definitive expression of this sex-mysticism; let his views stand for the type. *Lady Chatterley's Lover* is a religious book. That assessment didn't occur to people of my generation who, before laying hands on the book, assumed its title was *Lady Chatterley's Lovers,* and settled down for the inevitable orgy. The book offered, however, religion in a very traditional sense, for religion consists of some sort of experience of sacred power perceived in contest with other powers. The sacred grips the subject as overwhelming, alluring, and mysterious, and eventually orders the rest of life for the person or community so possessed. (Exodus 3, for example, describes the contest between Yahveh, God of the Jews, and the power of the Pharaoh. God liberates his people from Egypt and orders their life at Mt. Sinai; God prevails.)

Just so, the novel focuses on a woman who experiences in her own being a contest of the powers—those opposingly symbolized by Lord Clifford, her husband, and Oliver Mellors, her husband's gamekeeper. Her husband possessed those several powers which the English highly prized—status, money, and talent. He was at once an aristocrat, an industrial captain, and an author—an ironmonger and wordmonger. He wielded economic power and word power. Leaving such a man for his gamekeeper would utterly confound the commitments of Lady Chatterley's class. Lord Clifford's only trouble, his fatal trouble, however, was a war wound that left him dead from the waist down, a state of affairs which was but the natural issue of the kind of destructive power which he wields. Lady Chatterley discovers in the gamekeeper and in the grove where he breeds pheasants, a different kind of power, a growing power in the pheasant and the phallus, and this power prevails.

Lawrence's novel celebrates not random sex but a sex-mysticism. The grove where Lady Chatterley and Oliver meet serves as a sacred precinct removed from the grimy, profane, sooty, industrial midlands of England where men like Lord Clifford ruled. Lawrence explicitly uses the coronation Psalms of Israel to describe the act of sexual intercourse. "Open up, ye everlasting gates, and let the king of glory enter in." In using royal language, Lawrence advocated not sexual promiscuity, as the hungering undergraduates of my generation supposed. Far from it! Lawrence disdained the merely casual affair: he exalted sexual union into a sacred encounter. Tenderhearted sex is the closest we come to salvation in this life. It provides contact with all that nurtures and fulfills. Americans in the 1950s relied on a sentimental marital version of this religious expectation. As the song of the times put it, "love and marriage go together like a horse and carriage." In the oft-called "age of conformity" one tended to look to the sanctuary of marriage to provide respite from the loneliness and pressures of the outer world to which one conformed but which one found unfulfilling.

Sex as Casual

W. H. Auden once observed that the modern liberal offended Lawrence more than the Puritan. The Puritan mistakenly viewed sex as an outsize evil, but the liberal made the even greater mistake of reducing sex to the casual—to one of the many incidental goods that in our liberty we take for granted. Some have called this the drink-of-water theory of sex.

This casual attitude toward sex reflects a liberal industrial culture that prizes autonomy above all else, that reduces nature to raw material to be manipulated and transformed into products of man's own choosing, and that correspondingly reduces the body to the incidental—not to the prison house of the dualists, or to the Lord's temple of the monotheists, or to the sacred grove of the mystics, but to a playground pure and simple.

Some observers argue that this third attitude toward sexual experience dominates our time. Is not D. H. Lawrence, despite his flamboyance, actually somewhat quaint and old-fashioned, the reverse side, if you will, of the Victorian prude? Don't both the prude and the romantic make the mistake of taking sex too seriously? One elevates sex into the satanic, and the other celebrates it as divine. Have we not succeeded in desacralyzing sex and reducing it now to the casual?

This third and apparently prevailing theory of sex today, the so-called new sex ethic, takes two forms. First and most notoriously, its earlier, male chauvinist version converts sex into an instrument of domination. It reduces sex to the casual, by converting women into bunnies and by replacing heterosexuality with a not so latent male orientation. In its magazine formula, it condemns women, flatters the young male, and lavishes on him advice on how to dress, talk, choose his cars, and handle his women—all without involvement. The women's movement has shown proper contempt for this view.

The second version of the new sex ethic avoids the more obvious criticisms of the woman's movement; indeed, it seeks to join it by offering easy access, easy departure, and no long-term ties, but with equal rights for both partners. One of our entertainers best summarized this casual, tentative, experimental attitude toward sex and marriage by referring to his decision to do the "marriage bit"—a phrase from show biz. It suggests that marriage offers a role one chooses to play rather than a relationship by which one is permanently altered—not necessarily a one-night stand, but then not likely, either, to run as long as "Life With Father."

This reading of the social history of our time—from the religious to the secular—only apparently persuades. We are not quite as casual about sex as this analysis would suggest. Our popular magazines—men's and women's—may have evangelized for a cool attitude toward sex; but they would not have sold millions of copies if, underneath it all, in the steamy depths of our desires, we could toy with it that easily.

Denis de Rougemont neatly skewers our irrepressible fascination with sex in *The Devil's Share,* a book that included chapters on such topics as the "Devil and Betrayal," the "Devil and War," and the "Devil and Lying." His first sentence in his essay on the "Devil and Sex" reads, in effect: "To the adolescent amongst my readers who have turned to this chapter first . . ." I read de Rougemont's book when I was 32, but the age makes little difference. There one is—young or old—caught red-handed, eyes riveted, imagination stirred, ready for fresh rivulets of knowledge on that most fascinating of topics. Casual curiosity? Yes. But the lure of mystery as well. Elements of the religious as well as the casual characterize our attitude toward the subject.

Sex as a Nuisance

So far, this essay has covered three views of sex; symmetry alone would demand a fourth to complete two sets of paired attitudes. Dualists inflate sex into a transcendent evil; mystics view it as a transcendent good; and casualists reduce it to a trivial good. The demands of symmetry, then, would posit the existence of a fourth group composed of those prosaic folk who dismiss sex as a minor evil, a nuisance. Comic writers have rounded up this particular population and located them in Great Britain under the marquee: "No Sex, Please. We're British." Copulation is, at best, a burdensome ritual to be endured for the sake of a few lackluster goods. One has visions therewith of an underblooded, overarticulate clutch of aristocrats in whom the life force runs thin.

But a report in one of the most popular of American syndicated newspaper columns (in the *Washington Post,* June 14 and 15, 1985) suggests that the number of people occupying the quadrant of petty pessimists may be surprisingly large. Ann Landers asked her reading audience to send a postcard or letter with a reply to the question: "Would you be content to be held close and treated tenderly and forget about 'the act'? Reply YES or NO and please add one line: 'I am over (or under) 40 years of age.' No signature is necessary.'' Even discounting for the fact that the disgruntled find more time to write than the contented, the percentage of those replying to Landers' inquiry who deemed themselves to be sexually burdened was impressive. More than 70 percent replied YES and 40 percent of those affirmatives were under 40 years of age. Clearly the people who find sex to be a burden transcend the boundaries of the British Isles. Over 90,000 letters poured in from the U.S. and other places where Landers' column appears (in Canada, Europe, Tokyo, Hong Kong, Bangkok, Mexico). This outpouring has exceeded every inquiry that Landers has directed to her readers, except for the pre-fab letter to be sent to President Reagan on the subject of nuclear war. ''This sex survey beats . . . the poll asking parents, 'If you had to do it over again, would you have children?' '' (Seventy percent said NO.) (Some astute historians of religion have argued that Manichaeaism persists as the ranking heresy in the West.)

Critics of the Landers report have warned that her results are not scientific. Her respondents are self-selective and her question tips the responses negatively. By placing the term for intercourse in quotation marks and calling it ''the act,'' she tends to separate the sex act from tenderness. Still, the grammar of her question does not force an either/or response: tenderness or sex. However parsed, Landers uncovers a great deal of dissatisfaction amongst women . . . ''it's a burden, a bore, no satisfaction . . .'' Her letter-writers largely blame men for this state of affairs, but her survey and the ensuing discussion leave untouched the question as to whether the male failure to satisfy reflects a deeper masculine version of the experience of sex as a nuisance. One thinks here not of the occasionally impotent male who is agonizingly aware of sex as a nuisance, but, of the robust stallion who prides himself on his efficient performance but who finds foreplay, afterplay, tenderness, and gratitude an incomprehensible and burdensome detail.

Theological Interpretation

Since I am a trained Protestant theologian, not a social commentator, I will close with a few comments about each of these four attitudes on the basis of the biblical tradition. In these matters I don't think I stray too far from what my colleagues in the rabbinate and priesthood might say.

1. Whatever criticisms the biblical tradition might deliver against the casualist approach to sex, that approach has an element of truth to it. Not all sexual encounter should carry the weight of an ultimate significance. Sometimes sex is merely recreational, a way to fall asleep, a *jeu d'esprit,* to say nothing of a *jeu de corps*. But at the same time, the interpretation of a particular episode should not exhaust the full meaning of the activity. At first glance, the ideology of the casualist seems virile, optimistic, and pleasure-oriented. But a latent melancholy pervades it. The fantasy of transient pleasure as an interpretation of the full meaning of the act requires a systematic elimination of everything that might shadow the fantasy. The sacred grove trivializes into a playpen. Hugh Hefner's original policy of never accepting a story for his magazine on the subject of death

betrays the pathos of the approach. The fact of human frailty and death shatters the illusion upon which Hefner's world depends. By comparison, a sturdy optimism underlies a tradition that invites a couple to exchange vows that can stretch across the stark events of plenty and want, sickness and health, until death parts them. Since life is no playpen, it lets the world as it is flood in upon the lovers in the very content of their pledge.

Further, the casual outlook tends to ignore the inevitable complications of most sexual relationships. It lapses into a kind of emotional prudery. We are inclined to apply the word prudish to those who deny their sexual being. The modern casualist, however, is an emotional prude; that is, he tries to deny those emotions that cluster around his sexual life: affection, but not affection alone, loneliness in absence, jealousy, envy, preoccupation, restlessness, anger, and hopes for the future. The emotional prude dismisses all these or assumes that sincerity and honesty provide a kind of solvent that breaks down chemically any and all inconvenient and messy feelings: You hope for the future? But I never promised you a future. Why complain? I am emotionally clean, drip-dry. Why not you? This antiseptic view overlooks the element of dirt farming in sex and marriage. Caesar ploughed her and she cropped. Put another way, this view overlooks the comic in sex; adopting the pose of the casual it lacks a comic sense. It overlooks the way sex gets out of control. Sex refuses to stay in the playpen. It tends to defy our advance formulae. It mires each side down in complications that need to be respected.

If sex is a great deal more important, complicated, and consequential for the destiny of each partner than the committed casualists are wont to pretend, then it may not be out of place to subject it to a deliberateness, to submit it to a discipline, to let sexual decisions be *decisions* instead of resolving sexual ties by the luck of the draw, opportunity, and drift. The Hebrew tradition emphasized and symbolized the element of deliberateness in sexual life when it imposed the rite of circumcision. The rite does not deny the natural (as castration does with a vengeance) but neither does it accept the natural vitalities without their conforming to purposes that transcend them. Human sexual life is properly itself only when it is drawn into the self's deeper identity. Thus, against those who reduce sex to the casual, the tradition says sex is *important,* and should be subjected to discipline like anything important and consequential in human affairs.

2. The approach of the dualists to sex, either those who elevate it to a transcendental evil or those who reduce it to a doggish burden, hold to an element of truth. Sometimes, sexual activity can be abysmally self-destructive and destructive of others; at other times, it is merely a burdensome obligation. But, from the biblical perspective, both approaches wrongly estimate sexual love: they confuse the abuse of an activity with the activity itself. Sexual love is a good rather than an evil. God created man in his own image, *male and female* created He them. Genesis provides quite an exalted theory of sexual identity. Not divine, but in the image of God.

This differing estimate of sexual love shifts dramatically the meaning and warrants for discipline in one's sexual life. The Manichaeans disciplined sexual activity in the sense that they sought to eradicate it altogether; they justified radical denial on the ground that sex is inherently *evil*. The Jew and Christian, on the other hand, justify discipline on the basis of the goodness of sexual power.

Unfortunately, most popular justifications of discipline, especially in the perspective of the young, rest on the evilness of an activity or a faculty. Discipline the child because he is evil. Renounce your sexuality because it corrupts. This is the Manichaean way.

We may need to recover the vastly more important warrant for discipline that we already recognize in education and that the biblical tradition largely supports. The goodness and promise of the human mind, not its evilness, justifies the lengthy discipline of an education. Because the child has worthwhile potentialities, we consider it worth our while to develop her to the maximum. Because the piano is a marvelously versatile and expressive instrument, we think it worth the labors of the talented person to realize the full potentialities of the instrument rather than trivialize its capabilities with "Chopsticks." Some sexual encounters are not so much wicked as trivial, less than the best.

3. Finally, the sex-mystics also have an element of truth on their side. The event of sexual intercourse does supply us with one of our privileged contacts with ecstasy—the possibility of being beside ourselves, of moving beyond ourselves, experiencing a level of energy and urgency that both suspends and restores the daily round. But when all is said and done, sexuality, though a good, is only a *human* good, not *divine* as such. Despite Lawrence's perorations on the subject of love and the mountains atremble for Robert Jordan and his mate in Hemingway's *For Whom the Bell Tolls,* the act of sexual intercourse falls short of Exodus–Mount Sinai, death-resurrection. Intercourse is not an event of salvation; neither is marriage another name for redemption.

Biblical realism requires us to acknowledge three ways of abusing sex—to malign it with the dualists, to underestimate it with the casualists, but also to overestimate it with the sentimentalists and therefore to get angry, frustrated, and retaliatory when it fails to transcend the merely human. As a sexologist, St. Augustine had his faults, but he recognized that people tend to engage in a double torture when they elevate the human into the divine—whether it be sex, marriage, children, or any other creaturely good.

First, they condemn themselves to disappointment; they torture themselves. If men and women look for the resolution to all their problems in marriage, if they look to it for salvation, they are bound to discover that neither sex nor marriage converts an ordinary human being into someone sublime. They let themselves in for a letdown. Second, one not only tortures oneself, one also tortures the partner to whom one has turned. One places on the mate too heavy a burden. Dostoevsky tells of a dream in which a driver flogs a horse, forcing it to drag an overloaded wagon until the horse collapses under too much weight. We similarly overburden another when we look to him for too much. We expect others to function as a surrogate for the divine. Thus parents drive their thwarted ambitions through their children like a stake through the heart. Some marriages break up not because people expected too little from marriage, but because they have expected too much.

This biblical realism need not produce the sort of pessimism that expects little of the world and savors even less. Indeed, it should free us a little for enjoyment. Once we free our relationships to others from the impossible pressure to rescue us or redeem us, perhaps we can be free to enjoy them for what they are. Specifically, we can enjoy without shame and with delight a sexual relationship for the pleasurable, companionable, and fertile human good that it is.

16.

Sexuality and Intimacy in Marriage

Evelyn E. Whitehead and James D. Whitehead

Evelyn and James Whitehead have written extensively on marriage and sexuality. Their work is a model of wisdom and balance. This essay weaves together multiple strands around the subject of intimacy and sexuality. It is an area of much confusion in our culture.

The authors provide an understanding of the psychological dynamics of intimacy as a way of illuminating the personal and religious meaning of marriage. We can get better at being married. It is a lifelong psychological process and a religious passage incorporating several stages, each with challenges and invitations to intimacy.

Intimacy is a hallmark of adult maturity. It is the ability to come up close not only as sexual partners, friends, and colleagues, but in the experience of conflict and compromise. The interpersonal skills of intimacy (self-disclosure, empathy, and confrontation) can enrich the mutuality of our relationships and offer access to a richer developmental married life.

Some questions for reflection follow:

1. *What are the major obstacles to intimacy in marriage?*
2. *How can we apply the metaphor of wrestling to the quest for marital intimacy?*
3. *Is intimacy an endowed capacity or an acquired skill?*

An in-depth exploration of the issues raised in this essay are in the Whiteheads' study Marrying Well *(New York: Doubleday, 1983).*

Sexuality is an enduring aspect of human life. Sexuality is not simply romance. As a resource of the personality, sexuality contributes warmth, empathy, and energy to the wide variety of ways we have of being with other people. Sexuality is not simply for the young. Studies of adult development and aging remind us

From *Chicago Studies* 18:3 (1979): 251–261. Reprinted with permission.

that sexual interest and sexual experience continue into the mature years as an important part of the life of many elderly. And sexuality is not simple. For many in American culture it stands as a puzzle, a perennial question of personal life and a problematic issue in social experience.

Our intention here is to discuss the multifaceted issue, sexuality, in relation to a larger psychological category, that of intimacy. Our goal is to provide an understanding of the psychological dynamics of intimacy that can assist in illuminating the personal and religious meaning of marriage. It is our hope that this understanding will serve as working knowledge, as information that can influence attitudes and shape behavior. To this end we include a discussion of the skills of mutuality that must accompany these psychological dynamics if a couple's capacities for intimacy are to be realized in a rich interpersonal experience in marriage.

Some clarifications may be useful as we start. The first concerns the relationship among sexuality, genital expression, and intimacy. Frequently the terms are used interchangeably, with sexuality taken to mean genital expression and intimacy functioning as a "polite" synonym. We will use the terms more precisely: genital sexuality and orgasm are part of the larger experience of sexuality in adult life. This larger category, sexuality, includes my awareness of myself as male or female, along with my experience of affection, emotional attraction, and physical responsiveness. Intimacy, as an even larger category in adult experience, refers to the many ways in which I am brought "up close" to other people—not only in romance and sexuality, but in friendship, in cooperation and competition, in planning and collaborative effort, in conflict and negotiation.

Marriage involves us with each other at all these levels. We come "up close" to each other as sexual partners, as friends, as colleagues in the daily efforts of family life. We also experience each other "up close" in the experiences of conflict, competition, and compromise that are—inevitably and expectably—a part of our life together. These latter, often negative, interpersonal experiences are as much a part of the intimacy of marriage as are cooperation and love-play.

Here our religious rhetoric of marriage sometimes does not serve us well. In ceremonies and sermons about marriage, it is upon images of unity and peace and joy that we dwell. These images of life together in Christian marriage are important and true, but partial. When, as a believing community, we do not speak concretely to the more ambiguous experiences in marriage—experiences of anger, frustration, misunderstanding—we can leave many married people to feel that their marriages are somehow deficient.

Conflict and hostility are not goals of marriage, to be sure. But neither are they an indication that one's marriage is "on the rocks." Conflict is a normal, expectable ingredient in any relationship—whether marriage, team work, or friendship—that engages people at the level of their significant values and needs. The challenge is not to do away with all signs of conflict or, worse, to refuse to admit that conflicts arise between us. Rather we can attempt to learn ways to recognize the potential areas of conflict and to deal with these issues and feelings in ways that strengthen rather than destroy the bonds between us.

Marriage as a Process

Historically we have thought of marriage as a state—"the holy state of Matrimony." Today we are more conscious of marriage as a process, a path pursued, a journey which includes both expected and unexpected events. Understood psychologically, the process of marriage involves two movements. There is the transition from "I" toward "we," as two persons move from the strengths of independent identity toward creating and holding a life-in-common. The challenge here is to create a "we" that is an expression of both "I"s where the identity of each is tested and expanded, but not destroyed. There are new expectations of mutuality today in this "we" of marriage. It is less and less acceptable among women (as, to be sure, among many men) that the "we" of marriage be achieved primarily through the absorption of the wife into her husband's identity and life-ambition. Today the process of marriage involves the more difficult—and more rewarding—effort to create a "we" that bears the stamp of both spouses, a "we" that moves beyond each into the larger reality of their life-in-common.

A second transition in contemporary marriage is from romantic love to committed love. Over the course of history and cultures, many criteria have been used to determine who shall marry whom: dynastic purity, political considerations, consolidation of property, the advice of matchmakers, the decisions of parents. The dominant norm of marriage in America today is self-selection based on the criteria of romantic love. The process of maturity in marriage requires the movement from the exhilarating but largely passive experience of "falling in love" to the experience of love as a chosen and cultivated commitment. This is the movement from romantic love to the love of mutual devotion, strong enough to sustain commitment through the strain and confusions that are inevitably associated with continuing close contact. Such mutuality endures only if each partner is capable of generous self-disregard. In his *Insight and Responsibility* (New York: Norton, 1964), psychologist Erik Erikson expresses a conviction shared by those who have been unsuccessful as well as those successful in love: "only graduation from adolescence permits the development of that intimacy, the selflessness of joined devotion, which anchors love in a mutual commitment" (p. 128).

Marriage as a Religious Passage

If marriage is a psychological process, it is also a religious passage. The religious passage of marriage involves several stages, each with important challenges and invitations to intimacy. Current efforts in the church to strengthen its pre-Cana ministry witness to our growing realization of the importance of this early stage. As two Christians become engaged—as they move beyond casual acquaintance and the early experience of dating to genuinely take hold of each other—an important period of growth in intimacy has begun. During this stage, significant information about each other—strengths and weaknesses, hopes and apprehensions—needs to be shared, in order to test the viability of a life together.

Increasingly Christian ministers to marriage preparation recognize that a ministry of exhortation, urging the couple to love one another, is more effective if it is complemented by a ministry to the skills of intimacy—assisting the couple to learn the practical ways in which they can support and challenge one another as they move toward a life-in-common. Efforts to recover the historical notion of betrothal—a period of serious but not final commitment—may add weight and focus to this crucial stage of growth into marriage. In recent Catholic history ministers have been concerned that the "engagement" of the couple not be sexual in nature; a more important concern today may be that it be more than sexual.

The passage into Christian marriage is celebrated in the Rite of Matrimony. Ideally this rite marks neither the beginning of the couple's movement toward intimacy not its completion. Rather it signals a stage of deepened commitment. The engaged couple, with the community, celebrate a life together that has given signs of being well begun; they pledge a commitment of love that will give life beyond themselves. More and more, couples are finding the need to allow a year or more at the beginning of this commitment for the strengthening of their intimacy before expressing this love in the generation of a child. In these first years of marriage the couple can begin to develop the patterns of mutual care and emotional resilience that will serve them well as their family grows in size and challenge.

If engagement is an early stage of the religious passage of marriage, and the first years of matrimony represent a second stage of committed intimacy, a third stage is initiated with the advent of the first child. Couples testify to the sudden and significant changes in their relationship when a child is born. The advent of another person, with definite and clamorous demands, abruptly alters the marriage relationship. We can expect this advent to challenge the couple's established patterns of intimacy and to demand a re-understanding of how they are to be for each other. Here the Christian ministry to marriage and to Baptism converge. The significant event of a child's Baptism can be graceful not only for the child but for the parents and their own relationship. An invitation to share their hopes for their child, to reflect together on the faith that they want to transmit to this child—such an invitation to the parents by a ministering community creates an opportunity for growth in religious intimacy in marriage. Hearing each other's hopes and apprehensions for this child and its future can profoundly deepen the parents' own intimacy and commitment.

In addition to these stages of intimacy in the earliest years, a new crisis of intimacy is emerging later in contemporary marriages. As recently as a hundred years ago, a married woman could expect her husband to die before the last child left home. With the increased life expectancy of the late twentieth century, a couple can now expect two decades or more of life together after the children have left home. This stage of marital intimacy, with new and different challenges, *has never existed before* in the history of the Christian tradition. This crisis and opportunity for significant growth in intimacy occurs for many women and men in their forties. As the burden of child-rearing decreases and the attention to a career or careers becomes less absorbing, a married couple may feel the need to re-examine their ways of being together. With more time for each other they may become aware of the current inadequacy of ways of communicating and expressing intimacy that had once been more satisfying. This critical time can and does, of course, lead to divorce for some couples. But it can also be an extraordinarily graceful time as the couple set aside earlier expectations and roles and learn new, more mature, modes of communication and intimacy. The emergence of this mid-life crisis of intimacy reminds us that marriage is a continuing process and that an effective ministry to marriage must respond to the different and specific challenges encountered in the life-long journey.

Sexual Maturity in Marriage

Sexual maturity in marriage is, likewise, a process more than a state. It is the process through which both partners learn to contribute to what is, for *them,* mutually satisfying shared sexual experience. And the patterns of this mutually satisfying experience can be expected to differ widely from couple to couple. Here, as in most other criteria of maturity, to be ''normal'' does not mean to fit some external criteria of performance. The patterns of a couple's sexual sharing—patterns of frequency, of time and place, of initiation and response, of affection and ardor, of seriousness and fun—all these take their most important clue from the couple's developing (and, possibly, changing) sense of what is appropriate for *them,* what works for *them.* The expanding literature of sexual functioning can assist this process of sexual maturity in marriage, not by giving an external norm against which a couple is to judge itself, but by providing information that can guide a couple's experimentation and choice.

Sexual maturity in marriage is reached as the couple's capacity for intimate sexual sharing is developed and stabilized. ''Such experience makes sexuality less obsessive,'' Erickson remarks in *Identity: Youth and Crisis* (New York: Norton, 1968). ''Before such genital maturity is reached, much of sexual life is the self-seeking, identity-hungry kind; each partner is really trying only to reach himself. Or it remains a kind of genital combat in which each tries to defeat the other'' (p. 137). The experience of sexual play and orgasm in marriage can contribute to my willingness to risk myself, to let down my defenses in the presence of another. Sexual intimacy thus opens out into a larger psychological resource for intimacy.

Sexual love is often used as a model and metaphor for the wider experience of human intimacy. The rituals of love-making highlight, in dramatic fashion, features common to other experience of intimacy as well—the exhilaration of being attracted to someone, the impulse to share something of myself, the anxious moment of self-disclosure, the affirmation of being accepted, the delight in the give-and-take of mutuality. Love-play and orgasm provide a vivid example of the mutual regulation of complicated patterns that psychologists note as the distinguishing characteristic of adult intimacy. If the ''mutual regulation of complicated patterns'' describes successful love-making, it can also describe success in the other intimacy experiences of marriage and family life—preparing the household budget, taking a family vacation, raising teen-age children.

Cooperation and Competition

Experiences of sexuality and devotion are crucial to the full development of intimacy in marriage. But these are not the only experiences that summon one's resources for mutual commitment. Cooperation involves us in joint action to accomplish a common goal. Competition puts us in some opposition to one another in our pursuit of a goal. Both experiences are to be expected in marriage. Both illustrate the essential elements of intimate relationships. To be a good "cooperator" I must be aware of my contribution to the common goal, both its strengths and its limitations. I must be secure enough that I can make these strengths available to our common task. I must be flexible enough to accept my contribution being modified by your ideas and plans. Obviously my strengths as a cooperator will contribute to our experience of intimacy in marriage. Competition, on the other hand, suffers a bad reputation. Many therapists, educators, and religious persons share the conviction that competition is to be, as much as possible, eliminated. This conviction is born of much experience with the negative effects of competition in the lives of individuals and in marriage and family life. While we do not deny these negative experiences of competition, we wish to invite a consideration of another aspect of an admittedly ambiguous phenomenon. Most psychologists today would attest that an ability to compete maturely is an important ingredient of the adult personality. And the psychological characteristics of the mature competitor are remarkably similar to those of the mature cooperator. To compete well—in sports, for example—I must have a realistic sense of my own abilities, with an awareness of both the strong and weak points in my game. Competition forces me to express these abilities, to expose them to the test of a concrete challenge, with the risk that they may not be sufficient. But it is only in taking that risk of failure that I can confirm and develop my strengths. The exchange of competition reveals much about each participant. In the contest I come to a better knowledge of myself and to the special awareness of my opponent. My success in the game is often dependent upon flexibility and creativity in modifying myself in response to what I learn about my rival.

These characteristics—awareness of self and other, a sense of self adequate to the demands of mutuality and to the possibility of failure, flexible response to the individuality of other persons—are not germane to sports alone. They are resources that enhance intimacy in marriage. They are valuable in the variety of experiences in family life that involve planning, conflict resolution, negotiation, and compromise.

Marriage: An Intimacy That Is Competitive

The embraces of intimacy in marriage are not always warm and tender. Some are competitive, inviting us to struggle and threatening us with injury. In the Old Testament, Jacob's wrestling with Yahweh (Genesis 32) appears to be such a competitive intimacy. Jacob is "embraced" by Yahweh in the middle of the night—but is it an embrace of love or harm? Jacob's apprehension and dismay are repeated in many or our embraces. Can I trust such a close encounter or should I withdraw to a safer distance? Will I survive this embrace? The metaphor of wrestling is attractive as an image of intimacy because it suggests an intense closeness, one in which our strengths and weaknesses are critically tested and exposed. Wrestling takes intimacy beyond romance and fantasy with its introduction of apprehension, vulnerability, and hurt. It reminds us that loving is a contact sport with distinct possibilities of injury and loss. Jacob is not only vulnerable (as we must be in order to love), he is wounded (as we can expect to be if we try to love). Jacob loses something in this encounter: his name is changed. But the loss is not a deprivation but a growth. His new name, Israel, describes him as one who prevails in this relationship with God. And the loss or surrender in this struggle is mutual. Yahweh is forced to yield, surrendering to Jacob a blessing.

The ambiguity of the embrace in these intimacy struggles abounds. Jacob is said to prevail yet he limps away from the encounter. Such wrestling is exhausting but exhilarating. The intimacy of marriage obviously involves such wrestling embraces: the sexual embrace is complemented by the other, often ambiguous, embraces that generate decisions about career, children, finances, or even the more mundane decision of housekeeping and daily schedule. These competitive encounters arise in marriage not simply because the spouses are immature or selfish. Such struggles are important and expectable events through which the intimacy of marriage is tested and deepened.

There is another way in which marital intimacy and competition can be seen to intersect—both demand trust. Psychologist George Valliant in his study of successful adult maturity, *Adaptation to Life* (Boston: Little, Brown, 1978), traces this nexus among maturity, trust, and play. ''It is hard to separate capacity to trust from capacity to play, for play is dangerous until we can trust both ourselves and our opponents to harness rage. In play, we must trust enough and love enough to risk losing without despair, to bear winning without guilt, and to laugh at error without mockery'' (p. 309). There are marriages in which people do not play fair. There are also marriages where the partners are no longer playing. Either wounded by past hurts or frightened of possible injuries, they choose not to ''contest'' with one another. Though the marriage may continue, it has become a stalemate, an unplayful relationship of stale mates. A marriage matures as its partners get better at the different, sometimes threatening, embraces of the union. If they occasionally hurt each other, they also invigorate each other as the play of their intimacy deepens.

Intimacy as a Strength in Marriage

Intimacy is, then, as we have discussed elsewhere (see our *Christian Life Patterns* [New York: Doubleday, 1979]), a capacity, an ability, an abiding competence of adult maturity. It is the strength which enables me to commit myself, not to humankind in general or to idealized movements, but to particular persons in concrete relationships—aware of the limitation and incompleteness that are involved. Intimacy resources are drawn upon again in living out these commitments. Relationships are not static. People change and relationships develop over time. Some developments will bring fulfillment; others will make demands for accommodation, for understanding, for tolerance, for forgiveness. A well-developed capacity for intimacy enables a person to sustain the adjustments and compromises of life with others, without jeopardy to one's own integrity. A flexible identity, an empathetic awareness of others, an openness to continued development of the self—these strengths of intimacy make possible the creative commitment of marriage.

Here we see the power of marriage as a sign and source of intimacy. Marriage can provide the frame within which two persons move toward full and mutual devotion. Its structure supports the initial risks of self-disclosure and confrontation. Its commitments protect the fragile figure of a developing life-in-common. Its promise of duration invites the open-ended investment that is required for creative (and procreative) activity together. The marriage of mature—or maturing—Christians can thus display "that combination (by no means easily acquired, nor easily maintained) of intellectual clarity, sexual maturity, and considerate love," which, as Erikson notes in *Insight and Responsibility,* anchors men and women "in the actuality of their responsibilities" (p. 129).

Skills of Intimacy

Our religious tradition is rich in a wisdom of generous love. This heritage can provide a sound basis for the un-self-centered devotion that must undergird marriage. But "right attitudes" are not enough to sustain the long-term commitments of marriage. Attitudes must be translated into appropriate behavior. We must be able to express, in the give and take of marriage, our desires to be close, to share, to cooperate. In addition to the attitudes of intimacy we must develop the skills of intimate living. It is possible to "get better" at being married. We can learn more effective ways to express and to receive the gift of self that is at the core of our intimacy in marriage. Interpersonal skills especially important to intimate living are self-disclosure, empathy, and confrontation (see Gerard Egan, *Intepersonal Living* (Monterey, CA: Brooks, Cole, 1976). Each involves both an attitude and a set of behaviors. To share myself with you I must be psychologically disposed, able to overcome the hesitancy suggested by fear or suspicion or shame. But these inhibitions overcome, I must be able actually to share—to disclose myself in a way that is appropriate for me and for our relationship. Appropriate self-disclosure can be complicated. But I am not, however, limited to my current level of success. I can become more skillful, learning better ways to express my needs, my ideas, my feelings, my values.

This is equally true in empathy. An essential psychological strength undergirds my ability to stand with another, emotionally and intellectually. But the basic strength may not be enough. My capacity for accurate empathy can be enhanced by my development of a range of behavioral skills. An accepting posture, attentive listening, sensitive paraphrasing—each can contribute to my effective presence with you.

Confrontation, too, makes a critical contribution to intimacy in marriage. We do not limit "confrontation" here simply to its negative and narrow connotation as interpersonal conflict (though, as we have noted earlier, the ability to manage interpersonal conflict is itself an important resource for marriage). An ability to confront involves the psychological strength to give (and receive) emotionally significant information in a way that leads to self-examination rather than simply to self-defense. To do this I need to be skillful in communicating nonjudgmentally, in dealing with anger in myself and in you, and in offering you emotional support even as we disagree.

It is true, to be sure, that if these behavioral skills are not informed by a psychological capacity for intimacy they remain hollow and ineffective. They may even be used manipulatively. It is equally true, however, that the impulse toward intimacy can be frustrated and the challenge of mutuality in marriage exaggerated by the lack of these basic skills of interaction.

In the course of this discussion we have noted the relationship between sexuality and intimacy. The sexual encounter is for married persons a significant instance of intimacy in its psychological as well as physical meanings. Sexuality is one of the "complicated patterns" that must be "mutually regulated" in the intimacy of marriage. Two other comments on the relationship between sexuality and psychological intimacy are useful here. The first draws attention to the obvious contribution that psychological intimacy can make to the human experience of sexuality. The experience of sexuality and its genital expression is enhanced when the physical encounter is part of a larger mutuality. For most adults genital sharing is enriched when it occurs as a part of a broader pattern of shared experience. In most relationships, even most sexual relationships, sex is not enough. Lovers want to and need to have more in common. An inability to develop this broader deposit of common interests can lead to the deterioration of the quality of the sexual experience as well.

The second comment concerns the importance to sexual intimacy of the development of interpersonal intimacy skills. The burgeoning literature of sexual dysfunction and therapy attests to the necessity of self-disclosure, empathy, and confrontation in the communication between lovers. Wives and husbands must be able to let each other know, not only their preferences in love-play and intercourse, but their feelings of love, their desires for affection, their sense of vulnerability. A reluctance to confront my spouse on an issue of household concern can carry over into a reluctance to make love. An effort to develop these skills of communication often enhances a couple's life together more than does improved sexual performance. And couples who describe their sexual relationship as most satisfactory are likely to mention their overall patterns of communication as sound. Communication may not be the whole of intimacy, but—as both theology and psychology show us—it is at love's heart.

17.

Can We Get Real About Sex?

Lisa Sowle Cahill

The subject of sexual morality is controversial and dangerous territory to tread for a contemporary Roman Catholic theologian. The controversies are shaped by at least four constituencies: the traditionalists, the revisionists, the skeptics, and the "alienated." Each manifest different and, sometimes, conflicting perspectives. This has given rise to acrimonious debates and sidetracked us to tangential issues.

Lisa Cahill, a professor of theology at Boston College, finds the terms of the current discussion on sexual morality in the Roman Catholic Church impotent and inadequate. She attempts to transcend the present limits of the discourse and recover the essential message about sexuality in the heart of the Catholic tradition. The tradition embodies a set of moral attitudes toward sexuality that are wise guides for our day. Three dimensions of sexuality are noted: (1) mutual physical pleasure, (2) intimacy in interpersonal commitment, and (3) receptivity to procreation/parenthood. It is the deep unity, integration, and mutual reinforcement of the three aspects that contains the essence of the Roman Catholic vision of sexuality.

The essay is an appropriate return to fundamentals and offers secure ground and a guiding vision to the generations.

Some questions suggested by the essay follow:

1. *In relation to the official teaching of your own church, are you a traditionalist, a revisionist, a skeptic, or a member of an "alienated" constituency on sexual morality?*
2. *Can the moral parameters of marriage define and do justice to all the forms and expressions of sexuality today?*
3. *What set of values do you judge are needed to anchor sexuality in enduring and rewarding relationships? Is Cahill's proposal credible and convincing?*

During my years as college professor (since 1976), lecturer, writer, and mother I have learned that the task of trying to make sense of Catholic teaching on sexu-

From *Commonweal* 117:15 (14 September 1990): 497–503.

ality has to be geared to different audiences with different life experiences. Each generation has its own questions. Those of us who were teenagers before Vatican II still carry on a struggle of "liberation" from a negative and restrictive picture of sexual dangers. But most younger adults and virtually all teens today face a different battle: to carve out some sense of sexual direction in a peer and media culture which presents sex as a sophisticated recreational activity for which the only moral criterion is mutual consent. I have finally learned that my invitations to appreciate the goodness and pervasiveness of sexuality sound not only redundant but even naive to audiences hungry for a solid answer to shallow or cynical versions of precisely that same message.

Currents of ethical and theological thought within the church manifest similar differences in perspective. Controversies over sexual morality are shaped by at least four constituencies, having different and perhaps incompatible agendas. Though this is to simplify, we may think of them as the traditionalists, the revisionists, the skeptics, and the alienated. These groups represent different responses to the exciting but tumultuous changes which beset the church as well as the culture in the 1960s. The traditionalists put conformity to magisterial teaching high on the list of Catholic identity markers. They stand behind the idea that Vatican positions on matters like premarital sex, birth control, abortion, homosexuality, and divorce can brook no "dissent." Holding a united front on these questions is perceived as essential to the continued strength of church authority. Traditionalists try to connect past teaching with the modern world by arguing that those who *experience* sexual intimacy and honestly examine it will agree that the relationship is a form of "mutual self-gift" (in a phrase of John Paul II), which intrinsically requires heterosexuality, commitment, permanency, exclusivity, and procreation.

A second group, the "revisionists," mostly grew to maturity before Vatican II, and remember vividly the revitalization the council brought. These Catholics, some parents of adolescents and young adults, see the church as their religio-cultural home, even as "mother" and "teacher," but they disagree that sexual and marital experience necessarily confirm all current church teaching. At least since the sixties, they have had serious doubts about whether the positions on contraception and divorce can really hold water. They continue to struggle within the church to find room and a voice for moderate reformulations of—and possibly a few exceptions to—the Catholic view that sex belongs in indissoluble marriage and leads to parenthood.

A third group might be called the "skeptics." Many, but not all, of them are younger adults who are less willing to take church credibility for granted. They tend to look on in disbelief as the church of their parents promotes teachings on sex which appear oblivious to the realities of human relationships, at least in the U.S. They observe the same phenomena as do the "revisionists"—the widespread acceptance of contraception, "living together," homosexuality, abortion; the threats of AIDS and marital breakdown. But although the skeptics still consider themselves "Catholic," they differ from the "revisionists." They openly assert that church teaching on sex as formulated primarily by celibate male clergy should be declared irrelevant to modern needs. Ready to relegate the traditionalists to the lunatic fringe, they smile at the earnest and rather dogged reinterpretive efforts of the revisionists, wondering when they will realize the impossibility of making headway toward change within present structures.

The fourth group, the "alienated," no longer feel any special tie to the Catholic church or any necessity to justify, struggle with, or refute its teachings. Roman Catholicism is not a resource to which they turn (at least not consciously) for guidance on sex or any other issue. Church sexual teachings, when considered at all, are written off as obsolete, oppressive, and outrageous. Even though "alienated" Catholics are in a sense no longer a "constituency" for church teaching, it is significant that that teaching evokes so negative a response in a group whose size is far from negligible.

This author would be best identified as a member of constituency two attempting to convince constituency three that there is something worthwhile about sex still to be mined in Catholic teaching—though I would have to concur in the quite legitimate impression that its practical value is not always easy to discover.

The Catholic tradition on sexuality has always defined its moral parameters in terms of *marriage.* More than a union of two individuals, marriage is set in the context of *family,* and especially of *procreation.* Augustine and Aquinas saw procreation in marriage as the only reason fully justifying sexual intercourse, and saw both procreation and marriage as especially important insofar as they contribute to the species, the society, and the kinship network or extended family. Although companionship and friendship of spouses were ideals, premodern authors, like the society around them, were unable to recognize the later ideal of "interpersonal union" both because they lacked our sense of the importance of individuals, and because women were considered inferior and subordinate to their male partners. Procreation as the primary purpose of sex was maintained as late as 1930 (in Pius XI's encyclical *Casti connubii*). A break-through occurred in Vatican II's *Gaudium et spes* (1965) and in the encyclical *Humanae vitae* (1968), when love and procreation were ranked equally as the purposes of marriage. This shift raising love to a level with procreation represented the influence of philosophical "personalism," and the emergent awareness of women's equality. Yet official church teaching continues to tie respect for these values very much to the physical act of sexual intercourse, not to the overall or long-range relationship of the couple. Today it teaches that both purposes must be present in "each and every act." That is, every act must be part of a permanently committed, heterosexual, love relationship; and every single act must be procreative, *in the sense that* the outcome of procreation must not be artificially prevented.

Contemporary experience and thought raise challenges to this specific presentation of the teaching that should not be under-estimated. "Sexuality" is now recognized as a basic dimension of the personality, and covers far more than genital acts designed for reproduction. The affective and interpersonal dimensions of sex, along with the occasions it offers for intimacy and reciprocal pleasure, have become far more important. The wrongness and harm of defining all or most nonprocreative sexual pleasure as "sinful" is evident. The feminist movement has sharply critiqued the distorted forms with which patriarchy has shaped both marriage and family, and has begun to reshape sexuality with a new appreciation of women's experiences of sex, spousehood, and motherhood. Delayed marriage for many young adults who pursue educational and vocational goals also means a longer period of sexual maturity and potential relationships before marriage. The responsibility of the marriageable to choose their own partners rather than relying on parental negotiations or social and religious similarity, along with the high incidence of divorce, has led to premarital sex and "trial marriages," which many see as prudent exploratory arrangements. And there are many single adult Catholics who may not have the opportunity or desire for marriage, but who yearn for intimacy and sexual expression, for which they may find occasions outside marriage. Many lesbian and gay persons see their sexuality as a gift to be valued both for personal identity and relationships. They call for church support of their efforts to live as faithful Christians and to gain protection of their civil rights. All of these considerations pose challenges to church teaching which, it must be admitted, the church has not adequately met. They also raise questions which I cannot pursue here, though I have advocated modifications (a group-two goal!) of church teaching on many of these points. Indeed, it sometimes seems that both "conservatives" and "liberals" (terms that in practice refer to the traditionalists and the revisionists, since neither the skeptics nor the alienated see the relevant intra-ecclesial debates as worth the investment) become unduly distracted from more fundamental issues by battles over the morality of sexual acts: permarital, contraceptive, homosexual, etc. If we could transcend the limits of such discussions, we might recover the essential message about sexuality which Roman Catholicism transmits.

What is that message? What is a credible, convincing, and helpful expression of Christian sexual values today? I think that message pertains to three dimensions of sexuality: (1) sex as a physical drive for pleasure; (2) sex as intimacy or love; (3) sex as procreative. It is the value of the third that is the most necessary and the most difficult to communicate to today's young adults.

Sex as a physical drive. In the past, there has been in Christianity (in Augustine, for instance) a deep suspicion of sexual drives or sexual desire. This suspicion was no doubt based on sex's undeniable tendency to break social and moral restraints, and to seek fulfillment in self-centered, manipulative, and even violent ways. Today this attitude might be revised into the recognition that the sexual drive has real limits as a guide to sexual relationships. The dominant "cultural message"—that sex is natural, enjoyable, good, and even recreational—has an obvious legitimacy in itself, and exponents of a Catholic Christian approach to sexuality should not appear grudging in their acceptance of the "joy of sex." But the message is incomplete and inadequate. Using sexual acts and relationships as an outlet for our physical drives or as a means of access to physical enjoyment is not *bad,* but it is *limited.* Media images aside, I doubt that many people really disagree with this point, however much some may be tempted to rationalize indiscriminate sexual behavior. Physical desire and enjoyment taken alone as motives for sex make sex unfulfilling, lonely, and perhaps ultimately boring. Although women seem to understand better than men that sexual intimacy naturally entails psychological intimacy, I doubt this difference is innate. Rather, it is a matter of women being socialized or socially encouraged to take intimacy more seriously. Intimacy adds to the fulfillment of both men's and women's sexuality. This leads to sex's second dimension.

Sex as love. Seeing sex as an expression of love seems to verge on the romantic, yet it is not all that foreign to most of our experiences and personal goals. Our culture is prone to cynicism about the trustworthiness of human relationships. But sexual intimacy can express and augment psychological intimacy, affection, reciprocal understanding and encouragement, partnership, companionship, compassion, and even commitment. One Catholic Christian value of sexuality is permanence: the love relationship established sexually between a woman and a man should be long-term and not transient. Sex with little or no commitment shortchanges sex's potential for intimacy. Unlike other animal species, humans have a deep capacity for friendship and interpersonal reciprocity, which, when expressed sexually, constitutes the most intense of human relationships. The appropriate moral context for complete sexual union is a commensurate level of interpersonal commitment. Unequal commitment between partners leads to manipulation, disappointment, and pain. Sex without commitment is unfaithful to the human potential of sexuality.

The psychological and personal aspects of sexual union are complemented by the relation of the sexual couple to the family and society. Our sexuality is not simply an individualistic capacity but binds us with others in families, that is, in some of the most rewarding and most demanding of human relationships. A man and woman bring to their union links with and commitments to other persons, including their respective families and friends and, eventually, the children their union may produce. Although not all sexual couples give birth to children, the procreative potential of sex is always a part of that relationship. The prospect of pregnancy and parenthood is not always intentional and dominant in the relationship, but it is nonetheless a latent and morally important possibility. Obviously a faithful commitment between parents is the best context for the nurture of children.

Like many group-two Catholics, I have often considered the common practice of "living together" with some misgivings along with the feeling that there might be in it something to be learned about the nature of sexual commitment. Some of my intuitions came together upon hearing a Ugandan bishop observe that, in his culture, marriage was a *progressive* reality, which did not come into being in an instant during a single ceremony, but which developed through a process of negotiation, visiting, and gift-giving among bride, groom and their families. Although at some time during the process the couple might have sexual relations and even bear children, there was actually no one "point" before which a marriage did not exist and after which it did. Perhaps many couples in our culture are making a similar statement about their growing trust in and love for each other. However, the shortcoming of "progressive marriage" as we see it in the U.S. is the isolation of the unmarried sexual couple from the social support and accountability that accompanies formal marriage. In Uganda, the whole family has an investment in the growing relationship and expects the couple to make it work, persuading, admonishing, and supporting them as need be. This system also has clear understandings about how children will be cared for within the families if the couple subsequently parts. In other words the African form of gradual marriage carries with it at every stage an increasing level of personal, familial, and social weight and responsibility. One thinks also of ancient Israelite betrothal and marriage, reflected in Matthew's and Luke's stories of the premarital pregnancy of Mary. The sexual relations permitted before marriage occurred in a context of religiously and socially specified conditions. Though the provisional sexual relationships common in our culture may represent a valid insight about the development of commitment, they still lack the social forms which would make them accountable to the genuine personal and communal significances of sex. One of these is parenthood, which leads us to the next point.

Sex as procreation. In modern Western cultures, the value of procreative sex may be harder to ''sell'' than pleasurable sex and loving sex. We are in an era in which procreation has been reduced to an incidental meaning of sex, usually to be avoided, and certainly to be accepted only if freely chosen. The deep associations and mutual reinforcements of sex, love, and parenthood are missing—partly because the church's teaching authority itself has narrowed their reciprocity to an experientially unintelligible focus on reproductive genital acts taken as separate events. But it is the unity of sex, love, and parenthood in this broader sense that is probably the major message Roman Catholicism has to offer today's young adults, who more easily see that sex expresses love than that sexual love leads to permanent commitment, parenthood, and family.

A better expression of this link, one we should aim to attain, lies in contemporary thought's repudiation of dualism, and its insistence that the body and the spirit or psyche form an integrated reality, not two uneasily aligned ''components.'' We no longer tolerate a sexual ethics that sees the body as ''bad,'' and to be repressed, while only our spiritual side is ''holy.'' But a nondualistic view of sex requires that we premise a sexual morality on the goodness of sexual acts and sexual pleasure. It also requires that we look at these acts and their reproductive potential as an integrated whole or process. This is not to say that it is always wrong to interfere with conception as an outcome; but that moral analysis starts with a presumption in favor of the conduciveness of sex to shared parenthood. In other words, physical satisfaction or pleasure, interpersonal intimacy, and parenthood are not three separate ''variables,'' or *possible* meanings of sex which we are morally free to combine or omit in different ways. Sex and love as fully *embodied* realities have an intrinsic moral connection to procreativity and to the shared creation and nurturing of new lives and new loves.

A more flexible and experientially adequate way to express this unity is not in terms of acts, but of *relationships.* Certain basic human relationships come together through our sexuality and link us not only to our partners but to the wider community, through the social relationships of marriage and family. These basic relationships are the woman-man and the parent-child relationships. Spousehood and parenthood are linked in the long-term commitment of the couple, sexually expressed. Both are not only intersubjective, but also embodied relationships. *Spousehood* is embodied through the shared material conditions of economic and domestic life, and through sexuality, which can give rise to a shared physical relation to the child. *Parenthood* is embodied through the shared material conditions of family life, again through the genetic link, and through the fact that the physical relation of spousehood is that which gives rise to parenthood.

To recapitulate, the Catholic tradition yields a set of moral attitudes toward sexuality, even before the point of dealing with concrete moral dilemmas or moral norms and prohibitions. The tradition can encourage respect and appreciation for sexuality as mutual physical pleasure, as intimacy, and as parenthood (or at least receptivity to it). These three relationships come together in the ongoing relationship of a couple. In all its dimensions sex is both a psycho-spiritual experience and an embodied one, and both aspects contribute to its moral character.

Having said this, however, one also realizes that human circumstances sometimes arise in which not all three values (sex itself, commitment or love, and parenthood) can be realized simulaneously. In their sexual lives, as elsewhere, human beings are often confronted with moral comflicts, in which no choice is free of ambiguity. Of the three values, it is certainly love which is the *sine qua non,* the primary value in the triad. Since personhood is the most distinctive quality of the human being, it is the most personal aspect of sexuality which is *most* morally important. In unusual or difficult circumstances, the other two (sexual intercourse and procreation) can be subordinated to the love relationship of the couple as long as they are still given significant practical recognition. For example, in the use of contraception, procreation is temporarily set aside, but it still can be realized in the total relationship of the couple. In some infertility therapies, sex itself is set aside as the means to conception, but certainly the relationship of the couple which is both loving and procreative is otherwise given sexual expression. On the other hand, couples who are absolutely intolerant of the prospect of parenthood, perhaps resorting even to abortion as a means of birth control, do not give adequate moral recognition to the relation of sex, love, and procreation. Similarly, couples so desperate to conceive that they are willing to set aside the unity of their spousal-sexual-parental bond by using donor sperm or a surrogate mother to create a reproductive union between one spouse and a third party are also less than faithful to the values which sexuality represents. Although marriages as human realities sometimes fail, a tragedy that church teaching on "indissolubility" may not have met satisfactorily, the asset of the tradition is that it holds up an ideal of permanency. The meaning of "love" in the sexual triad goes far beyond romantic affections. It means a commitment to build a mutually satisfying relationship, to mutual respect, to understanding and support. It entails persistence, repentance, and forgiveness. However justified divorce may sometimes be, the very high incidence of divorce seems to be due to cultural forgetfulness that the commitment to marital partnership requires both ongoing personal dedication and strong social supports.

Since at least the 1960s, interpersonal values have moved to center stage in the Catholic picture of sex, just as more attention has been paid to the experience of actual sexual relationships. At the same time, the inclusion of love and commitment as central along with parenthood has been sidetracked by acrimonious exchanges over contraception and other issues. Such debates have drained energies from the real task of reappropriating the essentials of Catholic teaching for the next generation. What our culture most needs to hear is an effective critique of individualist, materialist, and transient sexual relationships—not lists of specific transgressions which are ''against church teaching.'' The Catholic ''message'' is that the interdependence of sex's pleasurable, intimate, and parental aspects can anchor our sexuality in some of the most enduring and rewarding human relationships. That message will be heard only if it is addressed honestly to the real sexual experiences of young adult Catholics, and only if the messengers can listen to and even learn from their audience's response.

V

Communication, Conflict, and Change

18.

How People Change

Allen Wheelis

Allen Wheelis's reflections on how people change contains some ideas important for those seeking the right marriage partner. He points out convincingly how our actions eventually make us a particular kind of person, whether we want to admit it or not. Thus there is a lot more to any person we meet than that person's initial attractiveness. That person's behavior has over time made him or her a particular kind of person. If you want to find out who a person is, look to the person's behavior.

This way of looking at people is actually quite hopeful, because, within certain limits, people can change their behavior and with it, even the kind of person they are. Some study questions follow:

1. *"We are what we do." In what ways is the author's argument convincing or unconvincing to you?*
2. *Freedom as the ability to choose alternatives is contingent on awareness and consciousness. What does this statement mean in dealing with a person whose behavior needs to change?*
3. *What implications does Wheelis's essay have for people who are dating?*
4. *At the end Wheelis explains how we can look at any life in terms of causes or choices. If the life is one's own or that of someone we love, we will emphasize choice. What implications does this position have for dealing with a potential or actual marriage partner. What, for example, would you say to someone who claims, "If I marry so and so, I will be able to change him or her through my love"?*

We are what we do . . . Identity is the integration of behavior. If a man claims to be honest we take him at his word. But if it should transpire that over the years he has been embezzling, we unhesitatingly discard the identity he adopts in words and ascribe to him the identity defined by his acts. ''He claims to be honest,'' we say, ''but he's really a thief.''

From *Commentary* (May 1969): 57–58, 63, 66. Reprinted with permission.

One theft, however, does not make a thief. One act of forthrightness does not establish frankness; one tormenting of a cat does not make a sadist, nor one rescue of a fledgling a savior. Action which defines a man, describes his character, is action which has been repeated over and over, and so has come in time to be a coherent and relatively independent mode of behavior. At first it may have been fumbling and uncertain, may have required attention, effort, will—as when one first drives a car, first makes love, first robs a bank, first stands up against injustice. If one perseveres on any such course it comes in time to require less effort, less attention, begins to function smoothly; its small component behaviors become integrated within a larger pattern which has an ongoing dynamism and cohesiveness, carries its own authority. Such a mode then pervades the entire person, permeates other modes, colors other qualities, in some sense is living and operative even when the action is not being performed, or even considered. A young man who learns to drive a car thinks differently thereby, feels differently; when he meets a pretty girl who lives fifty miles away, the encounter carries implications he could not have felt as a bus rider. We may say, then, that he not only drives a car, but has *become* a driver. If the action is shoplifting, we say not only that he steals from stores but that he has *become* a shoplifter.

Such a mode of action tends to maintain itself, to resist change. A thief is one who steals; stealing extends and reinforces the identity of thief, which generates further thefts, which further strengthens and deepens the identity. So long as one lives, change is possible; but the longer such behavior is continued the more force and authority it acquires, the more it permeates other consonant modes, subordinates other conflicting modes; changing back becomes steadily more difficult; settling down to an honest job, living on one's earnings, becomes ever more unlikely. And what is said here of stealing applies equally to courage, cowardice, creativity, gambling, alcoholism, depression, or any other of the myriad ways of behaving, and hence of being. Identity comprises all such modes as may characterize a person, existing in varying degrees of integration and conflict. The greater the conflict the more unstable the identity; the more harmonious the various modes the more durable the identity.

The identity defined by action is present and past; it may also foretell the future, but not necessarily. Sometimes we act covertly, the eye does not notice the hand under the table, we construe the bribe to have been a gift, the running away to have been prudence, and so conceal from ourselves what we are. Then one day, perhaps, we drop the pretense, the illusion cracks. We have then the sense of an identity that has existed all along—and in some sense we knew it but would not let ourselves know that we knew it—but now we do, and in a blaze of frankness say, "My God! I really am a crook!" or "I really am a coward!" We may then go too far and conclude that this identity is our "nature," that it was writ in the stars or in the double helix, that it transcends experience, that our actual lives have been the fulfilling of a pre-existing pattern.

In fact it was writ only in our past choices. We are wise to believe it difficult to change, to recognize that character has a forward propulsion which tends to carry it unaltered into the future, but we need not believe it impossible to change. Our present and future choices may take us upon different courses which will in time comprise a different identity. It happens, sometimes, that the crook reforms, that the coward stands to fight.

. . . *And may do what we choose*. The identity defined by action is not, therefore, the whole person. Within us lies the potential for change, the freedom to choose other courses. When we admit that those "gifts" were bribes and say, "Well, then, I'm a crook," we have stated a fact, not a destiny; if we then invoke the leopard that can't change his spots, saying, "That's just the way I am, might as well accept it," we abandon the freedom to change, and exploit what we have been in the past to avoid responsibility for what we shall be in the future.

Often we do not choose, but drift into those modes which eventually define us. Circumstances push and we yield. We did not choose to be what we have become, but gradually, imperceptibly became what we are by drifting into the doing of those things we now characteristically do. Freedom is not an objective attribute of life; alternatives without awareness yield no leeway. I open the door of my car, sit behind the wheel, and notice in a corner of vision an ant scurrying about on the smooth barren surface of the concrete parking lot, doomed momentarily to be crushed by one of the thousand passing wheels. There exists, however, a brilliant alternative for this gravely endangered creature: in a few minutes a woman will appear with a picnic basket and we shall drive to a sunny, hilltop meadow. This desperate ant has but to climb the wheel of my car to a safe sheltered ledge, and in a half hour will be in a paradise for ants. But this option, unknown, unknowable, yields no freedom to the ant, who is doomed; and the only irony belongs to me who observes, who reflects that options potentially as meaningful to me as this one to this ant may at this moment be eluding my awareness; so I too may be doomed—this planet looks more like a parking lot every day.

Nothing guarantees freedom. It may never be achieved, or having been achieved may be lost. Alternatives go unnoticed; foreseeable consequences are not foreseen; we may not know what we have been, what we are, or what we are becoming. We who are the bearers of consciousness but of not very much, may proceed through a whole life without awareness of that which would have meant the most, the freedom which has to be noticed to be real. Freedom is the awareness of alternatives and of the ability to choose. It is contingent upon consciousness, and so may be gained or lost, extended or diminished.

Personality is a complex balance of many conflicting claims, forces, tensions, compunctions, distractions, which yet manages somehow to be a functioning entity. However it may have come to be what it is, it resists becoming anything else. It tends to maintain itself, to convey itself onward into the future unaltered. It may be changed only with difficulty. It may be changed from within, spontaneously and unthinkingly, by an onslaught of physiological force, as in adolescence. It may be changed from without, again spontaneously and unthinkingly, by the force of unusual circumstance, as in a Nazi concentration camp. And sometimes it may be changed from within, deliberately, consciously, and by design. Never easily, never for sure, but slowly uncertainly, and only with effort, insight, and a kind of tenacious creative cunning.

Personality change follows change in behavior. Since we are what we do, if we want to change what we are we must begin by changing what we do, must undertake a new mode of action. Since the import of such action is change, it will run afoul of existing entrenched forces which will protest and resist. The new mode will be experienced as difficult, unpleasant, forced, unnatural, anxiety-provoking. It may be undertaken lightly but can be sustained only by a considerable effort of will. Change will occur only if such action is maintained over a long period of time.

The place of insight is to illumine: to ascertain where one is, how one got there, how now to proceed, and to what end. It is a blueprint, as in building a house, and may be essential, but no one achieves a house by blueprints alone, no matter how accurate or detailed. A time comes when one must take up hammer and nails. In building a house the making of blueprints may be delegated to an architect, the construction to a carpenter. In building the house of one's life or in its remodeling, one may delegate nothing; for the task can be done, if at all, only in the workshop of one's own mind and heart, in the most intimate rooms of thinking and feeling where none but one's self has freedom of movement or competence or authority. The responsibility lies with him who suffers, originates with him, remains with him to the end. It will be no less his if he enlists the aid of a therapist; we are no more the product of our therapists than of our genes: we create ourselves. The sequence is suffering, insight, will, action, change. The one who suffers, who wants to change, must bear responsibility all the way. "Must" because as soon as responsibility is ascribed, the forces resisting change occupy the whole of one's being, and the process of change comes to a halt. A psychiatrist may help perhaps crucially, but his best help will be of no avail if he is required to provide a kind or degree of insight which will of itself achieve change.

Should an honest man wish to become a thief the necessary action is obvious: he must steal—not just once or occasionally, but frequently, consistently, taking pains that the business of planning and executing thefts replace other activities which in implication might oppose the predatory life. If he keeps at it long enough his being will conform to his behavior: he will have become a thief. Conversely, should a thief undertake to become an honest man, he must stop stealing and must undertake actions which replace stealing, not only in time and energy, and perhaps also excitement, but which carry implications contrary to the predatory life, that is, productive or contributive activities.

Of two equally true accounts of the same life the one we choose will depend upon the consequences we desire, the future we intend to create. If the life is our own or that of someone who has come to us for help, if it involves suffering and there is desire to change, we will elect a history written in terms of choice; for this is the view that insists upon the awareness of alternatives, the freedom to make one's self into something different. If the life in question is one we observe from a distance, without contact or influence, for example a life which has ended, we may elect a history written in terms of cause. In reconstructing a life that ended at Auschwitz we usually ignore options for other courses of individual behavior, locate cause and responsibility with the Nazis; for our intent is not to appraise the extent to which one person realized existing opportunity, but to examine and condemn the social evil which encompassed and doomed him. In considering the first eighteen years in the life of Malcolm X few of us would find much point in formulating his progress from delinquency to rackets to robbery to prison in terms of choice, holding him responsible for not having transcended circumstance; most of us would find the meaning of his story to lie in the manner in which racism may be seen as the cause of his downward course.

Conflict, suffering, psychotherapy—all these lead us to look again at ourselves, to look more carefully, in greater detail, to find what we have missed, to understand a mystery; and all this extends awareness. But whether this greater awareness will increase or diminish freedom will depend upon what it is that we become aware of. If the greater awareness is of the causes, traumas, psychodynamics that ''made'' us what we are, then we are understanding the past in such a way as to prove that we ''had'' to become what we are; and, since this view applies equally to the present which is the unbroken extension of that determined past, therapy becomes a way of establishing why we must continue to be what we have been, a way of disavowing choice with the apparent blessing of science, and the net effect will be a decrease in freedom. If, however, the greater awareness is of options unnoticed, of choices denied, of other ways to live, then freedom will be increased, and with it greater responsibility for what we have been, are, and will become.

19.

Anger Defused

Carol Tavris

Carol Tavris's research on anger offers helpful clues in understanding how anger can help or hinder our ability to communicate. As a way of communicating, anger can be important in any relationship, including intimate ones. While most people realize that anger is an important issue in interpersonal relationships, they tend to lack a way of thinking about anger and of directing their anger constructively, instead of destructively. Read the following pages carefully for some important insights on this often misunderstood emotion. Some questions follow:

1. *What does Tavris mean by the ventilationist approach to anger, and what problem does she find in it?*
2. *What evidence does she offer to show that aggression inflames anger?*
3. *Why do you accept or reject the statement, "Talking out anger doesn't reduce it; it rehearses it."*
4. *Tavris claims nothing she says makes the case for keeping quiet when you are angry. If not, then what is her main point?*
5. *Why does she advocate different rules on anger for women and men?*
6. *What is the reappraisal method of dealing with anger?*

My husband, my teenage stepson, and I were enjoying a lavish brunch with friends one August day, when a neighbor of the hosts dropped in to visit. She is a journalist, and when she heard that I was working on a study of anger her curiosity was whetted. I was reluctant to talk about it, which aroused her interest all the more.

"Is it about women and anger?" she asked.

"Not specifically," I said.

"Is it about work and anger?"

"Not entirely."

"Then is it a sociobiology of anger?"

"No," I said curtly, trying to discourage her.

"Is it political?"

"You could say so." But I was failing.

"Is it a clinical analysis of anger in intimate relationships?"

"NO!" shouted my stepson, slamming his hand on the table in mock fury. The woman visibly jumped, and then all of us laughed. The interrogation and my tension were over, and Matthew had demonstrated one of my major points. Anger has its uses.

The social perspective on anger, I believe, explains the persistence and variety of this emotion far better than reductionistic analyses of its biology or its inner psychological workings. Anger is not a disease, with a single cause; it is a process, a transaction, a way of communicating. With the possible exception of anger caused by organic abnormalities, most angry episodes are social events: They assume meaning only in terms of the social contract between participants.

This is a minority view in an era that celebrates medical and psychological models of the emotions. Research on the brain and hormones, after all, promises exciting possibilities for "cures" of emotional abnormalities, and of course many people are accustomed to hunting for the origins of their emotional conflicts within themselves, rummaging around in their psyches. It is not that I think that these approaches are entirely wrong; rather, they are insufficient. An emotion without social rules of containment and expression is like an egg without a shell: a gooey mess.

Our contemporary ideas about anger have been fed by the Anger Industry, psychotherapy, which too often is based on the belief that inside every tranquil soul a furious one is screaming to get out. Psychiatric theory refers to anger as if it were a fixed amount of energy that bounces through the system. If you pinch it in here, it is bound to pop out there—in bad dreams, neurosis, hysterical paralysis, hostile jokes, or stomachaches. Therapists are continually "uprooting" anger or "unearthing" it, as if it were a turnip. Canadian psychiatrist Hossain B. Danesh, for example, writes that his profession has "succeeded in unearthing" anger that is buried in psychosomatic disorders, depression, suicide, homicide, and family problems. Yale psychiatrist Albert Rothenberg is not so sure about this success:

> In depression we look for evidence of anger behind the saddened aspect; in hysteria we experience angry seductiveness; in homosexuality and sexual disorders we see angry dependency; in marital problems we unearth distorted patterns of communication, particularly involving anger. We interpret the presence of anger, we confront anger, we draw anger, we tranquilize anger, and we help the working through of anger. . . . We operate on the basis of a whole series of assumptions, none of which has been clearly spelled out.

Anger-Ins, Anger-Outs

One of the assumptions most prevalent in the anger business is that the physical or verbal ventilation of anger is basically healthy, and suppressed hostility medically dangerous. "There is a widespread belief that if a person can be allowed, or helped to express his feelings, he will in some way benefit from it,'' writes psychiatrist John R. Marshall.

> This conviction exists at all levels of psychological sophistication. Present in one or another form, it occupies a position of central importance in almost all psychotherapies. . . . The belief that to discharge one's feelings is beneficial is also prevalent among the general public. Friends are encouraged to ''get if off their chests,'' helped to ''blow off steam,'' or encouraged to ''let it all hang out,'' Sports or strenuous physical activities are lauded as means of ''working off'' feelings, particularly hostility, and it is accepted that there is some value in hitting, throwing, or breaking something when frustrated.

But is there? It seems to me that the major effect of the ventilationist approach has been to raise the general noise level of our lives, not to lessen our problems. I notice that the people who are most prone to give vent to their rage get angrier, not less angry. I observe a lot of hurt feelings among the recipients of rage. And I can plot the stages in a typical ''ventilating'' marital argument: precipitating event, angry outburst, shouted recriminations, screaming or crying, the furious peak (sometimes accompanied by physical assault), exhaustion, sullen apology, or just sullenness. The cycle is replayed the next day or next week. What in this is ''cathartic''? Screaming? Throwing a pot? Does either action cause the anger to vanish or the angry spouse to feel better? Not that I can see.

Training Male and Female Responses

By looking at what happens, physically and psychologically, when people ''let anger out,'' we can see that the ways that we express anger actually affect how we feel. The decision of whether or not to express anger rests on what you want to communicate and what you hope to accomplish, and these are not necessarily harmonious goals. You may want to use anger for retaliation and vengeance, or for improving a bad situation, or for restoring your rights. Your goals determine what you should do about anger. For example, consider the popular notions that aggression is the instinctive catharsis for anger and that talking anger out reduces it. Jack Hokanson, a social psychologist who is a veritable Sherlock Holmes of psychological sleuthing, has been tracking catharsis theories for the past 20 years, using clues from one experiment to pose questions for the next. In the early 1960s, for example, Hokanson found that aggression *was* cathartic: The blood pressure of angry students would return to normal more quickly when they could retaliate against the man who had angered them. But Hokanson noticed that this was true only when the man who had angered them was a fellow student; when he was a teacher, retaliation had no cathartic effect. Naturally. In those days, teachers were still regarded as legitimate authorities, and one did not snap back at them in a cavalier manner.

So, after a flurry of studies on the target of catharsis, the first modification of catharsis theory appeared: Aggression can be cathartic only against your peers and subordinates; it does not work when the target is your boss, another authority, or an innocent bystander. If the reservoir-of-energy model were correct, though, ventilating anger against anyone or anything would result in reduced tension. It does not.

Then Hokanson noticed something that most of his fellow psychologists were ignoring in the 1960s: women. They were not behaving like the men. When you insulted then, they didn't get belligerent, they generally said something friendly to try to calm you down. When Hokanson wired them up to a physiological monitor to see whether they were secretly seething in rage, he discovered that one man's meat was a woman's poison. For men, aggression was cathartic for anger; for women, *friendliness* was cathartic. For them, any aggression, even toward a classmate, was as arousing and upsetting as aggression toward authority was for the men.

This difference gave Hokanson the idea that aggressive catharsis is a learned reaction to anger, not an instinctive one. People find characteristic ways to try to handle obnoxious persons, he concluded. In the case of sex differences, what works has a lot to do with the requirements of one's role. When women react to attack or threat by smiling, being friendly, or making gestures of accommodation, they traditionally assuage the other person's anger (often, however, at the cost of their own rights); when they act angrily or aggressively, they typically provoke a critical reaction from others ("what a bitch"), which in turn increases their arousal and anxiety. The reverse is true for men: Aggression and anger typically bring respect and results, whereas friendly accommodation is a sign of "caving in" ("what a weakling").

But sex roles are not straitjackets, and Hokanson's next experiment was prescient of the women's movement.

Imagine that you are seated at a console, facing an array of impressive gizmos and gadgets. Your partner, a fellow student of your gender and age, is sitting at an identical console nearby. You will communicate by pressing one of two buttons marked "shock" and "reward." You are probably in a friendly mood—"Only an hour of this," you think, "and I'll be done with my psychology requirement"—when suddenly, ZOT! Your partner has sent you an unpleasant shock.

If you are male, you are instantly irritated. "What the hell did the SOB do *that* for?" you mutter. "I'll show him he can't do that to me." So you press your shock button to give him a taste of his own medicine. Relief surges through your veins, according to that damned machine you're hooked to.

If you are female, you are instantly puzzled. "I wonder why she did that?" you murmur. "Maybe she hit the wrong button. Maybe she got a bad grade on the intro exam. I'd better be kind." So you press your reward button, which awards her a point in whatever mysterious game the experimenters have up their sleeve. You feel so good about your forgiving response that relief surges through your veins, according to that interesting machine you're hooked up to.

This mechanical conversation between you and your partner goes on for 32 rounds, and half the time the partner responds to whatever you do with a shock and half the time with reward. (The shocks and rewards, of course, are really controlled by the experimenters.) If you are a man, you notice the shocks. "Why, the bastard," you say to yourself. "He persists in attacking me." If you are a woman, you notice the rewards. ''She must be feeling better,'' you think. ''She's not sending me as many shocks.'' In either case, your blood pressure is down and your spirits are up.

Little do you know that those 32 rounds are only to inform the experimenters about your typical reactions. Now you go 60 rounds with new rules. If you are a man, your partner will reward you every time you are friendly in response to his shock. If you are a woman, your partner will reward you every time you are aggressive in response to her shock. In short, a woman can stop being treated unfairly by being aggressive, and a man can shake the bully by being friendly.

Hokanson and his associates observed that two things happened as a result of this new situation. First, the women rapidly learned the value of being aggressive and so became aggressive in response to shock more frequently, whereas the men learned the value of a generous reaction to insult. Second, the traditional form of catharsis for each sex was reversed. Women showed catharsis-like reduction in blood pressure when they responded aggressively and had a slow vascular recovery when they were friendly. The opposite was the case for men: Catharsis now followed friendliness, not belligerence.

Aggression Escalates Anger

Some schools of therapy, such as Alexander Lowen's bioenergetics, recommend any form of aggressive anger release that comes to mind, or foot: shouting, biting, howling, kicking, or slapping (anything short of assault and battery). Such aggression is supposed to get us ''in touch'' with our feelings. But aggression frequently has precisely the opposite effect of catharsis: Instead of exorcising the anger, it can inflame it.

One of the first studies to quarrel with the Freudian ventilationist position was conducted by Seymour Feshbach in 1956. Feshbach gathered a group of little boys who were not aggressive or destructive and encouraged them to play with violent toys, kick the furniture and otherwise run amok during a series of freeplay hours. This freedom did not ''drain'' any of the boys' ''instinctive aggression'' or ''pent-up anger;'' what it did was lower their restraint against aggression. On later occasions, the boys behaved in much more hostile and destructive ways than they had previously.

When you permit children to play aggressively, they don't become less aggressive, as the catharsis theory would predict; they become more aggressive. Indeed, aggressive play has no cathartic value at all. In a study in which third-grade children were frustrated and irritated by another child (whom the experimenters had enlisted in their cause), the children were given one of three ways of ''handling'' their anger: Some were permitted to talk it out with the adult experimenters; some were allowed to play with guns for ''cathartic release'' or to ''get even'' with the frustrating child; and some were given a reasonable explanation from the adults for the child's annoying behavior. What reduced the children's anger? Not talking about it. Not playing with guns; that made them more hostile, and aggressive as well. The most successful way of dispelling their anger was to understand why their classmate had behaved as she did (she was sleepy, upset, not feeling well).

The same principles of anger and aggression apply to adults. Murray Straus, a sociologist in the field of family violence, finds that couples who yell at each other do not thereafter feel less angry but more angry. Verbal aggression and physical aggression were highly correlated in his studies, which means that it is a small step from bitter accusations to slaps. Leonard Berkowitz, who has studied the social causes of aggression for many years, likewise finds that ventilation by yelling has no effect on anger. "Telling someone we hate him supposedly will purge pent-up aggressive inclinations and will 'clear the air,'" he says. "Frequently, however, when we tell someone off, we stimulate ourselves to continued or even stronger aggression."

Talking It Out Rehearses Anger

Like most people I know, I have always been a firm believer in the talk-it-out strategy. Talking things over makes you feel better. That's what friends are for. That's what bartenders are for. But that's not what the research shows. Talking out an emotion doesn't reduce it, it rehearses it.

Emotions are social constructions: The physiological arousal that we feel depends on cues from the environment to provide a label and a justification. Talking to friends is one way to find that label—to decide, for example, that you feel angry instead of hurt, or more sad than jealous. Sympathetic friends who agree with your self-diagnosis, or provide a diagnosis for you, are aiding that process of emotional definition.

The belief that talking it out is cathartic assumes that there is a single emotion to be released, but clinically you seldom find "pure" emotions. Most are combinations that reflect the complexity of the problem and of our lives: hurt and jealousy, rage and fear, sadness and desire, joy and guilt. Ventilating only one component of the mix, therefore, emphasizes it to the exclusion of the others. If you are upset with your spouse and you go off for a few drinks with a friend to mull the matter over, you may, in talking it out, decide that you are really furious after all. You aren't ventilating the anger; you're practicing it.

Three psychologists conducted a field experiment that showed just how this process works. Ebbe Ebbesen, Birt Duncan, and Vladimir Konečni were working in San Diego when the local aerospace-defense industry had to lay off many of its engineers and technicians. It was the right time to study anger. The employees were irate, and legitimately so, for they had been promised a three-year contract and the layoffs came after only one year. If this happened to you, at what or with whom would you be angry? Fate? The economy? The company? Your supervisor? Yourself?

The researchers seized this opportunity to interview 100 of the engineers who had been fired, comparing them with 48 who were voluntarily leaving the company at the same time. Birt Duncan conducted an exit interview with each man, during which he directed his questions—and the men's expressions of hostility—

in one of three ways: toward the company (In what ways has the company not been fair with you? Are there aspects of the company you don't like?); toward the company supervisors (What action might your supervisor have taken to prevent you from being laid off? Are there things about your supervisor that you don't like?); or toward the man himself (Are there things about you that led your supervisor not to give you a higher performance review? What in your past performance might have been improved?). Some men in each group were asked neutral questions only, such as their opinion of the technical library.

Notice that Duncan was not telling the men what they *should* be angry about, but asking them what they *did* feel angry about. When the interview itself was over, he asked each man to fill out a report that included questions about what he now felt about the company, the supervisors, or himself. "Ventilation" of the anger during the interview did not act as a catharsis in any way. On the contrary, the men became more hostile toward the company or their supervisors if they had taken an angry public stance against them in conversation. And their anger increased only toward the target they had discussed, not the others. The men who were asked to criticize themselves, however, did not blame themselves more later—a result I was happy to see, since clearly the engineers' ability was not at issue in the layoffs, and the men knew it. But neither did the self-criticizers get as angry at the company or their supervisors as the men who outspokenly blamed these targets.

The simple act of pinpointing the cause of your anger makes you more likely to repeat the explanation, even at the risk of harmful consequences (in this case, not getting rehired). As you recite your grievances, your emotional arousal builds up again, making you feel as angry as you did when the infuriating event first happened, and, in addition, establishing an attitude about the source of your rage. Friends and therapists, of course, often do just what Birt Duncan did with the laid-off engineers: ask probing questions, innocent or intentional, that direct us to a particular explanation of our feelings ("In what ways was Sheila not fair to you?" "Could you have done something to keep Herb from leaving?").

Learning How to Think About Anger

The psychological rationale for ventilating anger does not stand up under experimental scrutiny. The weight of the evidence indicates precisely the opposite: Expressing anger makes you angrier, solidifies an angry attitude, and establishes a hostile habit. If you keep quiet about momentary irritations and distract yourself with pleasant activity until your fury simmers down, chances are that you will feel better, and feel better faster, than if you let yourself go in a shouting match.

Now, none of this is to make a case for keeping quiet when you are angry, as some people seem to think whenever I talk about these research findings. The point is to understand what happens when you *do* decide to express anger, and to realize how our perceptions about the causes of anger can be affected just by talking about them and deciding on an interpretation. Each of us must find his or her own compromise among talking too much, expressing every little thing that irritates, and not talking at all, passively accepting injustices.

Silent sulking is a lousy and deadly weapon. Few people are magnanimous enough to never bear a grudge, nurse an indignity, or express their anger in devious ways—such as "forgetting" to do something they had promised to do, holding back sexually, or sulking irritably around the house. I argue, though, that these are not examples of "not talking"; they are examples of talking to yourself. You know that you feel angry, but in the guise of magnanimity or self-righteousness, or because you are afraid to announce your feelings, you pretend that everything is fine. Meanwhile, you are muttering imprecations to yourself and holding elaborate conversations in your head.

This is not going to reduce your feelings of anger either. The purpose of anger is to make a grievance known, and if the grievance is not confronted, it will not matter whether the anger is kept in, let out, or wrapped in red ribbons and dropped into the Erie Canal.

Of course, emotional release can feel awfully good. Telling off someone whom you believe has mistreated you is especially satisfying. Publishing the true story of how you were victimized by the bigwigs makes you feel vindicated. These cathartic experiences feel good not because they have emptied some physiological energy reservoir, but because they have accomplished a social goal: the redemption of justice, reinforcement of the social order. Further, some expressions of emotional release are morally and politically necessary. Bureaucracies, hospitals, and other large institutions have plenty of subtle and not-so-subtle ways to "cool out" the legitimate anger of mistreated customers and patients (by making them feel that they, the mistreated customers, are crazy, misguided, or stupid). People who sustain their anger under such circumstances often are taking a lonely but heroic course.

To know whether and how catharsis gets rid of anger, therefore, you have to know what you're angry about, and what the outside circumstances are. When you let out an emotion, it usually lands on somebody else, and how you feel—relieved, angrier, depressed—is going to depend on what the other person does. For example, the calm, nonaggressive reporting of your anger (those "I messages" that so many psychologists recommend) is the kindest, most civilized and usually most effective way to express anger, but even this mature method depends on its context.

Consider the predicament of Margie B., a 40-year-old mother of two, and a part-time decorator, who told me that, to her dismay, an abusive rage is the only way that she can convince her husband, a 32-year-old construction worker, that she is really angry. Calm discussion of her feelings, she said, may be the preferred middle-class way, but it gets her nowhere:

"The next day [after a furious argument], he ran around, and mowed the lawn, and cleaned the gutters, and washed the kitchen floor. And he said he was sorry he was so lazy. Then he said, "You'll never get me to do anything unless you get flat-out mad, like you did, which puts me in an awkward position because I hate getting angry.

"Usually our fights end in frustration all the way around and in a very empty feeling. For the next day or so I feel definitely terrible. I'm tempted to totally abase myself just to get rid of the feeling, the horrible emptiness. We don't make connection in our fights at all, and when they are over I feel so dead, so lonely.

"They say that expressing your anger is so good for you, so good for all concerned, but I don't think it has very good results as all. My therapist was always saying to me, 'Well, can't you just say, in a calm but irate tone of voice, "You know, it's driving me crazy that you're doing this"?' But I've never found that using that tone does any good, cuts any ice. Either people ignore you, or else you get a little *too* angry and they get angry back and you get in a fight that settles nothing."

There are no simple guidelines to determine when talking is better than yelling, because the choice depends as much on the receiver of the message as on the sender. Many of us know this intuitively, and tailor our communications to fit the purpose. A 42-year-old businessman, Jay S., described how his eyes were opened when he overheard his usually even-tempered boss on the phone one afternoon:

"I've never heard him so angry. He was enraged. His face was red and the veins were bulging on his neck. I tried to get his attention to calm him down, but he waved me away impatiently. As soon as the call was over, he turned to me and smiled. 'There,' he said. 'That ought to do it.' If I were the guy he'd been shouting at, let me tell you, it would have done it, too. But it was all put on."

The catharsis of anger, like its creation, depends on mind and body, if you want to "let go" of anger, you have to rearrange your thinking, not just lower your pulse rate. The most successful therapeutic methods for helping people who are quick to rage, therefore, take mind and body into consideration. Practitioners of yoga learn techniques to relax the heart rate and slow breathing, and they learn techniques to calm distracting, worrying, or infuriating thoughts. Western psychologists are catching on. For example, Ray Novaco, who teaches at the University of California at Irvine, works with people who have problems with chronic anger, teaching them two things: how to think about their anger, and how to reduce tension. Novaco reasons that anger is fomented, maintained, and inflamed by the statements we make to ourselves and others when we are provoked—"Who does he think he is to treat me like that?" "What a vile and thoughtless woman she is!"—so he teaches people how to control anger the same way, by showing them how to reinterpret a supposed provocation: "Maybe he's having a rough day"; "She must be very unhappy if she would do such a thing." (This is what people who are slow to anger do naturally: They empathize with the provoker's behavior and try to find justifications for it.) And then, because anger may have been an effective strategy for easily provoked people, they often have to learn new and gentler reactions that will bring them results as effectively as anger.

The reappraisal method is being used with people who are exposed to constant provocations as part of their jobs (or who believe that they are constantly provoked), such as police officers and bus drivers. New York City bus drivers, for example, may now see a film in which they learn that passengers who have irritating mannerisms may actually have hidden handicaps. Repeated questions ("Driver, is this 83d Street?") may indicate severe anxiety, which the passenger cannot control, apparent drunkenness may actually be cerbral palsy, mild epileptic seizures can make a passenger seem to be deliberately ignoring a driver's orders. "[The film] makes you feel funny about the way you've treated passengers in the past," says a bus driver from Queens. "Before I saw this film, if a passenger rang the bell five times, I'd take him five blocks to get even. Now I'll say, 'Maybe the person is sick.'"

There are, though, some events that cannot be reappraised. A gorgeous woman I know who has multiple sclerosis was asked to leave her elegant health club because she was, the manager said, not one of the "beautiful people," and beautiful people don't want to exercise with the disabled. (The extent of her disability at the time was a cane and brace, and, in fact, other women in the club were extremely helpful to her.) When she told me of this appalling experience, I found I had no desire to worry about the manager's personal life and woes to reduce my feeling of outrage.

But ventilating anger directly is cathartic only when it restores your sense of control, reducing both the rush of adrenaline that accompanies an unfamiliar and threatening situation and the belief that you are helpless or powerless. So the question is not "Should I ventilate anger?" or even "How should I ventilate anger?" but instead, "How should I behave in this situation to convince Harry that I'm angry—and get him to do something about it?"

A charming (if psychologically flawed) example of how to rethink a provocation to make yourself feel less angry comes from screenwriter Larry Gelbart. A friend of Gelbart's, who worked at a movie studio for a tyrannical employer, was irate because the boss had chewed him out for some insignificant matter once too often.

"I will not be treated like a worm," he raged to Gelbart. "I'm going to punch him out the next time he shouts at me like that. How dare he?"

"Hold on, hold on," Gelbart said, "I think I see the problem. You're not Jewish, are you?"

"What the hell has that got to do with anything?" his friend replied irritably. "We're talking about common courtesy, damn it."

"Listen," Gelbart said. "Relax. What any Jewish person would know is that he's not yelling at *you,* he's yelling for *himself.* Next time he shouts at you, this is what you do. You lean back in your chair, fold your arms, and let his screams wash over you. Tell yourself: 'Oy, such good it's doing him to get it out of his system!'"

The friend says that this advice works wonders. The mogul won't feel any better for yelling (in fact, as we now know, he's bound to feel worse), but Gelbart's friend will surely feel better for thinking that he does.

20.

What Does It Mean to Be Assertive?

Robert E. Alberti and Michael L. Emmons

In this reading, Alberti and Emmons deal with our responses to others in situations of conflict. Though they sidestep entirely the important matter of anger, keeping the problem of anger in mind when reading the following pages will help one to decide for oneself when it is possible to be assertive instead of aggressive. Alberti and Emmons offer us a way of avoiding anger by learning to communicate our needs clearly to others without demeaning these others. They refuse to see conflict situations in terms of someone winning and someone else losing. Instead of winning, they recommend a focus on communicating clearly and respectfully. Being able to do so is an important skill to have in dealing with our intimates. Some questions follow:

1. *What are some typical situations in which you act aggressively?*
2. *In what situations do you typically act nonassertively?*
3. *Describe a recent occasion when you responded to conflict assertively.*
4. *What would be some examples of situations where boyfriends or girlfriends might need to communicate more assertively?*
5. *Is assertiveness more of a problem for men or for women?*

We are all controlled by the world in which we live. . . . The question is this: are we to be controlled by accidents, by tyrants, or by ourselves?
—*B. F. Skinner*

Chapters 4 and 5 of *Your Perfect Right: A Guide to Assertive Behavior*, 6th ed. (San Luis Obispo: Impact Publishers, 1990), pp. 25–45.

Understanding the Terms

Assertiveness is not a simple characteristic. The fact is, there is no general agreement on a definition of the term. Here we will examine several approaches to the concept of assertiveness.

Assertive behavior promotes equality in human relationships, enabling us to act in our own best interests, to stand up for ourselves without undue anxiety, to express feelings honestly and comfortably, to exercise personal rights without denying the rights of others.

Let's examine those elements in greater detail:

To promote equality in human relationships means to put both parties on an equal footing, to restore the balance of power by giving personal power to the "underdog," to make it possible for everyone to gain and no one to lose.

To act in your own best interests refers to the ability to make your own decisions about career, relationships, life style and time schedule, to take initiative starting conversations and organizing activities, to trust your own judgment, to set goals and work to achieve them, to ask help from others, to participate socially.

To stand up for yourself includes such behaviors as saying no, setting limits on time and energy, responding to criticism or put-downs or anger, expressing or supporting or defending an opinion.

To express feelings honestly and comfortably means the ability to disagree, to show anger, to show affection or friendship, to admit fear or anxiety, to express agreement or support, to be spontaneous—all without painful anxiety.

To exercise personal rights relates to competency as a citizen, as a consumer, as a member of an organization or school or work group, as a participant in public events to express opinions, to work for change, to respond to violations of one's own rights or those of others.

To not deny the rights of others is to accomplish the above personal expressions without unfair criticism of others, without hurtful behavior toward others, without name-calling, without intimidation, without manipulation, without controlling others.

Thus, assertive behavior is a positive self-affirmation which also values the other people in your life. It contributes both to your personal life satisfaction and to the quality of your relationships with others.

Studies show that, as a direct result of gains in self-expressiveness, individuals have improved their self-esteem, reduced their anxiety, overcome depression, gained greater respect from others, accomplished more in terms of their life goals, increased their level of self-understanding, and improved their capacity to communicate more effectively with others. We can't promise those results for you, of course, but the evidence is impressive!

Assertive, Nonassertive, and Aggressive Behavior

The way we live in the late twentieth century presents some mixed messages about appropriate behavior. A typical example is found in common attitudes and teachings about human sexuality. While restraint is the sexual norm of the middle-class family, school and church, the popular media bombard audiences with a different view.

Aggressiveness is highly valued in male sexual behavior: the "lover" is glorified in print, on the screen, and by his peers. Paradoxically, he is cautioned to date "respectable" girls, and warned that sexual intercourse is allowable only after marriage. Women are given similar mixed messages. On one hand, they are expected to be sweet and innocently non-assertive; whereas on the other hand, they are rewarded for being sultry, seductive, and sensual.

Examples of such conflicts between *recommended* and *rewarded* behavior are evident in many other areas of life. Even though it is typically understood that one should respect the rights of others, all too often we observe that parents, teachers, business, and government contradict these values in their own actions. Tact, diplomacy, politeness, refined manners, modesty, and self-denial are generally praised; yet to get ahead it is often acceptable to "step on" others.

The male child is carefully coached to be strong, brave, and dominant. His aggressiveness is condoned and accepted—as in the pride felt by a father whose son gets in trouble for socking the neighborhood bully in the nose. Ironically, and a source of much confusion for the child, the same father will likely encourage his son to "have respect for adults," to "let others go first," and to "be polite."

. . . We believe that you should be able to *choose for yourself* how to act in a given circumstance. If your "polite restraint" response is too strong, you may be unable to make the choice to act as you would like. If your aggressive response is over-developed, you may be unable to achieve goals without hurting others. Freedom of choice and self-control are made possible by developing assertive responses for situations in which you have previously responded nonassertively or aggressively.

Contrasting assertive with nonassertive and aggressive actions will help to clarify these concepts. The chart (p. 235) displays several feelings and consequences typical for the person (sender) whose actions are nonassertive, assertive, or aggressive. Also shown, for each of these actions, are the likely consequences for the person toward whom the action is directed (receiver).

As the chart shows, a *nonassertive* response means that the sender is typically denying self-expression, and is inhibited from letting feelings show. People who behave nonassertively often feel hurt and anxious since they allow others to choose for them. They seldom achieve their own desired goals.

The person who carries a desire for self-expression to the extreme of *aggressive* behavior accomplishes goals at the expense of others. Although frequently self-enhancing and expressive of feelings in the situation, aggressive behavior

Nonassertive Behavior	Aggressive Behavior	Assertive Behavior
Sender	*Sender*	*Sender*
Self-denying	Self-enhancing at expense of another	Self-enhancing
Inhibited	Expressive	Expressive
Hurt, anxious	Deprecates others	Feels good about self
Allows others to choose	Chooses for others	Chooses for self
Does not achieve desired goal	Achieves desired goal by hurting others	May achieve desired goal
Receiver	*Receiver*	*Receiver*
Guilty or angry	Self-denying	Self-enhancing
Deprecates sender	Hurt, defensive, humiliated	Expressive
Achieves desired goal at sender's expense	Does not achieve desired goal	May achieve desired goal

hurts other people in the process by making choices for them and by minimizing their worth.

Aggressive behavior commonly results in a *putdown* of the receiver. Rights denied, the receiver feels hurt, defensive, and humiliated. His or her goals in the situation, of course, are not achieved. Aggressive behavior may achieve the sender's goals, but may also generate bitterness and frustration which may later return as vengeance.

. . . Appropriately *assertive* behavior in the same situation would be self-enhancing for the sender, an honest expression of feelings, and will usually achieve the goal. When you choose for yourself how to act, a good feeling typically (not always) accompanies the assertive response, even when your goals are not achieved.

Similarly, when the consequences of these three contrasting behaviors are viewed from the perspective of the person receiving the action, a parallel pattern emerges. Nonassertive behavior often produces feelings ranging from sympathy to confusion to outright contempt for the sender. Also, the receiver may feel guilt or anger at having achieved goals at the sender's expense. The receiver of aggressive actions is often hurt, defensive, put-down, or perhaps aggressive in return. In contrast, assertion enhances the feeling of self-worth of both parties, and permits both full self-expression and achievement of goals.

In summary, then, it is clear that the sender is hurt by self-denial in nonassertive behavior; the receiver (or even both parties) may be hurt by aggressive behavior. In the case of assertion, neither person is hurt, and it is likely that both will succeed.

The series of example situations which appear in the following section will help to make these distinctions more clear.

It is important to note that assertive behavior is *person-and-situation-specific*, not universal. That is, what may be considered assertive depends upon the persons involved and the circumstances of the situation. . . .

Cultural Differences in Self-Assertion

While the desire for self-expression may be a basic human need, assertive behavior in interpersonal relationships is primarily characteristic of western cultures (although by no means limited to the United States).

In Asian cultures, group membership (family, clan, workgroup) and ''face'' are highly valued. How one is seen by others tends to be more important to an individual than is self-concept. Politeness is a key virtue, and communication is often indirect so as to avoid confrontation or offending one another. Assertiveness, in the Western sense of direct self-expression, is generally not considered appropriate by those who value tradition. Many young people, however, and those whose business activities include considerable contact with the United States and Europe, have developed more direct, informal, and assertive styles.

Many individuals and subcommunities in Latin and Hispanic societies have emphasized the notion of ''machismo'' to the point that assertiveness—as we have defined it—seems rather tame, especially for men. In those cases, a greater display of strength is the norm for self-expression.

Yet, people from those cultures where self-assertion traditionally has not been valued may be just those who most need its benefits. People of some other lands have tended to express themselves in ways we might consider nonassertive or aggressive. While for some cultures those styles represent thousands of years of tradition, it may be that current and future international relations require more open and direct communication, and a greater sense of equality—expressed on both sides of the table.

''But Isn't Aggression Just Human Nature?

Assertion and aggression are clearly different. Aggression and violence are often excused on the grounds that they are innate in the human organism, and cannot be avoided. Not so, say the most distinguished scholars who have researched the subject. The ''Seville Statement''—written in 1986 by 20 distinguished social and behavioral scientists from 12 nations, and endorsed by the American Psychological and American Anthropological Associations—says: ''It is *scientifically incorrect* to say . . .

• . . . we have inherited a tendency to make war from our animal ancestors. Warfare is a peculiarly human phenomenon and does not occur in other animals. War is biologically possible, but it is not inevitable. . . .

• . . . that war or any other violent behavior is genetically programmed into our human nature. Except for rare pathologies the genes do not produce individuals necessarily predisposed to violence. Neither do they determine the opposite.

• . . . that humans have a 'violent brain.' While we do have the neural apparatus to act violently, there is nothing in our neurophysiology that compels (such behavior).

• . . . that war is caused by 'instinct' or any single motivation. . . .

We conclude that biology does not condemn humanity to war, and that humanity can be freed from the bondage of biological pessimism. Violence is neither in our evolutionary legacy nor in our genes. The same species (that) invented war is capable of inventing peace."

Classifying Behavior: "A Rose, By Any Other Name . . ."

"I told my father-in-law not to smoke his cigar in my house. Was that assertive or aggressive?"

Members of assertiveness training groups and workshops often ask trainers to classify a particular act as "assertive" or "aggressive." What criteria do make the important difference?

We have suggested that assertive and aggressive behavior differ principally in that the latter involves hurting, manipulating, or denying others in the course of expressing oneself.

Practitioners and writers with a psychoanalytic orientation have proposed that *intent* must be considered. That is, if you intended to hurt your father-in-law, that's aggressive; if you simply wanted to inform him of your wishes, you were acting assertively.

Many psychologists insist that behavior must be measurable according to its *effects*. Thus, if your father gets the assertive message and responds accordingly—by agreeing not to smoke—your behavior may be classified as assertive. If he pouts in a corner, or shouts, "Who do you think you are?" your statement may have been aggressive, as described by this criterion.

Finally, as we have noted, the *social-cultural context* must be taken into account in classifying behavior as assertive or aggressive or nonassertive. A culture, for example, which regards honoring one's elders as one of its ultimate values may view the request as clearly out of line and aggressive, regardless of the behavior, response, or intent.

There are no absolutes in this area, and some criteria may be in conflict. A particular act may be at once assertive in *behavior* and *intent* (you wanted to and did express your feelings), aggressive in *response* (the other person could not handle your assertion), and nonassertive in the social *context* (your culture expects a powerful, put-down style). It's not always easy to classify human behavior!

A specific situation may vary considerably from the examples we discuss here. In any event, the question "Is it assertive or aggressive?" is not one which may be answered simply! Each situation ultimately must be evaluated on its own. The labels "nonassertive," "assertive," and "aggressive" themselves carry no magic, but they may be useful in assessing the appropriateness of a particular action.

Our concern is not with the labels, but with helping you to choose for yourself how you will act, and with helping you know that you have the tools you need to succeed.

Social Consequences of Assertion

While it is our purpose here to teach you skills so that you may improve your own ability to express yourself appropriately and responsibly, we believe strongly that self-expression must be modulated by its context. Just as freedom of speech does not convey the right to yell "fire" in a crowded theater, so the form of self-expression we advocate is one which considers its consequences.

The perfect right you have to say "no" exists alongside the other person's right to say "yes." And your desire to accomplish your goals through self-assertion must be weighed against the needs of the larger society. Speak out or write about any idea you choose to support, but recognize the other person's right to do the same. And be prepared to pay some dues—perhaps in jail—if your expression goes beyond words and includes civil disobedience. Just as there are taxes for those who accumulate wealth, there is a price to pay for freedom of expression.

While you have the perfect right to advocate a viewpoint, everyone else has the same right—and your views may conflict. Keep this in mind on your journey toward greater personal assertiveness.

Ten Key Points About Assertive Behavior

To summarize this section, here is a list of ten key qualities of assertive behavior.

Assertive behavior is:

1. Self-expressive;
2. Respectful of the rights of others;
3. Honest;
4. Direct and firm;
5. Equalizing, benefitting both self and relationship;
6. Verbal, including the content of the message (feelings, rights, facts, opinions, requests, limits);
7. Nonverbal, including the style of the message (eye contact,voice, posture, facial expression, gestures, distance, timing, fluency, listening);

8. Appropriate for the person and the situation, not universal;
9. Socially responsible;
10. Learned, not inborn.

Now you have a better idea of what it means to be assertive, and you are probably ready to begin taking steps toward increasing your own assertiveness.

The following section provides many examples of life situations calling for assertive action. It's likely you'll find youself nodding in recognition as you read some of these!

Examples of Assertive, Nonassertive, and Aggressive Behavior

A look at some everyday situations will improve your understanding of the behavioral styles we've discussed. As you read the examples in this chapter, you may wish to pause and think about your own response before reading the alternative responses we have presented. These examples are oversimplified, of course, so we can demonstrate the ideas more clearly.

Dining Out

Adam and Evelyn are at dinner in a moderately expensive restaurant. Adam has ordered a rare steak; but when the steak is served, he finds it well done. His action is:

NONASSERTIVE: Adam grumbles to Evelyn about the ''burned'' meat, and vows that he won't patronize this restaurant in the future. He says nothing to the waitress, responding ''Fine!'' to her inquiry, ''Is everything all right?'' His dinner and evening are spoiled, and he feels angry with himself for taking no action. Adam's estimate of himself and Evelyn's estimate of him are both deflated by the experience.

AGGRESSIVE: Adam angrily summons the waitress to his table. He berates her loudly and unfairly for not complying with his order. His actions ridicule the waitress and embarrass Evelyn. He demands and receives another steak, this one more to his liking. He feels in control of the situation, but Evelyn's embarrassment creates friction between them and spoils their night out. The waitress is humiliated and angry for the rest of the evening.

ASSERTIVE: Adam motions the waitress to his table. Noting that he had ordered a rare steak, he shows her the well-done meat. He asks politely but firmly that it be returned to the kitchen and replaced with the rare-cooked steak he originally requested. The waitress apologizes for the error, and shortly returns with a rare steak. Adam and Evelyn enjoy dinner and Adam feels satisfaction with himself. The waitress is pleased with a satisfied customer and a generous tip.

Something Borrowed

Helen is an airline flight attendant, bright, outgoing, and a good worker liked by customers and peers. She lives in a condo with two roommates, and is looking forward to a quiet evening at home one Friday when her roommate Mary asks a favor. Mary says that she is going out with a special man, and wants to borrow Helen's new and quite expensive necklace. The necklace was a gift from her brother, with whom Helen is very close, and it means a great deal to her. Her response is:

NONASSERTIVE: She swallows her anxiety about loss or damage to the necklace. Although she feels that its special meaning makes it too personal to lend, she says "Sure!" She denies herself, rewards Mary for making an unreasonable request, and worries all evening.

AGGRESSIVE: Helen is outraged at her friend's request, tells her "Absolutely not!" and rebukes her severely for even daring to ask "such a stupid question." She humiliates Mary and makes a fool of herself too. Later she feels uncomfortable and guilty. Mary's hurt feelings show all evening, and she has a miserable time, which puzzles and dismays her date. Thereafter, the relationship between Helen and Mary becomes very strained.

ASSERTIVE: Helen explains the significance of the necklace to her roommate. Politely but firmly, she observes that the request is an unreasonable one since this piece of jewelry is particularly personal. Mary is disappointed but understanding, and Helen feels good for having been honest. Mary makes a big hit with her date just by being herself.

The Heavyweight

Barry and Madalynne, married nine years, have been having marital problems recently because he insists that she is overweight and needs to reduce. He brings the subject up continually, pointing out that she is no longer the woman he married (she was 25 pounds lighter then). He keeps telling her that being overweight is bad for her health, that she is a bad example for the children, and so on.

Barry teases Madalynne about being "chunky," looks longingly at thin women while commenting how attractive they look, and makes reference to her figure in front of their friends. Barry has been acting this way for the past three months and Madalynne is highly upset. She has been attempting to lose weight for those three months, but with little success. Following Barry's most recent rash of criticism, Madalynne is:

NONASSERTIVE: She apologizes for her weight, makes feeble excuses, or simply doesn't reply to some of Barry's comments. Internally, she feels both hostile toward her husband for his nagging, and guilty about being overweight. Her feelings of anxiety make it even more difficult for her to lose weight and the battle continues.

AGGRESSIVE: Madalynne goes into a long tirade about how her husband isn't any great bargain anymore, either. She brings up the fact that at night he falls asleep on the couch half the time, is a lousy sex partner, and doesn't pay enough attention to her. She complains that he humiliates her in front of the children and their close friends, and that he acts like a "lecherous old man" by the way he eyes other women. In her anger, she succeeds only in wounding Barry and driving a wedge between them by her counterattack.

ASSERTIVE: Approaching her husband when they are alone and will not be interrupted, Madalynne says that she feels that Barry is right about her need to lose weight, but she does not like the way he keeps after her about the problem. She points out that she is doing her best and is having a difficult time losing the weight and maintaining the loss. He admits that his harping is ineffective, and together they work out a plan in which he will systematically reinforce her for her efforts to lose weight.

The Neighbor Kid

Edmond and Virginia have a two-year-old boy and a baby girl. Over the last several nights, their neighbor's son, who is 17, has been sitting in his car, in his own driveway, with his car stereo blaring loudly. He begins just about the time their two young children go to bed on the side of the house where the boy plays the music. They have found it impossible to get the children to bed until the music stops. Edmond and Virginia are both disturbed and decide to be:

NONASSERTIVE: They move the children into their own bedroom on the other side of the house, wait until the noise stops (around 1 A.M.), then transfer the children back to their own rooms. Then they go to bed much past their own usual bedtime. They quietly curse the teenager, and soon become alienated from their neighbors.

AGGRESSIVE: They call the police and protest that "one of those wild teenagers" next door is creating a disturbance. They demand that the police put a stop to the noise at once. The police do talk with the boy and his parents, who get very angry as a result of their embarrassment about the police visit. They denounce Edmond and Virginia for reporting to the police without speaking to them first, and resolve to have nothing further to do with them.

ASSERTIVE: Edmond goes over to the boy's house and tells him that his stereo is keeping the children awake at night. Edmond suggests they try to work out an arrangement which allows the boy his music, but does not disturb the children's sleep. The boy reluctantly agrees to set a lower volume during late hours, but he appreciates Edmond's cooperative attitude. Both parties feel good about the outcome, and agree to follow up a week later to be sure it is working as agreed.

The Loser

Russell is a 22-year-old college dropout who works in a plastics factory. He lives alone in a small one-room, walk-up apartment. Russell has had no dates for the past fourteen months. He left college after a series of depressing events—academic failures, a ''Dear John'' letter, and some painful harassment by other students in his residence hall. He has been in jail overnight for drunkenness on two recent occasions.

Yesterday, Russell received a letter from his mother inquiring about his well-being, but primarily devoted to a discussion of his brother's recent successes. Today, his supervisor criticized him harshly for a mistake which was actually the supervisor's own fault. A secretary at the plant turned down his invitation to dinner.

When he arrived at his apartment that evening, feeling particularly depressed, his landlord met him at the door with a tirade about ''drunken bums'' and a demand that this month's rent be paid on time. Russell's response is:

NONASSERTIVE: He takes on himself the burden of the landlord's attack, feeling added guilt and even greater depression. A sense of helplessness overcomes him. He wonders how his brother can be so successful while he considers himself so worthless. The secretary's rejection and the boss' criticism strengthen his conviction that he is ''no damn good.'' Deciding the workd would be a better place without him, he begins to think about how he will commit suicide.

AGGRESSIVE: The landlord has added the final straw to Russell's burden. He becomes extremely angry and pushes the landlord out of the way in order to get into his room. Once alone, he resoves to ''get'' the people who have been making his life so miserable recently: the supervisor, the secretary, the landlord, and possibly others as well. He remembers the guns he saw in the pawn shop window yesterday.

ASSERTIVE: Russell responds firmly to the landlord, noting that he has paid his rent regularly, and that it is not due for another week. He reminds the landlord of a broken rail on the stairway and the plumbing repairs which were to have been accomplished weeks earlier. The following morning, after giving his life situation a great deal of thought, Russell calls the local mental health clinic to ask for help. At work, he approaches the supervisor calmly and explains the circumstances surrounding the mistake. Though somewhat defensive, the supervisor acknowledges her error and apologizes for her aggressive behavior.

Recognizing Your Own Nonassertive and Aggressive Behavior

The examples given here help to point out what ''assertiveness'' means in everyday events. Perhaps some of the situations ''rang a bell'' in your own life. Take a few minutes to listen honestly to yourself describe your relationships with others who are important to you. Carefully examine your contacts with parents, peers, co-workers, classmates, spouse, children, bosses, teachers, salespeople, neighbors, relatives. Who is dominant in these relationships? Are you easily taken advantage of in dealings with others? Do you usually express your feelings and ideas openly? Do you take advantage of or hurt others frequently?

Your responses to such questions provide hints which may lead you to explore in greater depth your asssertive, nonassertive, or aggressive behavior.

21.

Two Views of Marriage Explored: His and Hers

Daniel Goleman

In the next three brief articles, Daniel Goleman reports recent research on marriage, some of which has important implications for dating and engagement. In the first piece, Goleman describes the different notions held by women and men of what makes for emotional intimacy. Men and women tend to interpret the signs of intimacy differently. Also, men, without intending to deceive, tend to pay more attention during courtship to their mates' needs and less attention after marriage. Understanding these attitudes and tendencies in oneself and one's partner can help communication toward greater intimacy. Some questions follow:

1. *What are the differing ways women and men interpret the signs of intimacy?*
2. *Why is it that in adulthood women tend to be less comfortable with separateness and men less comfortable with intimacy?*
3. *Why is it claimed that marriage is harder on women than on men?*
4. *From what you read, why might you agree or disagree with the statement: "Men have more trouble with closeness and intimacy than do women"?*

Every marriage, researchers are discovering, is actually two marriages: his and hers. The differences suffuse even the happiest of relationships between husbands and wives.

Some of the new research suggests that, paradoxically, the differences need not be divisive, but can be sources of marital growth.

Psychologists say that couples who openly acknowledge these differences improve their chances of avoiding strife. And those who seek to free their mar-

From *The New York Times* (1 April 1986): pp. C1, C11.

riages of such male-female differences are better able to do so if they are aware of how powerful, though largely hidden, the differences are.

One of the great gaps between husbands and wives is in their notions of emotional intimacy and how important they feel it is in a marriage. For many men, simply doing such things as working in the garden or going to a movie with their wives gives them a feeling of closeness. But for their wives that is not enough, according to Ted Huston, a psychologist at the University of Texas at Austin who has studied 130 couples intensively.

"For the wives, intimacy means talking things over, especially talking about the relationship itself," Dr. Huston said. "The men, by and large, don't understand what the wives want from them. They say, 'I want to do things with her, and all she wants to do is talk.'"

While women expect more emotional intimacy than their husbands do, "many men seem to feel they've fulfilled their obligation to the relationship if they just do their chores," said Robert Sternberg, a psychologist at Yale University who has studied couples. "They say, 'I took out the garbage; now leave me alone.'"

In courtship, Dr. Huston has found, men are much more willing to spend time talking to women in ways that build a woman's sense of intimacy. But after marriage, and as time goes on, the men tend to spend less and less time talking with their wives in these ways, and more time devoted to their work or with buddies. The trend is strongest in marriages that follow traditional patterns, and most of the current research suggests the traditional patterns are still prevalent despite two decades in which these conventional attitudes and mores have been assailed for the stereotypes they breed.

"Men put on a big show of interest when they are courting," Dr. Huston said. "But after the marriage their actual level of interest in the partner often does not seem as great as you would think, judging from the courtship. The intimacy of courtship is instrumental for the men, a way to capture the woman's interest. But that sort of intimacy is not natural for many men."

As with all such differences, there is a lesson to be learned. The starkness of the husband's apparent change in behavior after marriage can lead to disappointment, demands and acrimony—in short, a relationship in trouble. Dr. Huston suggests that in more successful marriages there is a middle ground in which the couple share experiences that naturally lead to more intimate conversation.

"You can't force intimacy," Dr. Huston said. "It has to arise spontaneously from shared activities."

Husbands' and wives' differing stances toward intimacy signify deeper disparities between the sexes, in the view of Carol Gilligan, a psychologist at Harvard University. Dr. Gilligan says young boys take pride in independence and are threatened by anything that might compromise their autonomy, while young girls tend to experience themselves as part of a network of relationships and are threatened by anything that might rupture these connections.

"Boys, as they mature, must learn to connect, girls to separate," said Kathleen White, a psychologist at Boston University, who, with her colleagues, is the author of an article on intimacy in marriage in *The Journal of Personality and Social Psychology*.

In adulthood this means women tend to be uncomfortable with separateness, while men are wary of intimacy. Some psychologists have proposed that one lesson teen-age boys learn from their girlfriends is how to be emotionally intimate, a lesson that can extend into marriage, particularly for those who never really master it.

The more comfortable a husband is with intimacy, Dr. White's research shows, the more satisfied with the marriage is the wife likely to be.

Changing View Toward Parents

Another telling finding is that marriage typically makes a woman draw closer to her parents, while a man often becomes more distant from his. For a woman, closeness to her parents ranks among her most important expectations, the new research shows, while husbands tend to rank a warm association with either set of parents comparatively low.

For the men, the marriage evidently supplants earlier closeness to parents. What the men often seem to be saying, according to Dr. White, is, "I don't need my parents anymore; I have my wife."

But for women, Dr. White has found, the marriage seems to offer a crucial footing from which they can set aside earlier rebelliousness and make peace with their parents, particularly their mothers, and develop a new warmth.

Other factors may be at work, too. "Because many husbands focus their lives outside the marriage, on their work or their friends, many wives have a sense of being abandoned," Dr. Huston said. "They turn to their mothers for an intimate involvement they do not get from their husbands."

Other research has found that wives place more emphasis than their husbands do on preserving ties with both sets of parents, not just their own. Some experts say a couple that blends the stances of the wife and the husband toward their parents can find a healthy balance in which independence and family ties coexist.

The disparities between husband and wife in such areas as intimacy and family ties are part of a wide range of differences that men and women bring to marriage, according to many experts.

Differing Sets of Values

Some of the starkest evidence of these differences has come from a study by Dr. Sternberg of people 17 to 69 years old, some of whom have been married for as long as 36 years. The men in the study rated as most important in their marriage their wives' ability to make love and the couple's shared interests. But the wives listed marital fidelity and ties to family and friends as important. The wives were particularly concerned with how well the husband handled both sets of parents, especially hers, and also how well the couple got along with each other's friends.

Dr. Sternberg's study found evidence that the double standard still holds for many couples. Wives said fidelity was very important for both spouses; husbands said it was more important for wives than for husbands.

Perhaps the most dramatic difference is in evaluating the relationship. "The men rate almost everything as better than do the wives," Dr. Sternberg said. Men have a rosier view of love-making, finances, ties with parents, listening to each other, tolerance of flaws and romance.

"The only thing the women rate better than men is the couple's degree of fidelity," Dr. Sternberg said.

Wives generally complain more about the state of their marriage than husbands do. Jesse Bernard, a sociologist, has proposed in *The Future of Marriage* (Yale University Press) that this is because marriage is harder on women than on men. Dr. Bernard, citing a wide range of studies, says the "psychological costs of marriage seem to be considerably greater for wives than for husbands."

Other marital experts, though, say wives appear to suffer more than husbands because women are more willing than men to admit their problems. Men often feel it is "unmanly" to admit depression or anxiety.

According to John Gottman, a psychologist at the University of Illinois who has studied happy and unhappy couples, wives are more willing to complain about problems in marriage than men are, and this is particularly so among unhappy couples. Men generally try to avoid conflicts in marriage, he has found, while women are more willing to confront them.

"Men and women have different goals when they disagree," Dr. Gottman said. "The wife wants to resolve the disagreement so that she feels closer to her husband and respected by him. The husband, though, just wants to avoid a blowup. The husband doesn't see the disagreement as an opportunity for closeness, but for trouble." Dr. Gottman believes this is because men are more vulnerable than women to physical stress from emotional confrontations, at least as he reads the research evidence.

The preferences that draw couples together also show a major, and perhaps unsurprising, difference between the sexes: women, by and large, place great importance on a man's earning potential, while men place great stress on a woman's attractiveness, according to Dr. White's study in *The Journal of Personality and Social Psychology*.

Some psychologists say that one lesson to be learned from the differences is that each partner in the psychological alliance that is marriage can be enriched by learning from the other. In this view, marriage offers a unique opportunity for psychological growth. "Men and women," said Dr. White, "need to take on each other's strengths."

The differences also point to what people mean when they say that it takes work to make a marriage work. Those who are complacent about the differences between spouses, who see no need to accommodate the partner's perspectives, may be putting the marriage at risk, the experts say.

"It's too easy to portray the expressive wife as good and the stoic husband as horrible; in fact, they are just different," said Carol Tavris, a social psychologist whose book *The Longest War* (Harcourt, Brace Jovanovich) reviews sex differences.

"While you want to understand the differences, it is probably futile to try to change your partner to be just like your best friend," Dr. Tavris said. "It's better for husbands and wives to develop a sense of humor and tolerance, and to accept their mates as they are."

22.

Marriage Research Reveals Ingredients of Happiness

Daniel Goleman

What makes for happiness in marriage? Researchers have been studying this question by talking with couples who say they are happy in their marriages. In the following pages Daniel Goleman looks at some ingredients of a happy marriage turned up by this research. An important finding: a couple's ability to talk over problems is more important in the long run than how "in love" the couple is. How happy the couple was before marriage is unrelated to how long the marriage in fact lasted. Study questions follow:

1. *What conclusions do you come to from the fact that so many happily married couples admitted to sexual problems?*
2. *Why do you suppose that women are more likely to spot trouble in a marriage than men?*
3. *In what ways might the tension between intimacy and separateness found in all marriages show up in engagements or even in dating?*
4. *Which of the marital-communication skills listed in this essay would be most helpful for premarriage communication, and why?*

In the psychological web that is marriage, there is no sure way to know which threads are crucial for happiness and which are illusory. But new research is revealing some of the hidden ingredients of happy marriages.

Such elements as a couple's ability to talk over problems effectively, the research shows, are more crucial to compatibility, and are often more telling in the long run, than even how much a couple is in love. Such findings are being used as the basis for educational programs for couples who are about to be married.

To be sure, there is no single formula for marital compatibility. For every rule

From *The New York Times* (16 April 1985), pp. C1, C4.

of thumb that psychological studies suggest, there are bound to be marriages that defy it.

But, although general prescriptions for marital happiness are elusive, the new research highlights broad patterns that distinguish between those marriages that thrive and those that wither. For example, long-range studies have found that one of the strongest predictors of success after five years of marriage is how well a couple communicated before they married. But, according to the same studies, how happy a couple were before marriage was unrelated to how long the marriage lasted. "What counts in making a happy marriage," said George Levinger of the University of Massachusetts, "is not so much how compatible you are, but how you deal with incompatibility."

Early marital research focused on finding the personality profiles of individuals who would fit together best. Over the years that effort was largely abandoned; researchers found no clear relationship between personality traits and marital happiness.

Although the prevailing view among some psychotherapists is that marital difficulties result from psychological problems in one partner or the other, the consensus among most marital researchers is that personality is less crucial to marital success than is the nature of the relationship itself. As the late Nathan Ackerman, a pioneering family therapist, put it, even two neurotics can have a happy marriage.

As long ago as 1938, Lewis Terman, a Stanford University psychologist who was better known for his studies of intelligence, published findings he hoped would bring clarity to what he called "the chaos of opinion on the determiners of marital happiness."

Dr. Terman discovered that most of the opinions of the day on what was required for a happy marriage were dead wrong. For example, his research found little or no relationship between the frequency of sexual intercourse and marital satisfaction.

That finding has held up over the years. In a 1978 study of 100 happily married couples, it was found that 8 percent had intercourse less than once a month, and close to a quarter reported having intercourse two or three times a month. Most couples reported having intercourse one to three times a week. Two of the couples had no intercourse, while for one couple the rate was daily.

Moreover, one-third of the men and two-thirds of the women reported a sexual problem, such as difficulty getting an erection in men and difficulty reaching orgasm in women. And both men and women complained of such difficulties as inability to relax and a lack of interest in sexual activity. Even though the couples with sexual complaints reported being dissatisfied with their sex lives, they still felt their marriage to be "working" and happy.

The precursors of later marital success or failure can be detected in the earliest stages of a marriage, according to Howard Markman, a psychologist at the University of Denver. Writing in *Marital Interaction* (Guilford Press), Dr. Markman reports on a study in which 26 couples were regularly observed while discussing their problems over a period of almost six years, beginning just before their marriage. By the end of the study, eight of the couples had separated or divorced.

The best predictor of the couples' satisfaction after five-and-a-half years of marriage was how well they communicated before the marriage.

"Private Language" of Couples

Other research has found that happily married couples seem to develop a "private language," a set of subtle cues and private words that have meaning only to them. Researchers at the University of Illinois found that in happy marriages husbands were much better than were strangers at understanding exactly what their wives meant. But in distressed marriages, strangers were as adept at understanding messages from wives as were their own husbands.

Likewise, happily married couples show a high degree of responsiveness to each other in sharing events of the day. The absence of such responsiveness can lead to heightened tensions, according to John Gottman, the leader of the research group at the University of Illinois. Dr. Gottman said he believes the friendship built up through such day-to-day exchanges makes couples willing to go through the difficulties of repairing their relationship when it becomes strained.

One of the striking differences between the satisfied and distressed couples in the University of Denver study was in how they viewed the way they talked over their problems. The wives were aware of stress even when it was not readily apparent to others. But the husbands weren't.

In a series of studies, Dr. Markman and his colleagues had objective observers and the partners themselves rate how "positive"—by which Dr. Markman means friendly—or "negative"—that is, hostile, the spouses were during discussions of problems.

In unhappy couples, husbands seemed oblivious to the hostility of their wives in some of these conversations, although objective observers noted it.

The wives were sure their husbands were hostile, although the observers said the husbands, for the most part, did not seem to be especially difficult during these encounters.

The husbands' seeming lack of hostility was consistent with the widely observed tendency among husbands to avoid confrontation while wives seem more ready to engage it.

But why did the wives in these unhappy couples see the husbands as hostile? Dr. Markham suggests that the wives tend to be more sensitive to trouble in a marriage than husbands. Evidently, they were reading the true stress in their marriage into their husbands' behavior even though the behavior did not overtly reflect it.

"This does not mean that wives are the cause or the victims of marital distress," said Dr. Markman, but rather that the wives are better barometers of problems in the marriage than are the husbands.

In general, women are more comfortable with confrontation because—in happy marriages, at least—they more readily can end the fight by switching from a hostile stance to a conciliatory one, according to Dr. Gottman.

Husbands, in Dr. Gottman's research—even happy ones—were less able to make this switch in the heat of an argument.

In unhappy marriages, the wives no longer seemed willing to play this role, according to Dr. Gottman.

Men Withdraw, Women Argue

As the emotional tone of a marriage becomes more negative, men are more likely to react by withdrawing, while women are more likely to escalate emotional pressure, becoming more coercive and argumentative. The result is often an escalating cycle of pressure from the wife and withdrawal by the husband, a cycle that happily married couples seem better able to prevent.

"All couples go through ups and downs in marriage," Dr. Markman said in an interview. "But it's those couples who don't communicate well whose marriage is more likely to be the victim of such a difficult period."

Couples planning to marry who were trained by the University of Denver group in a range of marital skills were, after three years, still as happy as they were just before their marriage. But other couples who did not receive the training showed "dramatic declines in overall satisfaction," Dr. Markman said.

"Inevitable Push-Pull in Marriage"

Occasional tensions, of course, are inevitable in any marriage. A psychoanalytic view of marriage holds that each partner unconsciously gravitates to someone whom they see as fulfilling a deeply held need. The two most prominent of these needs exist in a state of tension: the desire for intimacy, on the one hand, and the need to establish one's identity as a separate person on the other. Marriage thus becomes a forum for the negotiation of a balance between these conflicting urges.

"There's an inevitable push-pull in marriage," said Michael Kolevzon, a professor of social work at Virginia Commonwealth University. "As a couple's intimacy increases, you often see a corresponding increase in their desire for distance. How satisfied they will be in the marriage depends to a great extent on how they communicate these needs."

"Anger is one means this balancing act is negotiated in marriage," Dr. Kolevzon added in an interview. "Sometimes it's a way to ask for distance: I'll get angry with you so I can justify being alone for a while. Or with some people their anger is actually a plea for intimacy. They build up to an angry confrontation so it can resolve into intimacy as an affirmation that their spouse loves them despite their faults."

Programs Teach Marital Skills

The new research is leading to programs for couples who are about to be married. The programs are designed to strengthen those skills that seem to help couples weather the stresses and changes that challenge a marriage. These programs teach a range of marital skills that concentrate on communication, including these:

- The couples are taught to focus on one topic at a time and to make other preoccupations clear, such as "I may seem angry because I had a bad day at work."
- They are also trained to "stop the action" until the partners cool down when repetitive cycles of conflict begin. One of the major signs of distress in a couple is escalating hostility, often in the form of nagging that provokes an angry response. The escalation seems unstoppable once begun.
- By being specific in criticizing or praising a spouse's actions, the couples learn how to prevent nebulous complaints that often trigger arguments. Thus, a spouse might say, "When I see your coat on the floor, I feel you are not doing your share around the house and I feel taken advantage of," instead of "You're a slob."
- The couples also learn to edit what they say, so as to avoid saying things that would needlessly hurt a spouse. "It's especially unproductive to dredge up past events and old grudges during a fight," Dr. Kolevzon said.

23.

Want a Happy Marriage? Learn to Fight a Good Fight

Daniel Goleman

Which makes for a happy marriage? Knowing how to avoid arguments and "fights" or knowing how to use them as a means of greater closeness? Some in the first phases of "falling in love" think they should never have serious disagreements or that "being in love means never to have to say you're sorry" (Love Story). *The research reported here by Daniel Goleman shows the importance of conflict-resolution skills in a marriage. In a time when so much conflict is resolved with fists or even guns, the ability to move through conflict respectful of the other person is an important communication skill. Some questions follow:*

1. *What is the difference between productive and destructive arguments or fights?*
2. *What are the characteristics of a fruitful fight? Of a destructive one?*
3. *Why are fights predictable after the initial, romantic phase of marriage?*
4. *Why is paraphrasing so useful in a fight or argument?*
5. *Why is being specific so important in communicating complaints to your partner?*

What's the difference between a marriage that gets happier as time goes on and one that grows more miserable? In many cases it is fights, according to a new study that pinpoints exactly which kinds of arguments help a couple grow closer and which split them apart.

Ways of smoothing things over that seem to keep the peace in the short term, the study indicates, can undermine the relationship in the long run. This is particularly true of ignoring deep disagreements or pretending that they don't exist.

The conventional view that a couple's satisfaction with their marriage predicts how happy they will be in years to come was also challenged by the study. Paradoxically, those couples who were unhappy, but fought well, tended to have become much happier by the time they were contacted again three years later.

Although the findings may seem self-evident, they took the researchers themselves by surprise because they contradict the prevailing wisdom among professionals and are prompting more research.

There are those couples who are so well attuned that they rarely, if ever, fight over their differences. And there are other couples who simply do not fight, despite their grievances. In other research, psychologists have found that such couples are typically composed of partners who both are agreeable in all spheres of life, marriage among them.

But the findings on fights apply to the large majority of couples who have some degree of conflict in their relationship.

The study of marital arguments contradicts longstanding findings that had shown that couples who were more prone to arguments were the least satisfied with their marriage. The studies that led to those findings, however, failed to distinguish among the kinds of fights that couples have or to follow the course of the marriages to see whether they got better or worse.

The new study shows that certain kinds of fights can improve some marriages, and draws a clear distinction between the kinds of arguments that nurture a relationship and those that sink it. Arguments in which one or the other partner becomes defensive or stubborn, or whines or withdraws, are particularly destructive. Those fights in which the partners freely express their anger while not letting the intensity escalate out of control bode well for the future.

The new findings, published in the February [1989] issue of the *Journal of Consulting and Clinical Psychology,* are from the first study to analyze closely the specific emotional maneuvers during actual arguments between couples and then to track down those same couples three years later to see how satisfied they were with the marriage.

"We had assumed, like everyone else, that what made a couple happy now would make them happy in the future; we were surprised by what we found," said John Gottman, a psychologist at the University of Washington in Seattle who conducted the study with Lowell Krokoff, a psychologist at the University of Wisconsin.

"We were puzzled to find that the patterns that made some couples complain they were dissatisfied led to improvements in the relationship as time went on," Dr. Gottman said. "I thought at first there was an error in our techniques—it went against all prevailing wisdom. Marital therapists rarely recommend that couples have a fight, and rarely saw couples who wished to."

Fights with Value

The most fruitful fights, the study showed, were those in which the partners felt free to be angry with each other, felt they made themselves understood to their partner, and finally came to a resolution involving some degree of compromise. Such fights, according to Dr. Gottman, give a couple the strong sense that they can weather conflict together.

But few fights follow this pattern; many couples rely on emotional ploys that are destructive to the relationship. The study found that among the most frequent destructive maneuvers during fights were the following:

- Defensiveness or making excuses instead of taking responsibility for a problem.
- Telling partners what one wants the other to stop doing as opposed to what the partner might do more of.
- Erroneously attributing blameworthy thoughts, motives or feelings to the other.
- Stubbornness.
- Contemptuous remarks or insults.
- Whining complaints.

But what of couples who never seem to fight? Other research has found that in marriages where both partners have little hostility toward the world in general, fights rarely if ever occur. When such couples are brought into a laboratory for research, investigators have found it virtually impossible to provoke an argument, according to a study by Timothy Smith, a psychologist at the University of Utah.

But in couples where either partner tends to be cynical or hostile toward the world in general, fights are easily provoked, Dr. Smith found. In such cases the more hostile partner is able to trigger an angry response even in a generally easygoing spouse.

Coming to the Surface

Fights are predictable during the second phase of marriage, after the initial romantic period, according to Harville Hendrix, a marital therapist in New York City who is the author of *Getting the Love You Want,* published by Henry Holt & Company. "During that period, the conflicts surface as couples negotiate unresolved childhood issues and needs they bring to marriage," he said. "If they have the tools to resolve those issues, then the conflicts fade during the next phase."

But in some marriages, the absence of fights may augur poorly for a couple." In some marriages where there is an agreement not to fight, things are fine as long as their lives go well," said Dr. Gottman, "but if something bad happens, they are too brittle to handle the problem."

Dr. Gottman added, "Couples who have healthy fights develop a kind of marital efficacy that makes the marriage stronger as time goes on." In recognition of the usefulness of healthy fighting, a project at the University of Denver, led by Howard Markman, a psychologist, has been training couples in the skills that help them handle conflict well.

In keeping with Dr. Gottman's results, Dr. Markman found that those couples who learned how to argue productively were less happy at first, as their differences were aired, but became progressively more satisfied. After six years, the divorce rate was half that of a group of comparable couples who had not had the training.

Wives tend to have a special role in orchestrating a couple's fights, Dr. Gottman's study found. "Wives are the emotional managers of most marriages," said Dr. Gottman. "The wife is usually the one who brings up disagreements and makes a couple confront their differences. Our results suggest that a wife's anger is a valuable resource in a marriage."

This role is a delicate one, Dr. Gottman points out. "She has to be able to express her anger, but do it in such a way that it doesn't drive her husband away—make him withdraw, or defensive, for instance."

The best response for a husband whose wife is starting a fight, Dr. Gottman said, is to "let her know he's listening, show respect for the disagreement, and acknowledge that there's something there that should be dealt with," along with being mad right back, if that is what he is feeling.

Even though anger can be productive, couples need to keep it within bounds. "Fights in which tempers or feelings like fear and sadness get out of hand bode poorly for a couple," Dr. Gottman said.

"I suspect that wives whose marriages improve are careful not to let the argument get out of hand; they keep the lid on," said Dr. Gottman.

One of the ways partners can de-escalate a fight in progress, Dr. Gottman said, is to paraphrase what a partner has said and look for a solution rather than disagreeing. Another is to suggest a compromise.

The idea that couples should be encouraged to learn to fight by certain rules in order to strengthen their relationship was popular during the 1970s, part of the movement that also fostered "encounter groups." But that idea has become passé among marital therapists, according to Richard Simon, a marital and family therapist and editor of the Family Therapy Networker.

"Most marital therapists downplay the idea of having better fights these days," Dr. Simon said. "They tend to see fights as symptomatic of something else in the relationship. Many therapists try to shift the couple to a more positive focus, finding solutions. The basic attitude is that encouraging fights simply makes it more likely a couple will fight."

Still, some therapists welcome the findings. Dr. Gottman's study "fits our clinical observations—we find that certain kinds of fights help couples clarify and resolve their differences," said Aaron Beck, a psychiatrist at the University of Pennsylvania medical school who has written *Love Is Not Enough,* published by Harper & Row. "Stating what you're mad about and feeling that what you've said has been heard gives you the feeling you will be responded to by your mate."

"The worst kinds of fights are those in which partners resort to character assassination and blame; it just leads to a dead end," Dr. Beck added. "But when you state a concrete, specific complaint, then there is a good chance it will lead not just to a resolution, but to an improvement."

For example, it is unhelpful to make the charge that a spouse is "a slob" when what is meant is "it bothers me when you leave the cap off the toothpaste."

Dr. Beck finds that the note a fight ends on is particularly important. "If the fight leads to an escalating cycle of retaliation, it leaves the partners with a sense of hopelessness. But if it ends at a good point, it leaves them optimistic about their marriage and each other."

24.

On Hostile Ground

James Check, Neil Malamuth, Barbara Elias, and Susan Barton

If you asked most men what their basic attitude was toward women, they would tell you they loved *women. The same is true for most women: they like men and find them attractive. Yet underneath such attractions can be lurking latent hostility toward the opposite sex. Examining such possible hostility in ourselves can help us rethink our way of relating with men and women. The following questionnaire on hostility will give students a lot to think about and talk about.*

Yes, there does seem to be a war between the sexes. In our research on sexual aggression, it has become readily apparent that a good deal of hostility exists between men and women. In a study of 305 college men at the University of Manitoba, for example, we found that more than one-third basically agreed with the statement, "Women are usually sweet until they've caught a man, but then they let their true self show." And the hostility is not all one-sided. Almost 40 percent of a sample of 278 college women agreed with the statement, "A lot of men talk big, but when it comes down to it, they can't perform well sexually."

Given the implications of these statements (taken from the Adversarial Sexual Beliefs scale developed by our colleague Martha Burt at the Urban Institute in Washington, D.C.), we decided to investigate hostility between the sexes and developed scales to help us measure it in both sexes.

How hostile are you toward members of the opposite sex? To find out, complete the appropriate scale on the following pages. Carefully read each statement in the sex-appropriate scale and circle "T" (true) if it applies to you or if you agree with it. Circle "F" (false) if the statement does not apply to you or if you disagree with it. A self-scoring key and an indication of what your score means follow, with a discussion of what we are learning about hostility between the sexes.

Hostility Toward Women

To Be Completed by Men

1. I feel that many times women flirt with men just to tease them or hurt them. T F
2. I feel upset even by slight criticism by a woman. T F
3. It doesn't really bother me when women tease me about my faults. T F
4. I used to think that most women told the truth but now I know otherwise. T F
5. I do not believe that women will walk all over you if you aren't willing to fight. T F
6. I do not often find myself disagreeing with women. T F
7. I do very few things to women that make me feel remorseful afterward. T F
8. I rarely become suspicious with women who are friendlier than I expected. T F
9. There are a number of women who seem to dislike me very much. T F
10. I don't agree that women always seem to get the breaks. T F
11. I don't seem to get what's coming to me in my relationships with women. T F
12. I generally don't get really angry when a woman makes fun of me. T F
13. Women irritate me a great deal more than they are aware of. T F
14. If I let women see the way I feel, they would probably consider me a hard person to get along with. T F
15. Lately, I've been kind of grouchy with women. T F
16. I think that most women would not lie to get ahead. T F
17. It is safer not to trust women. T F
18. When it really comes down to it, a lot of women are deceitful. T F
19. I am not easily angered by a woman. T F
20. I often feel that women probably think I have not lived the right kind of life. T F
21. I never have hostile feelings that make me feel ashamed of myself later. T F
22. Many times a woman appears to care but just wants to use you. T F
23. I am sure I get a raw deal from the women in my life. T F
24. I don't usually wonder what hidden reason a woman may have for doing something nice for me. T F
25. If women had not had it in for me, I would have been more successful in my personal relations with them. T F
26. I never have the feeling that women laugh about me. T F
27. Very few women talk about me behind my back. T F

28. When I look back at what's happened to me, I don't feel at all resentful toward the women in my life. T F
29. I never sulk when a woman makes me angry. T F
30. I have been rejected by too many women in my life. T F

Hostility Toward Men

To Be Completed by Women

1. Sometimes men bother me by just being around. T F
2. At times I have a strong urge to do something harmful to a man. T F
3. Lately, I've been kind of grouchy with men. T F
4. I think that most men would lie to get ahead. T F
5. It is safer not to trust men. T F
6. When it really comes down to it, a lot of men are deceitful. T F
7. Men are responsible for most of my troubles. T F
8. I think nearly any man would lie to keep out of trouble. T F
9. I do not blame a woman for taking advantage of a man who lays himself open to it. T F
10. I used to think that most men told the truth, but now I know otherwise. T F
11. If you aren't willing to fight, men will walk all over you. T F
12. Many times a man appears to care, but really just wants to use you. T F
13. Men today don't deserve the breaks they get. T F
14. I can easily make a man afraid of me, and sometimes do just for the fun of it. T F
15. I am sure I get a raw deal from the men in my life. T F
16. I commonly wonder what hidden reason a man may have for doing something nice for me. T F
17. My motto is "Never trust a man." T F
18. Once in a while I cannot control my urge to harm a man. T F
19. When a man is bossy, I take my time just to show him. T F
20. I know that men tend to talk about me behind my back. T F
21. Most men make friends because friends are likely to be useful to them. T F
22. Men always seem to get the breaks. T F
23. I can't help feeling resentful when I think of how easy men have it in life. T F
24. There are certain men I dislike so much that I am inwardly pleased when they get into trouble for something they have done. T F
25. When I look back at what's happened to me, I can't help feeling mildly resentful toward the men in my life. T F

26. I don't seem to get what's coming to me in my relationships with men.	T	F
27. At times I feel I get a raw deal with the opposite sex.	T	F
28. I don't blame women for trying to get everything they can from men nowadays.	T	F
29. Men irritate me a great deal more than they are aware of.	T	F
30. If I let men see the way I feel, they would consider me a hard person to get along with.	T	F

Scoring Keys

Following is the scoring key for each of the two scales. Please score yourself by simply counting the number of items that you answered according to the key.

Hostility Toward Women Scale:	**Hostility Toward Men Scale:**
1. T	1. T
2. T	2. T
3. F	3. T
4. T	4. T
5. F	5. T
6. F	6. T
7. F	7. T
8. F	8. T
9. T	9. T
10. F	10. T
11. T	11. T
12. F	12. T
13. T	13. T
14. T	14. T
15. T	15. T
16. F	16. T
17. T	17. T
18. T	18. T
19. F	19. T
20. T	20. T
21. F	21. T
22. T	22. T
23. T	23. T
24. F	24. T
25. T	25. T
26. F	26. T
27. F	27. T

	28. F	28. T
	29. F	29. T
	30. T	30. T
TOTAL SCORE:		

Hostility: How You Rate

Men

Your score on the Hostility Toward Women scale will range from 0 to 30. In our latest sample of 305 college men, the average score was 8.79. If you scored between 0 and 5 on this scale, you are below the 25th percentile, meaning that at least 75 percent of the men in our sample scored higher—more hostile—than you. If you scored between 6 and 11, you have an average score (the middle 50 percent of our sample). If you scored 12 or more, you rank above the 75th percentile, meaning that you scored higher than 75 percent of our sample. A score between 0 and 3 is considered low, between 3 and 11 is medium and 12 to 30 is high.

Women

Your score on the Hostility Toward Men scale will range from 0 to 30. In our latest sample of 278 college women, the average score was 7.57. If you scored between 0 and 2 on this scale, you are below the 25th percentile, meaning that at least 75 percent of the women in our sample scored higher—more hostile—than you. If you scored between 3 and 11, you have an average score (the middle 50 percent of our sample). If you scored 12 or more, you rank above the 75th percentile, meaning that you scored higher than 75 percent of our sample. A score between 0 and 3 is considered low, between 3 and 11 is medium and 12 to 30 is high.

A Precautionary Note

It is important to understand that our findings with respect to hostility between the sexes are quite new and will have to be confirmed in studies during the next several years. Accordingly, we caution you not to overinterpret your score on our hostility scale. The relationships we have found so far are not perfect predictors, so it would be unwise, for example, to assume that you are overly hostile or sexually aggressive just because you scored 14 on the hostility scale. Instead, you should examine your own beliefs and behavior and decide for yourself whether you have a problem in this area. The scales are to be used as guidelines, and we hope that our findings will stimulate you to think seriously about the important problem of hostility between the sexes.

Interpretations

Male Hostility toward Women

We have learned a great deal about male hostility toward women from our studies of convicted rapists, men from the general population and college students who have been rated on the hostility scale and other measures. In general, we find that age has little effect on hostility toward women, that men of higher socioeconomic status tend to score slightly lower and that married men usually score lower than do single men.

We also found that hostility toward women is closely related to certain attitudes and beliefs about women and sexuality. One of our most consistent findings, for example, is that men who score high on the hostility scale tend to believe that men and women are essentially adversaries in their sexual relationships with each other. As you might imagine, these adversarial attitudes can lead to feelings of loneliness, and we and our colleague Daniel Perlman of the University of British Columbia have found that men who score high on the hostility scale do tend to be quite lonely and depressed and have low self-esteem. Why? We suspect it may be because lonely people usually are overly sensitive to signs of both friendship and rejection. A lonely man, therefore, might come on too strong with a woman who is only trying to be polite and cause her to shy away. The lonely guy might then overreact to this rejection, get angry and bring about further rejection by the woman. Thus, we have a vicious cycle in which the man becomes more and more hostile toward the women he seeks companionship from and so gets lonelier and lonelier.

Another consistent finding is that men who score high in hostility toward women tend to have traditional sex-role beliefs. That is, they believe that a woman should get married and rear a family, that a wife should never contradict her husband in public, that a woman should be a virgin when she marries and that men should pay for dates. Men with high hostility levels also tend to believe in various rape myths, such as the false belief that women enjoy being raped and that they are responsible for their rapes. Such attitudes, we believe, can lead to very negative behavior during sexual interactions with women. And some of the high-hostility men we have studied admit that they use various levels of force in their attempts to get women to have sex with them, and that they would be likely to do so again. Similarly, in our laboratory studies of aggression, we find that some high-hostility men get very angry when women reject them and subsequently use high levels of physical punishment when given the opportunity to retaliate.

Our research also suggests that hostility toward women might be related to the viewing and enjoyment of violent pornography by men. Although our results are preliminary, we have found that some men who score high on the hostility scale report that they find explicit depictions of rape sexually arousing and rate sexually violent videotapes more stimulating and entertaining than do men who score low on the scale. We and our colleague Arthur Gordon at the Regional Psychiatric Center, a federal hospital-prison in Saskatoon, Saskatchewan, have found that the relationship between hostility toward women and sexual arousal by depictions of rape is particularly strong in convicted rapists.

Our search for the causes of male hostility toward women has just begun, but one puzzling finding has emerged: Men who have been sexually abused as children score high on the hostility scale and are somewhat lonely. Since most research shows that the vast majority of those who sexually abuse children are men, it is not clear why men who experienced sexual abuse as children should be more hostile toward women. What we need to find out is whether these men were victimized by a man or a woman. This is only one of the many questions that we will address in our future research.

Female Hostility toward Men

Our study of female hostility toward men is in its initial stages, so we know less about it than we do about male hostility toward women. However, several of our findings with women parallel those with men. For example, there is little if any relationship between hostility and age or socioeconomic status, but married women, like married men, do tend to score lower on the hostility scale than do singles.

One of the more telling findings in this research comes from our colleague Susan Dickens at York University. In her study, the Hostility Toward Men scale was administered to various groups of Toronto women ranging from 18 to more than 60 years old. In all groups, the women who were most hostile toward men were those who most feared being raped and who were likely to believe that the average man would rape a woman if he could be assured of not being caught and punished. These fears and beliefs might explain why we have found that women, like men, who score high in hostility also endorse adversarial sex beliefs and are somewhat lonely and depressed.

And finally, as with men, we find that women with low self-esteem score higher on the hostility scale. This is not surprising, since previous studies have suggested that people of either sex who are low in self-esteem tend to be lonely, somewhat hostile and to have problems in their relationships with others.

In conclusion, the problem of hostility between the sexes seems to be a serious one, affecting both men and women. In light of our findings implicating childhood sexual abuse as a potential cause of such hostility—at least among men—we are currently undertaking a large-scale study of adolescents who have been sexually victimized. And, based on the work of Dianne Garrels at York University, we already have a clue to what the link might be between childhood sexual abuse and later hostility. She has found that children who report that they have been abused often experience as much or more trauma from the aftermath of the disclosure as they did from the actual abuse. The family is thrown into turmoil, and the child may experience severe punishment from the abuser. It may well be that this trauma and turmoil may be at the root of the sexual hostility that subsequently develops.

It is our hope that we will see more research designed to deal with this most disconcerting problem.

25.

Communication and Conflict

Evelyn E. Whitehead and James D. Whitehead

Enriching this section on Communication, Conflict, and Change is the following essay by Evelyn and James Whitehead. Like so many of the other writers in this section, they situate conflict within the context of intimacy. In other words, conflict can be a means for greater intimacy—or, if handled poorly, for greater distance and alienation. The Whiteheads state clearly that any couple "can get better at being married" by getting better at the skills of intimacy. One of these is skill with confrontation and conflict.

This essay may well be a classic for its helpful way of explaining conflict's connection with communication and intimacy. It deserves careful study. Some questions follow:

1. *What are the skills of intimacy and which of them seems to you most crucial?*
2. *Why is premature judgment so dangerous in confrontation?*
3. *How common is it for people to grow up "assuming love does away with conflict, that love and conflict are mutually exclusive"?*
4. *Do you think most people accept that "conflict is a normal, expectable ingredient in any relationship," including marriage?*

In marriage we see intimacy in both its most inviting and its most challenging face. Our daily patterns of life—living together, working together, sleeping together—these are the substance of intimacy. Here we feel ourselves being tested and getting better at being "up close." To live well these "up close" patterns of marriage we need the resources of psychological maturity: a sense of who I am, an openness to others, a capacity for commitment, some tolerance for the ambiguity both in myself and in other people. But these resources may not be enough. Aptitudes for intimacy must be expressed in behavior. We must be able, in the give-and-take of our life together, to develop a lifestyle that is mutually

From *Marrying Well* (New York: Doubleday Image Books, 1983), pp. 309–319.

satisfying. Our desire to be close must be expressed in the way we act toward one another. It is encouraging to know that we can get better at being married. We can learn more satisfying ways *for us* to be close; we can learn more effective ways to give and receive the gift of ourselves that is at the core of our married love. And among the most valuable resources for this growth in marriage are the skills of intimate living.

The Skills of Intimacy

Over the past two decades there has been much interest in psychology and other disciplines in understanding better what happens in communication between people. As a result we are more clearly aware today of both what helps and what frustrates understanding in close relationships. Values and attitudes are important in our ability to live up close to others, but so, especially, is our behavior. There are skillful—that is, effective—ways to be with and behave toward one another. Interpersonal skills that are especially important to the intimate life of marriage include empathy, self-disclosure and confrontation. Each involves both attitudes and behaviors; each can contribute significantly to marriage for a lifetime.

Empathy enables me to understand another person from within that person's frame of reference. Empathy begins in an attitude of openness which enables me to set aside my own concerns and turn myself toward you. But this basic openness is not always sufficient. My capacity for empathy can be enhanced by my developing a range of behavioral skills. An accepting posture, attentive listening, sensitive paraphrasing—each of these can contribute to my effective presence to you.

My posture can give you important information about who you are to me and how important I judge your communication to be. If I appear distracted or edgy, if I keep glancing at my watch or rush to take an incoming phone call, I am likely to let you feel that you are not very important to me now. In the midst of the hectic schedules of most married couples today, it is often necessary to take steps to ensure the postures of presence: taking the time to sit down together to talk, finding ways to give each other some undivided attention, learning when to hold a personal concern until later and when to "stop everything" in order to deal with an issue now.

Learning to listen well to each other can be the most important skill of our marriage. To listen well is to listen actively, alert to the full context of the message—the words and silences, the emotions and ideas, the context in which our conversation takes place. To listen is to pay attention: paying attention is a receptive, but not a passive, attitude. If I cannot pay attention, it will be difficult for me to hear; if I do not listen, it will be difficult for me to understand and to respond effectively to you. The skills of active listening are those behaviors which enable me to be aware of your full message. This includes my being alert to your words and their nuances. But equally and often even more important are the non-verbal factors involved. Your tone of voice, your gestures, the timing, the emotional content—these may tell me more than the words between us. To listen actively, then, calls for an awareness of the content, the feelings and the context of our communication.

Sensitive paraphrasing is a skill of empathy as well. I show you that I understand you by saying back to you the essence of your message. To paraphrase is not merely to "parrot"—to repeat mechanically what you have just said. Rather I want to show you that I have really heard *you,* that I have been present not just to your words but to their deeper meaning for you. I go beyond the simple assurance that "I understand" by offering you a statement of what I have understood. You can then confirm that, in fact, I have understood you—or clarify your message so that my understanding may be more accurate. In either case, I demonstrate my respect for you and for your message. It is important to me that I understand what you say, and it is to you that I come to check my understanding.

Empathy, then, is my ability to understand your ideas, feelings and values from within your frame of reference. The goal of empathy is to understand; as such, it precedes evaluation. Empathy does not mean that I will always agree with you; it does not require that I accept your point of view as my own or even as "best" for you. I may well have to evaluate your ideas. We may well have to discuss and negotiate as we move toward a decision we can share. But these movements of evaluation and judgment come later in our communication. My first goal is to accurately understand you and what you are trying to say to me. Judgment and decision are not secondary in our communication but they are subsequent to accurate understanding.

Empathy is the practical ability to be present to another person. Its exercise is a discipline: if I am distracted by fatigue or agitated by fear I cannot be present to my spouse. As virtuous behavior, empathy depends on a (relatively) strong sense of my identity and vocation. I do not have to defend myself: being aware of and comfortable with who I am, I can give my full attention to another person. Empathy thus is the stuff of intimacy. Without some skill, some virtue here, it will be difficult for me to express my love for my partner. Finally, to speak of empathy as both a skill and a virtue is to remind us again that we can get better at it. A Christian spirituality or asceticism of marriage will include these efforts to learn to be more effectively present to those we most love.

The open stance of empathy does much to enhance communication in marriage. But communication involves more than receptivity. I must be able to speak as well as to listen; to initiate as well as to understand. Self-disclosure thus becomes an essential skill of intimacy. To share myself with you I must be able to overcome the hesitancy suggested by fear or doubt or shame. But these inhibitions overcome, I must be able to act in a way that gives you access to my mind and heart, in a way that is fitting for me and for our relationship. Appropriate self-disclosure can seem complicated. But I am not limited to my current level of success. I can become more skillful, learning better ways to express my values and needs, my ideas and feelings.

Self-disclosure begins in self-awareness. I must *know* what I have experienced, what I think, how I feel, what I need, what I want to do. This knowledge is not likely to be full and finished; an unwillingness to speak until I am completely sure of myself can be a trap in communication. Self-awareness is rather an ability to know where I am now, to be in touch with the dense and ambiguous information of my own life. Beyond knowing my own insights, needs and purposes, I must value them. This need not mean that I am convinced that they are "the best." It means rather that I take them seriously as deserving of examination and respect, from myself and from others as well. My feelings, my perceptions of myself and of the world—these have worth and weight. By valuing them myself I contribute to the possibility that they can be appreciated by others as well. My needs and purposes exist in a context of those of other people, to be sure. But a conviction that my own ideas and goals are of value is basic to mature self-disclosure.

An important skill of self-disclosure is my ability to speak concretely. I must be able to say "I," to acknowledge my own ideas and feelings. Self-disclosure can be thwarted by a retreat into speaking about "most people"; "everybody knows . . ." instead of "I think that . . ."; "most people want . . ." instead of "I need . . ."; "people have a hard time . . ." instead of "it is difficult for me to . . ." Beyond this willingness to "own" my experience, I can learn to provide more specific details about my actions and emotions. To share myself with you in our marriage I will need, for example, a well-nuanced vocabulary of feelings—one that goes well beyond "I feel good" and "I feel bad." To tell you that "I feel good" is to share some important information about myself but not yet very much. What does this mean for me? Is this good feeling one of confidence? or affection? or physical vigor? Does it result from something I have done or something that has been done for me? Are you an important part of this good feeling for me or are you really incidental to it? My self-disclosure becomes more concrete when I can name my feelings more precisely and when I can describe the events and actions that are part of them for me.

Confrontation, too, makes a critical contribution to intimacy in marriage. For most of us the word "confrontation" implies conflict. And, as we shall see shortly, the ability to deal well with conflict between us is an important skill of marriage. But we use the word "confrontation" here in a meaning that goes beyond its narrow and, most often, negative connotation as interpersonal conflict. The ability to confront involves the psychological strength to give (and to receive) emotionally significant information in ways that lead to further exploration rather than to self-defense. Sometimes the emotionally significant information is more positive than negative. To say "I love you" is to share with you emotionally significant information. And many of us know how confrontive it is to learn of another's love for us. Similarly, to give a compliment is to share emotionally significant information, and there are people who defend themselves against this "good news" as strongly as others of us defend ourselves against an accusation of blame. But most often, to be sure, when confrontation becomes necessary and difficult in our marriage, it is because there is negative information we must share with our spouse. It may be some practical issue of daily life that we must face—the use of the automobile, our bank balance, plans for next summer's vacation. The issues, however, may be more sensitive—the way you discipline the children, my parents' influence in our home, how satisfying is our sex life.

Skills of confrontation are those behaviors that make it more likely that our sharing of significant negative information in these instances will lead us to explore the difficulty between us rather than to defend ourselves against one another. My ability to confront effectively is enhanced when I am able to speak descriptively rather than judgmentally. To tell you that I missed my meeting because you came home late is to *describe;* to call you a selfish and inconsiderate person is to *judge*. While both may be hard for you to hear from me, one is more likely to escalate into a quarrel than is the other. As we have noted before, judgment is not irrelevant in marriage, but premature judgment is likely to short-circuit the process of exploration and mutual understanding. Perhaps there are extenuating circumstances that caused you to be late; perhaps you are genuinely sorry that you inconvenienced me and want to do something to make amends. My attack on your selfishness is not likely to leave room for this kind of response on your side. It is more likely to lead you to defend yourself against my accusation, perhaps by calling up instances of my own selfishness, perhaps by leaving the scene altogether. In neither case has communication between us been furthered.

There are other behaviors that make our confrontation more effective, that is, more likely to further communication between us. These include the ability to accept feelings of anger in myself and in you and the ability to show respect for you even as I must disagree with you or challenge your position. These skills become especially important in dealing with conflict in marriage.

Conflict and Love

Conflict is an aspect of Christian marriage about which our rhetoric can be misleading. In ceremonies and sermons about marriage, it is upon images of unity and peace and joy that we dwell. These images of life together in Christian marriage are important and true, but partial. When, as a believing community, we do not speak concretely to the more ambiguous experiences in marriage—experiences of anger, frustration, misunderstanding—we can leave many married people feeling that their marriages are somehow deficient.

Conflict and hostility are not goals of marriage, to be sure. But neither are they an indication that our marriage is "on the rocks." Conflict is a normal, expectable ingredient in any relationship—whether marriage, teamwork or friendship—that brings people "up close" and engages them at the level of their significant values and needs. The challenge in close relationships is not to do away with all signs of conflict or, worse, to refuse to admit that conflicts arise between us. Rather we can learn ways to recognize the potential areas of conflict *for us* and to deal with these issues and feelings in ways that strengthen rather than destroy the bonds between us.

Conflict is normal in interpersonal exchange; it is an expectable event in the intimate lifestyle of marriage. Whenever people come together in an ongoing way, especially if significant issues are involved, we can expect that they will become aware of differences that exist between them. Sometimes these differences will be simply noted as interesting. But often they will involve disagreements, misunderstanding and discord. It is here that the experience of conflict begins.

Marriage engages each of us at a level of our most significant values and needs. My sense of who I am, my convictions, my ideas and ideals, what I hope to make of my life—in our marriage all these are open to confirmation or to challenge and change. In addition, every marriage is a complex pattern of interaction and expectation. We develop our own way of being together and apart; we come to our own understanding of what each of us gives and receives in this relationship. The process through which we develop the patterns of our own marriage is ongoing. We can expect times of relative stability when the rhythms of our life together seem to fit especially well. We can also expect periods marked by significant adjustment and change. The process of marriage includes this continuing exploration, even trial and error, as we attempt to learn more about ourselves and our partner. It is these normal and even inevitable experiences of personal challenge and mutual change in marriage that set the stage for conflict.

Conflict is a response to discrepancy or disparity. "Things are not as I expected or as I want them to be." In interpersonal conflict the other person is seen as somehow involved in, or responsible for, this discrepancy. "You are not as I expected; it is your fault that things are not as I want them to be." Marriage brings us together in so many ways, as friends and lovers as parents and householders, in cooperation and competition, in practical decisions about our money and our time. These overlapping issues give us many opportunities both to meet and to fail each other's expectations. Thus discrepancy and conflict are predictable.

This predictability of conflict in marriage is not simply a cause for concern. Conflict is not "all bad." Its effects in intimacy are not simply or necessarily destructive. As many marriage counselors know, conflict is as often a sign of health in a relationship as it is a symptom of disease. The presence of conflict between us indicates that we are about something that is of value to us both. Conflict thus marks a relationship of some force. This energy can be harnessed; it need not always work against us. A marriage in which there is nothing important enough to fight about is more likely to die than one in which arguments occur. Indifference is a greater enemy of intimacy than is conflict.

Many of us have grown up assuming that love does away with conflict, that love and conflict are mutually exclusive. But this romantic view of love is challenged by our experience of tension in our own marriage. We come to know conflict as a powerful dynamic in our relationship and one with ambiguous effect. For most of us, it is the negative effects of conflict that we know best and fear. Conflict feels bad and seems to have bad results. To be in conflict seems like a move away from intimacy. I am angry or hurt, you feel rejected or resentful. And most often my own past experience reinforces the unpleasant conclusion that conflict leads to the disintegration of relationships. Sometimes the relationship ends immediately; sometimes it continues, but with a burden of bitterness and unhealed grievances that ultimately leads to its death. In the face of this negative sense of the power of conflict, the evidence that it is expectable and even inevitable in our marriage is likely to strike us with alarm.

Conflict Can Be Constructive

But these negative results of conflict do not give the full picture. Conflict can make a constructive contribution to our marriage. It can bring us to a more nuanced appreciation of who each of us is; it can test and strengthen the bonds that exist between us; it can deepen our capacity for mutual trust.

The experience of conflict points to an area of discrepancy between us. I am uncomfortable with the way you discipline our oldest child; you don't like me to let my new job interfere with our weekends together as a family; I no longer want to be "the perfect housewife and mother," though this is the way you have always seen me; you no longer want to be "the strong and self-sufficient male," "though you know it frightens me to see your weakness.

If we are willing to face the conflict, we may be able to learn from the experience of discrepancy that is at its root. Exploring this discrepancy—between what I want from you and what you are able to give, or between who I am and who you need me to be, or between our differing views about money or privacy or sex or success—we can come to know one another more fully. We can grow toward a greater and more respectful mutuality, based on a greater awareness and respect for who each of us really is.

Conflict is not necessarily a part of every development in marriage. Some changes are accompanied more by a sense of fulfillment than frustration. Some couples are open to the processes of mutual exploration in such a way that there is little sense of discrepancy and little experience of conflict between them. But change is frequently a source of confusion and conflict, even if only temporarily. A relationship that cannot face at least the possibility of conflict will soon be in trouble. In order to ensure the conflict will not arise between us, we may decide that our relationship should touch us only minimally, in areas where we are not vitally concerned. Or we may believe that we must be willing to disown our response to the concerns that do matter to us. But to disown conflict does not strengthen a relationship. It tends instead to have us look away from part of the reality that exists between us. But the reality that is there—the troublesome concern that stands beneath the conflict—does not go away. The discrepancy remains, unattended, as a likely source of more serious trouble between us in the future.

We may know from previous experiences that conflict, faced poorly, can lead to resentment and recrimination. But not to face it does not ensure that our marriage will be free of these negative emotions. A more useful stance involves our willingness to face the conflicts that may arise between us, aware of their ambiguous power both to destroy and to deepen the love we share. This willingness to accept conflict as inevitable and even as protentially valuable need not mean that we find it pleasant to be at odds. But it does mean that we are willing to acknowledge and even tolerate this discomfort that conflict brings, in view of the valuable information it provides about our relationship and ourselves.

The experience of facing together the conflicts that arise between us can give greater confidence, an increased security in the strength and flexibility of the commitment between us, since we have seen it tested and found it sufficient to the test. Conflict can have this positive effect in a relationship but it remains a powerful and ambiguous dynamic. Just as the presence of conflict does not necessarily or automatically signal a relationship in trouble, neither does it necessarily or automatically result in new learning or growth. Whether the expectable event of conflict in our marriage will have positive or negative effect is due in large part to how we respond to it. To deal well with the ambiguous power of conflict we must first appreciate that conflict can be more than just negative between us. We must believe that the benefits of working through our conflict are worth the trouble and discomfort that attend. We must both have the resources of personal maturity that enable us to face strong emotion and to look at ourselves anew and possibly change. And we must have the skills that enable us to deal effectively with one another even in the heat of our disagreement.

26.

Communication

Thomas N. Hart and Kathleen Fischer Hart

Thomas and Kathleen Hart's essay on communication ends this section on communication, conflict, and change. They begin, unsurprisingly, with a tale of conflict in Tony and Sue's marriage. The conflicts they are going through sound like they could be those of many couples dating or even engaged.

Behind their eleven hints for communication lies an assumption that we can change our basic patterns of communicating if we are willing to try and if there are others to help us. One of the hardest lessons of life comes in realizing that nobody can force another person to change. You can encourage the person, but change can only come from the person him- or herself. This means that we ourselves have to face our own willingness to change so as to become a more trustworthy and loving gift to others.

The eleven techniques, or "hints," offered here need to be practiced. A person who wants to develop these as skills might ask a close friend to help. They could practice them together. This essay, like so many others in this book, is meant to be shared with those we love and with whom we are trying to improve communication. Some questions for reflection follow:

1. *The authors stress that in marriage two people pledge to each other "the gift of the self," which "always involves the disclosure of feelings," including ones we may find difficult to communicate. This raises the question of whether the disclosure of feelings may be just as difficult before a marriage—say, during dating or engagement—as afterward. Which feelings would you say are most difficult to communicate in dating? Why are these particular feelings so difficult to express?*
2. *This essay offers eleven hints to help foster the art of communication in intimate relationships. If you were to pick a small number of these—say, five or fewer—that are more important than the others, what would they be and why do you pick them?*
3. *The Harts seek to make us much more aware of how we actually go about communicating to others. Considering that our patterns of communicating are learned as habits in childhood interaction with parents, what*

Chapter 2 of *The First Two Years of Marriage* (New York: Paulist, 1983), pp. 18–29.

chances do you think we have of changing them? For example, how do you react to the common defense: "This is how I am; I can't change; you have to accept me as I am"?

When Tony and Sue came in for counseling, they had been married five years and had two children. But things were not going at all well. At a recent workshop for couples, the two of them had been asked to recall one or two of the peak experiences of their marriage, and Tony could not think of any. He wondered if he was just so angry that his recall of anything good was blocked. As we talked, the roots of the anger were gradually uncovered. Tony summed up the problem this way: "Sue never understands me, and so she always reacts in the wrong way. But you know, the reason she doesn't understand me is that I have never really let her know me." Sue chimed in and said: "Everything he said is true. What's worse, he doesn't understand me either. And that's my fault, because I've held back too. So you can imagine what our guessing-game interaction is like." They went on to say that it had been like this from the beginning.

This clear analysis of their problem, offered by the couple themselves, set the agenda for counseling. The task was to assist them to begin to talk to one another about what was going on deep inside, to open out their inner worlds to one another. The heart of that was to get them expressing how they felt about themselves, each other, and the myriad situations of their lives. Tony admitted that the reason he held himself in was that he did not like himself very much, and he found it much easier to remain crouched behind his wall sniping at Sue than to come out from behind it and let her see who he was. To do the latter would be to make himself vulnerable, to admit hurt, weakness, inadequacy, and need, and to put the truth in Sue's hands for her to deal with. That required a degree of courage he had long been unable to muster.

We talk in Christian terms of the gift of the self. In marriage, that is what two Christians pledge to one another. It is *the* great gift of love. Many married people stay with one another and serve each other in many ways. But this is not yet the gift of the *self,* which can only be given if one is willing to open one's heart to the other. One can do many external deeds of love and still hold back the really precious gift, the inner self. This gift can be given only through communication. It costs, like all of the better gifts. But union between two persons is hardly possible if they have not let each other into their inner worlds. This always involves the disclosure of feelings.

Some feelings are harder to communicate than others. Some people, for some reason, find it next to impossible to say "I love you." Many people, especially men, find it difficult to admit their fears, their sense of failure, or their sadness. Some find it hard to affirm others, to give positive feedback. Husbands stop telling their wives that they are beautiful, saying, "She already knows that," or "It would go to her head." We have yet to meet someone who does not need to be told the good things over and over (and even then still doubts it), or whose head is in much danger of an over-swell from too much affirmation. But there is another reason why people hold in their feelings in marriage. They care about their mates and do not want to hurt them. So they do not express dissatisfaction, irritation, or any other uncomfortable feelings.

What happens in an intimate relationship when unconfortable feelings are held in? Terrible outbursts from time to time, when feelings reach the breaking point. A note on the kitchen table announcing the divorce. Or, in milder forms, sarcasm, silence, and various forms of subtle punishment. One thing is certain. If people cannot deal with their anger, they cannot be intimate either. You either have a relationship in which there are angry exchanges at times and warm closeness at others (often shortly after the angry exchange), or you have a relationship in which all is smooth on the surface but the psychological distance is unbridged. In these latter relationships there are several forbidden subjects and an abundance of silence.

In intimate relationships, communication is the foundational skill. There is none more basic. It is the indispensable condition of union. It is the key to resolving conflict. It is the only way two people can continue growing together, or even living together.

The question is: How do you do it? It is an art, and it takes time to cultivate. The following are eleven hints to point the way.

1. Use I-statements rather than You-statements. Talk about yourself rather than your mate. Don't say things like, "*You* never care about anybody but yourself," or "*You* think you know everything all the time," or "*You* never do anything around here." Say instead, "*I* often feel lonely," or "Sometimes *I* feel put down by things you say," or "*I* feel overburdened with household chores and sometimes *I* resent it because it doesn't seem fair to me." Talk about yourself, in other words, and your feelings, in response to concrete behaviors for your mate. Do that instead of making judgments about your mate ("You're always flirting"; "You're so damn sure about everything") or giving commands ("Get out of here"; "Why don't you loosen up once in a while?").

The approach we are suggesting is risky because it exposes you. But it has many distinct advantages. It does not make your mate so defensive, and so it gives you a better chance of getting a hearing and an honest response. It lets your mate in on your inner world, and so reveals important information. It does not pronounce judgment about who is wrong, but leaves the question open. For instance, if I am bothered by the way you socialize at parties, it may indeed be your fault. But it could just as well be mine. Maybe I am very insecure, and cling too much, and am very easily threatened and jealous. Maybe I misinterpret what you are doing. If I find you overly emotional, it may be that you are. But it may also be that I am emotionally repressed and uncomfortable with the expression of feelings, or simply that I feel inadequate to meeting the needs you make known to me. When I stay with my own feelings, owning them and letting you know them, I let you know me and I leave the question open about who should do what. We can work on that together. "I feel uncomfortable around your dad" is not yet a comment about your dad, still less about you. So far, it is just an informative comment about me.

Talking about your mate is legitimate to this extent: As far as possible, tie your feeling statements to your mate's concrete *behaviors*. "When you come in without saying hello, I feel unloved." "I start to feel insecure when I see you having a good time with another man." You are talking here just about concrete *behaviors,* externally observable. You are not guessing at your mate's *feelings* or *intentions,* which are hidden from view.

There are four basic I-statements which carry most of the weight in an intimate relationship. They are: (1) I think, (2) I feel, (3) I want, and (4) I need. All of them are positive steps in self-assertion and indicate an underlying self-respect. All of them make me vulnerable to you. They do not state what is right, nor do they make a demand. They simply tell you who I am and what is going on with me right now. If you are willing to make a similar self-revelation, we have the materials for really learning to care for one another.

2. Express feelings rather than thoughts. Not that it is bad to express thoughts. There is plenty of scope for those too. But our feelings reveal more of who we are. One can sit for an entire course before many professors, and know a good deal of their thought and almost nothing of who they are. A wife expressed this eloquently once, saying that everything her husband said to her could be said on television. He was an engineer, and lived much more in the realm of thought than of feeling. He was not untalkative, but she was always left wondering what was going on inside. It is in expressing our feelings that we give the gift of the self. That was the gift she was still waiting for.

People sometimes hide their feelings behind their thoughts. "A woman's place is in the home" is an apparent statement of principle, but it may be a man's way of saying that he feels he has failed as a provider if his wife takes a job outside the home, or that he fears he will lose her if she has much occasion to be with other men. Those are feelings. "Should" statements can also be masks. "You should enjoy sex" is probably best translated "I enjoy sex," or "I feel inadequate as a lover when you don't seem to enjoy sex, or hurt when you turn me down." In more open communication, people just talk about their feelings, not about the eternal order of things (as they see it), or the commonsense truths embraced by all (but you). Most of our feelings, after all, come out of our cultural relativity.

Some people are not very aware of their feelings. You cannot communicate what you do not know. To develop a greater awareness of feelings, it is a helpful exercise to go inside yourself from time to time during the day, inquiring what you are feeling right now. Watch the variations in typical situations: talking with your child, talking with the boss, hearing the phone ring, approaching the front door at home, waking up in the morning, watching TV at night, going about your daily work, getting into bed at night. Move gradually from becoming more aware of feelings to becoming more expressive of them.

3. Listen attentively without interrrupting. Good communication requires more than good talking. It demands good listening too. Listening is difficult. It requires setting other things aside, even the ruminations of the mind. It is especially hard when we do not like what we are hearing, or when we think we have heard it all before. One of the things a marriage counselor does most frequently is stop married partners from interrupting each other, suggesting instead a three-step process which can revolutionize the way they talk to each other: (a) listen without interrupting; (b) say back what you heard, and check it out; (c) respond to it.

In poor communication, the listeners are working on their responses instead of listening, and cut in to make them as soon as they are ready. In the approach suggested here, you have to listen closely or you will not be able to say back what you heard. This is how you make sure you got the message. There are often surprises here, as the original speaker makes the necessary corrections. Then you can respond. This may seem cumbersome and time-consuming. But if you really get your mate's message, and respond to it rather than to something else, you end up saving time. And your mate has the gratifying feeling of being heard, even if you end up differing. If two people are in the habit of interrupting each other as they argue back and forth, the entire time is probably being wasted. Neither is listening. Neither is open. Nothing is being produced except more bad feeling.

4. Check out what you see and hear. Part of this is summarizing what you hear your mate saying and asking if that is the message. It keeps the conversation on track. But there are other parts to this checking out too. Listen for the feelings behind the words, and check those out. Listen for anger and frustration. Listen for loneliness. Listen for fear. And test what you think you hear, saying, for example, "You sound weary," or, "You sound as if this is really hard for you to tell me." This approach is especially useful with people who do not express feelings directly, but prefer to make statements of fact, pronounce judgments of good and bad, and give direct or indirect commands. You cannot get them to play by your rules. If you say, "Don't make judgments; tell me your feelings," you have given them a command instead of expressing your frustration and leaving them free. Even if they persist in their usual ways, you can listen for the feelings behind the words, and check those out.

Checking out can be useful outside of times of conversation. You come home, and seem tired, discouraged, or distant. But I am not sure. Mostly you are silent. If I want to relate to you appropriately—giving you space, encouraging you, or inviting you to unburden yourself—I have to know what you are feeling. I can ask the open question, "How are you doing?" Or I can check out my impressions: "You seem distant." Or, "You look as if you had a hard day."

Such an approach invites mates to express themselves. The ideal situation would be that they would volunteer this information, and ask for what they need. But the situation is not always ideal.

5. Avoid mind-reading. Mind-reading is the attempt to reach inside the sanctuary of the other person's psyche and declare what is going on there. You tell others what they are feeling or what their motives are. "You're saying that because you're jealous." "You're just telling me what you think I want to hear." This kind of statement almost always gets an angry reaction. It deserves it. The statement is a violation of the other's privacy, and what is alleged is often inaccurate besides. We have *impressions* of other people's feelings, and *hunches* about their motives. It is legitimate to inquire. It is also all right to voice our impressions, if we do it tentatively, with recognition of our uncertainty. Mind-reading is another matter. It is a violation of the person. It shows disrespect for that person's integrity, destroys trust, and invites retaliation.

6. Make your needs known. Sometimes those needs are general: "I need about half an hour's space when I get home from work before I can face any new challenges." "I need relief from child care at least one day a week." Sometimes they are particular: "I need a hug." "I need to get away some weekend soon."

A couple came for counseling. The problem they brought was that the husband was angry much of the time. What came to light was that he expected his wife to anticipate his needs and take care of them, and when she did not, he got angry. He expected her to know when he needed space, and not to talk to him then. He expected her to know when he needed affection, and to be affectionate then. When she guessed wrong and acted unsuitably, he was angry. This is an extreme form of a common fallacy: "If you really loved me, you would know what I am feeling and what I need." Not so. All of us are unfamiliar territory, often even to ourselves, certainly to others. Our only hope of getting our needs met is to be assertive in declaring them. They will not always get met, of course, because others have their needs too and some limitations in their ability to meet ours. But if needs are declared, they can at least be negotiated.

7. Learn your mate's language of love. All of us have a language in which we like to be told that we are loved. And one person's language differs from another's. What tells Randy that he is loved is a massage by Betty. But what tells Betty that she is loved is not a return message by Randy, but just being held by him. What told your mother that your father was sorry was a single red rose, but what tells *your* wife you are sorry is not a rose at all but an apology and an explanation of what was going on inside you at the time of the incident. A woman might express her love by keeping a very clean house, but what would actually speak love to her husband would be her relaxing more with him.

There is usually a lot of love in the first two years of marriage, but sometimes it is spoken in your own language rather than your mate's, and so it does not have much impact. The trick is to learn your mate's language and to speak that. One very ironic situation of our acquaintance was that of a couple who differed in how they wanted to be treated when they were sick. She liked people to come into her room, freshen the air, bring her some orange juice, ask her how she was, and leave some flowers behind. He liked to be left completely alone so he could sleep. So when she was sick, he left her alone. And when he was sick, she visited him often and did all kinds of nice things for him. It took them a while to learn each other's language. It usually does. The Golden Rule here is "Do unto others as they would have you do unto them."

8. Avoid the words "always" and "never." This is an easy one to understand, but a hard one to do. "Always" and "never" are very tempting words, especially when you are angry. "You *never* listen." "You're *always* complaining." "You *never* want to do anything but watch TV." Because they are exaggerations, they provoke anger and invite a quick denial. And so the point is missed. Wouldn't "sometimes" be a better word than "always," more accurate and easier for the other person to hear? Then there is "often," and, when you really want to be emphatic, "usually." Never use "never."

9. Avoid name-calling. Names usually come into the game in the heat of anger. They hurt (which is why they are used). They stick. That is the problem. The fight ends, you make up, and things are supposed to be all right again. But your mate cannot forget that name you called her. Did you really mean that? If you didn't mean it, why did you say it? You have unwittingly planted a weed, and weeds are very hard to eradicate.

A couple once agreed in marriage counseling to just two contracts with one another. They would not read each other's minds and they would not call each other names. Seventy-five percent of their wrangling dropped away.

10. Deal with painful situations as they arise. Have you had the experience of setting off an angry tirade by some simple slip-up, like being five minutes late to pick up your mate? There has probably been some gunnysacking going on, and the sack has just burst. You take it over the head not just for the present offense but for several others stretching back over weeks and months. You didn't know your mate was carrying all this around. You can't even remember the incidents they refer to.

If couples would be close, they must learn to deal with anger honestly and constructively. That means handling it by occasions, not allowing it to build. There is no point in trying not to hurt your mate with the bad news that you did not like something. You will harm your mate more in the long run if you hold these things back. You can soften the pain by telling the truth with love each time. That way you avoid the big outburst with white-hat anger, exaggerated statements, name-calling, and sometimes physical violence.

11. Make time for talk that goes beyond practical problems. Most couples manage to get the day-to-day problems solved. They communicate enough to get the bills paid, the food bought, the baby taken care of, the guests entertained, the car fixed. But many couples gradually neglect talking about themselves. The very thing that made the courting period such a deeply happy time, talking about you and about me and about us, gets pushed from the center to the periphery and sometimes dies out altogether. We make love less frequently, and I notice it, but I don't say anything. We do more things separately, and talk about them less. I go to work and muse a lot on the general drift of my life, but I keep these thoughts and feelings to myself. You go off to be with your parents. We keep solving the day-to-day problems, but we do not talk about *ourselves*. And both of us notice in our interaction a growing distance and irritability.

Marriage Encounter has a simple idea to keep marriages going and growing. Each day the couple write each other a short letter, no more than ten minutes' worth, on some subject that draws out feelings. Some couples accomplish the same purpose in other ways. They keep an agreement to do something together one night a week, to be by themselves and talk about things that are important to each of them. It may be a dinner, or it may just be a walk. Other couples commit themselves to a weekend away every few months. These exercises in deliberate cultivation of the relationship are vital. Often all that is needed is to remove the obstacles, the challenges that ordinary living throws in the way of deeper communication. If we can free ourselves from these regularly, the deeper currents can keep flowing and joining. There is an amazing power of resurrection in marital relationships if they are not neglected too long. The coals may seem a little quiet at times, but don't call the fire out. Couples who make a little time and tend the embers see some amazing things happen. The habit of doing this needs to be formed in the first two years.

Communication is the foundational skill, and the key to all the rest of the elements that build a marriage. It is learned over time. Doing some reading about it, and attending marriage enrichment events that foster it, are very helpful in making it grow. It is actually possible for two people to be open and honest with each other, to entrust each other with nothing less than the gift of the whole self, and to become increasingly one instead of increasingly two. It takes courage, but it is one of the most satisfying experiences that life offers.

VI

Getting Beyond Stereotypes

27.

Home and Work: Women's Roles and the Transformation of Values

Rosemary Radford Ruether

Work has come to center stage as a critical human and religious issue of our time. The revolution in work patterns and the shifting work roles in marriage have challenged our established categories and expectations.

Rosemary Radford Ruether, a leading Christian feminist, begins this section with an exploration of the current relationship between home and work. They are worlds split apart into stereotypical male and female realms. The result is alienated labor, the privatized home and lack of integration in our lives. Ruether chronicles the evolving relation of women and work, beginning with the preindustrial period and through the critical turning points of urbanization and industrialization. The lesson of history frees us from fixity and a paralyzing sense of destiny in male and female roles. The author's agenda points to resocializing the home by bringing access to work and political decision-making back into a more integral relationship to it.

Some questions for reflection follow:

1. *Where do you see the cult of the lady ("true womanhood") and the idealized home operative in society today?*
2. *What impact has the modern technological revolution had on home and work?*
3. *How would you assess Ruether's claim: "As long as women are economic dependents on men, marriage is a degraded exchange of sexual rights and domestic labor for economic support."*
4. *What can we learn from the kibbutz communal model about work, child-raising, political decision-making, and context for family life?*
5. *What distinctive values cultivated in the female sphere can lead to the recovery of meaningful work and a rediscovery of community? Are these values being brought into the work force by women today?*

From *Theological Studies* 36: 4 (1975): 647–659.

Women in Western societies are apt to identify the question of women's liberation with the "right to work." The discussion of the rights of women often involves heated controversy over how it is possible for women to "go out to work" and still "take care of the home" and "be mothers." The home-work dichotomy splits male and female on opposite sides of the economic system, locating men on the side of production, women as managers of the consumer support system. When women gain the right to enter a profession, it is still very hard for them to compete with men on an equal footing, since they are also presumed to be in charge of this domestic support system. Even the childless or unmarried woman is handicapped in relation to a married male on the job who has a wife who cleans his house, cooks, shops, and plans the household, thus freeing the man for full-time attention to the "job." In this system woman's work remains invisible and unpaid. It is this double bind that is the primary reason why so few women have been able to take advantage of work opportunities even when, theoretically, they are open to them in industrial societies.

People in modern society tend to assume that this role of women is static and primordial, that women were always "unproductive" members of society. The liberation of women focuses on the integration of women into paid work roles. However, in actuality, this split of home and work, with its consequent segregation of women from "productive" or exchange-value labor, is characteristic of industrialization. The real history of women and the changing structure of the family, the relationship of the home to the economic system, is concealed when we suppose that the way these appear today is the primordial role of the family and of women. If one were to ask an African woman in a traditional village if she would like to "leave the home" and "go out to work," she might have difficulty understanding what is meant. In such societies the home, embedded in the tribal community, is the unit of economic production. Here women do much of the productive labor. They are the chief agriculturalists and produce most of the handicrafts. They sow and harvest the fields, which often belong to the women. They command the transformatory processes that turn the raw into the cooked, herbs into medicines, raw materials into clothes, baskets, and pots. Often marketing is in their hands. Men protect the village from aggression, conduct war, clear and fence the fields, and make weapons. The grown men are freed by the work of women and youth in order to "palaver," to engage in the political and social discourse of village government. Women are the productive laborers of society.[1] Here there is no split between home and work, because the economy still has its original locus in the home.

Women and Work in Preindustrial Societies

The picture of women in many preindustrial societies is found in Proverbs 31: 10–31:

> She seeks wool and flax and works with willing hands. . . .
> She rises while it is yet night and provides food for her household and tasks for her maidens.

> She considers a field and buys it, with the fruit of her hands she plants a vineyard. . . .
> She perceives that her merchandise is profitable;
> Her lamp does not go out at night;
> She puts her hands to the distaff and her hands hold the spindle. . . .
> Her husband is known in the gates when he sits among the elders of the land.
> She makes linen garments and sells them;
> She delivers girdles to the merchant. . . .

In this picture we see women as the primary workers and managers of the economic realm, freeing men for the roles of political discourse "at the gates." This role did not disappear with the urban revolution, although women and men's roles became less equalitarian than they had been in village life and sharp class divisions appeared. Not only did women continue to be workers in peasant life, but the latifunda of the great landed families were often largely managed by the wife as an extension of a family-centered economy, while the men occupied themselves with war and politics. This was to a large extent true even of the plantations of the American South. Lacking an industrial base, plantations managed by the wives supported the economy that was squandered by their husbands in the Civil War.[2] As long as the economy was centered on the family, women had an important economic role and even an economic bargaining power in society, despite the patriarchal character of the political system that might define women as dependent and rightless.

The transition from rural to urban life was an important turning point in the history of women. The urban revolution originally affected only a small segment of society, however, while the rest of society remained agrarian and in a family-centered handicraft economy. But the urban revolution created a new elite group of males whose power was no longer based on the personal prowess of the hunter or warrior, but on an inherited monopoly of political power, land, and knowledge. The political sphere, which had already fallen into the hands of males in village society by and large, could now be monopolized by this elite to define women and lower classes in a dependent and inferior relation to themselves. Generally we find women excluded not only from political leadership (although they may be place holders for male heirs) but from those professional roles in culture and religion that buttress political power. Scribes and priests exclude women programmatically, although no biological differences would have prevented women from entering these fields on equal terms. It is from these classes that we also get those religious laws and ideologies that codify the doctrines of female inferiority in classical societies.

Yet various professions often remained in women's hands in classical societies which they were subsequently to lose. In general, we may say that roles remain open to women as long as they are based more on experience and folk knowledge. Once the training necessary to enter them becomes professionalized, women are excluded. The exclusion of women from education in classical societies becomes the chief means of excluding women from the entire process of the reflection upon and transmission of culture, as well as access to the training necessary for all the valued professional roles.

Medicine was often monopolized by women in earlier societies. Pharmacy was an extension of their role as cooks and gatherers of herbs. As mothers, they were also midwives and healers of injuries and diseases. Certain "wise women" often specialized in these healing arts in villages. As medicine became professionalized, sometimes a few women were allowed to participate in the early stages. There were women in the schools of medicine in Spain in the eleventh and twelfth centuries in medieval Europe, for example. But generally professionalization meant both the exclusion of women from the necessary training for the profession and a gradual proscription of their earlier exercise of it based on folk knowledge.[3] This is particularly true when scribal and priestly exclusions of women join together. Such an exclusion of women from the study of medicine is represented by the decree issued by the faculty of the University of Bologna in 1377 A.D.:

> And whereas woman is the fountain of sin, the weapon of the Devil, the cause of man's banishment from Paradise and the ruin of the old laws, and whereas for these reasons all intercourse with her is to be diligently avoided; therefore we do interdict and expressly forbid that any one presume to introduce into the said college any woman, whatsoever, however honorable she be. If this nonwithstanding anyone should perpetrate such an act, he shall be severely punished by the Rector.[4]

The effect of such an exclusion of women is sometimes a dual system. There is the trained doctor for the upper classes and the folk "wise woman" for the poor. The result in the Middle Ages was not always an improvement of medicine, because university medical training was highly theoretical, based on classical authorities without experimental verification, while the medicine of the "wise woman" was based on experience and practice. But there was also magic mixed up with it. The great persecution of old women as witches in the Late Middle Ages down to the eighteenth century had, as one aspect, the crushing of the wise woman as folk doctor and pharmacist. Soon after, the professionally trained doctor also displaces the woman as midwife as well.[5] This new male hegemony in obstetrics had the side effect of an outsider's approach to the woman in delivery as a "patient" who is ill, whose body is treated as an object, rather than as an active participant in a natural process.

In Europe in the seventeenth century women's traditional role in crafts meant that some guilds, especially those associated with weaving and clothmaking, were female. Women also were trained in many crafts as assistants to their husbands. The proximity of shops and homes and the family aspect of guild membership meant that a widow often continued her husband's craft. In the seventeenth century there was a concerted elimination of women in crafts through professionalization and new forms of organization and through licensing that specifically forbade woman's participation. For example, women traditionally had been brewers, but new laws forbade the granting of licensing for brewing to women. The tradition of women in skilled crafts, as entrepreneurs and owners of taverns and businesses, continued longer in colonial America, where the frontier situation made the working woman still valuable. The elimination of women in business and crafts that affected Europe in the seventeenth and early eighteenth centuries only became general in America at the end of the eighteenth century.[6]

Industrialization and the "Cult of True Womanhood"

Industrialization is a second critical turning point in the socioeconomic history of women. On the one hand, it added a new economic dependence of men to the legal dependence that had been imposed on women in classical patriarchy. This had already been the case of upper-class women, but working-class women could still be self-supporting as long as the economy was family-centered. Industrialization progressively removed all self-supporting functions from the home, refashioning the family as a sphere totally dependent on an economic system outside of it. Women's role was also refashioned from that of an active laborer in vital economic processes to that of a manager of consumption and an ornament to her husband's economic prowess.[7]

But industrialization also created new frustrations and contradictions for a larger mass of women, increasingly deprived of active participation in the life of society. This frustration made feminism a mass movement, rather than treatises written by a few educated women of the upper classes. Industrialization also forced many poor women out of the home into doubly oppressive conditions of the factories. The liberal doctrines of equal rights began to be taken up by women and by workers who had not been included in those declarations of the "rights of man" declared by the victorious bourgeois of the French and the American Revolutions. The efforts to press woman into her newly limited and intensified role in the home created new ideologies of women's "natural" difference from men. But the revolt against this stifling sphere also began the systematic challenging of the classical patriarchal status of women as property and political dependents on men.

Industrialization completed the reshaping of the role of the home and the ideologies of womanhood and childhood that had begun under bourgeois commercial society. The home is privatized as an intensive center of personal life and nurture. The retainers, servants, and other dependents that have lived with their masters, even in relatively modest households, are gradually thrust out and the nuclear family withdraws into itself. Many family functions, such as childbirth, that have been public occasions, withdraw into secrecy. The home loses its more open, public character. Bedrooms cease to be areas of public socializing, although the great halls of the aristocracy keep these traditions longer. As the family withdraws into intensified private life, the concept of childhood is reshaped into an increasingly extended period of nurture and shaping of a malleable being. Women's role, in turn, is defined by this intensified domesticity and increasingly prolonged concept of childhood dependency.[8]

The ideology of "true womanhood" or the "lady" shaped and reflected this intensified domestication of the middle-class woman. The cult of the lady and the idealized Home also played a crucial compensatory role in the new industrial society that was being formed. Although built on the earlier aristocratic traditions of courtly love, it was popularized in the nineteenth century as part of the middle-class reaction against secularization, social revolution, and industrial society, with their threats to traditional values. The Home and Womanhood were to be everything that the modern industrial society was not. Here in the home patriarchy and the natural aristocracy of ''birth'' still held sway in male-female relations, although democratic concepts were everywhere else challenging this concept. Here the religious world view of fixed certainties could be maintained in an age of growing unbelief. Here emotionality and intimacy held sway in a world dominated outside by unfeeling technological rationality. Here sublimated spirituality compensated for the outward capitulation to the fierce materialism of industrial competition. Here an Eden of beauty and peace walled the bourgeois at home off from the ugly work world of the factories. The home was, above all, the realm of nostalgic religiosity, to be cultivated by women, to which men could repair to escape the threatening outside world of doubt, insecurity, and social restiveness. Women were to remain precritical and insulated against this threatening world, in order to preside over a home where men could preserve their faith in those values in which they no longer believed but wanted to believe that they still believed. The almost religious veneration of the home and womanhood in Victorian society must be seen in this context of escape and compensation for the threats to all these traditional values posed by the industrial revolution.[9]

This idealization of woman in the home as effectively removed her from the ''real world'' of men and public affairs as had her earlier denigration as that ''devil's gateway, font of sin, and unsealer of the forbidden fruit.'' It is said that women are ''too pure,'' too noble, to descend into the base world of work and politics. To step out of her moral shrine to work or to vote, to attend universities with men, and mingle with them in the forums of power is to sully her virtue and destroy instantly that respect which accrues to her in the ''sanctuary'' of the home. This ''down from the pedestal'' argument became the chief tool by which social conservatives in the Church and society rebutted every effort of the rising women's movement to enlarge the public sphere of women. Much the same arguments are used today against the Equal Rights Amendment.

These arguments reveal the fundamental ambiguity of the male ideology of ''femininity.'' These characteristics are seen simultaneously as unchangeably rooted in woman's biological ''nature,'' and yet as something that can be lost instantly as soon as she steps out of her assigned social role. In the early twentieth century the Catholic bishops of the United States put themselves solidly on record against women's suffrage. An interview with Cardinal Gibbons reveals the line of the argument:

> ''Women's suffrage,'' questioned the Cardinal. . . . ''I am surprised that one should ask the question. I have but one answer to such a question, and that is that I am unalterably opposed to woman's suffrage, always have been and always will be. . . . Why should a woman lower herself to sordid politics? Why should a woman leave her home and go into the streets to play the game of politics? Why should she long to come into contact with men at the polling places? Why should she long to rub elbows with men who are her inferiors intellectually and morally?

> Why should a woman long to go into the streets and leave behind her happy home, her children, a husband and everything that goes to make up an ideal domestic life? . . . When a woman enters the political arena, she goes outside the sphere for which she was intended. She gains nothing by that journey. On the other hand, she loses the exclusiveness, respect and dignity to which she is entitled in her home.
>
> "Who wants to see a woman standing around the polling places; speaking to a crowd on the street corner; pleading with those in attendance at a political meeting? Certainly such a sight would not be relished by her husband or by her children. Must the child, returning from school, go to the polls to find his mother? Must the husband, returning from work, go to the polls to find his wife, soliciting votes from this man and that . . . ? Woman is queen," said the Cardinal, in bringing the interview to a close, "but her kingdom is the domestic kingdom."[10]

This split between woman's sphere in the home and the male world of work created a new ideological dualism which divided the feminine from the masculine, the private self from the public world, morality from facts. Religion, driven into the private realm by secularization, also participated in and was shaped by this split. This split partially reversed the older typologies of female "nature." Whereas classical Christianity unhesitatingly saw women as less religious, spiritual, and moral than men, nineteenth-century culture typically saw women as inherently more moral, spiritual, and religious than men. Whereas earlier culture had regarded women as more sexual than men, almost insatiably so, Victorian womanhood was regarded as almost asexual. The "true woman" is almost incapable of feeling sexuality, and sexual desire is banished from her mind. Carnality is ceded to the male nature, as part of his rough dealings with the "real world" of materialism and power. Religion likewise recedes into the "feminine" world of spirituality divorced from truth or power. The material world is now seen as the "real world," the world of hard, practical aggressivity, devoid of sentiment or morality.

Rationality is still located in the man and "his world," but it loses the quality of wisdom and becomes that functional rationality that is the tool for manipulating matter through science. Reason is split from morality, making reason "value-free" and morality sentimentalized. Religion especially falls victim to this sentimentalization of spirituality and morality. Morality and religion fall into the realm of "private man," in the home, *de facto* the realm of women. The ethical split between "moral man" (private man) and "immoral society" (public man) unconsciously is split along the lines of work and home, masculine and feminine. Christian virtue, agape, comes to be seen as peculiarly "feminine." Christ too in nineteenth-century religion comes to be seen typically as a "feminine" figure, no longer the Christ Pantocrator of Christendom. The Church and the clergy function, like women, to create a nostalgic place of escape and compensation for an evil public world. But this realm of Church, home, and women also is the tacit support of the secular realm of male power, by pacifying the private self in relation to it. Christian virtue is both politically conservative and yet apolitical. It is "feminine" in a way that makes it also "unrealistic" and "out of place" in the world of "manly men."

The Victorian cult of true womanhood was a class myth. Industry, together with a still existing servant class, made possible a new group of leisured middle-class women who displayed through their delicacy, elegance, and idleness the wealth of the new economic leaders.[11] But the myth of the lady also ignored the large numbers of working women driven into the factories to work long hours at pitiable wages. Its sublimated leisure culture of affluence was built on sexual and social oppression.[12] The asexual "purity" of the "good woman" had its underside in the proliferation of houses of prostitution.[13] Its affluence was built on the exploitation of factory labor. These two forms of oppression mingled in the poor woman, who, often unable to survive on her low factory wages, turned to prostitution. Work itself was seen as a kind of "fall" from purity, destroying that "femininity" and "purity" of the lady. Thus the class division between the lady and the working woman also fissured into the dual ideologies of the pure asexual feminine and evil carnal femaleness. Since the cult of true womanhood made the leisured woman normative, woman going to work could only be viewed as a downfall from the sanctity of home.

Nineteenth-century working women also developed their own political struggle and articulated their own needs, which differed radically from middle-class feminism. The vote, education, and professions were the class privileges of the sisters of those in power that did not speak to the condition of working women. Middle-class feminism often spurned working women or reached out to them only in the patronizing form of moral uplift for the fallen. Working women organized around their economic needs, better wages, and working conditions. But they generally found little help from their working brothers in the union movement. Women's work was either not taken seriously as real economic need or else the low wages of women were regarded as a threat to male wages. It was assumed that women work for "luxuries," not real support, despite the numbers of households headed by women. They are regarded as unreliable workers whose biology makes them irregular and who can be expected to stay on the job only until they get married. Doubtless the bad conditions often made these assumptions self-fulfilling prophecies.

Male unionists have seldom fought for equal pay for equal work for women, but instead have either ignored them or sought to segregate them in special types of low-paid work which did not threaten their own wages.[14] Fundamentally, women's work comes to be structured into job-support systems, such as stenography, which aid male work, or are used as a surplus labor force to be hired in times of added need, such as wartime, and fired when this need recedes. Despite the numbers of women in the work force today (about 40%), neither the ideology of womanhood nor the planned relations of home and work have been willing to adjust to this reality. Women at work still have to find *ad hoc* solutions for childcare, housekeeping help, performance of domestic work for themselves and their husband and children. "Women's work" in the home is still presumed to be theirs in a work world that makes no effort to adjust to the special reality of women. The world of work still organizes itself as though workers were male and have nonworking wives providing for their domestic needs.

The Socialist Critique of Women's Role

The utopian socialists of the early nineteenth century recognized that their critique of the family and private property involved a criticism of the role of women. Both sexual liberation and equal work roles for women were part of the program of the St. Simonians, Owenites, and Fourierites. Marx and Engels extended and deepened this connection between socialism and the liberation of women. Their experience with working women in English factories alerted them to the class character of the standard myth of the delicate lady, incapable of hard labor. There they saw women working ten and twelve hours a day under brutal conditions, only to return home to care for their domestic chores as well. But they also concluded that industrialization, despite its doubly brutal conditions for women, was creating the economic basis for the emancipation of women. Marx and Engels and subsequent Marxists have hinged their concept of women's emancipation upon the restoration of women to the world of production. Only when women have autonomous incomes from their own labor will they have the economic basis for personal equality with men. As long as women are economic dependents on men, marriage is a degrading exchange of sexual rights and domestic labor for economic support. However softened by custom, its reality remains that of a kind of slavery and economic bondage. Autonomous work and independent income are the bases for all other rights and dignities of women. Without it, all rights and dignities are extended to her on the sufferance of males, who still retain the title to them in their own hands. In the *Origin and History of the Family, Private Property and the State,*[15] Engels worked out their view of the relation between the subjugation of women to the rise of private property and women's deprivation of an autonomous role in production. This they saw beginning to be restored by industrialism, as far as the working-class woman was concerned.

Engels believed that communism would establish complete equality between men and women by integrating women into all spheres of work equally with men. Women would receive the same education and could enter any occupation. The economic independence gained from work would be the foundation of their personal independence. They would no longer have to sell their sexuality for economic security or have their income and property owned and managed by their husbands, be coerced into marriage, or kept in marriages grown cold, by economic need. Marriage could return to being what Engels believed it had been originally, before the rise of the patriarchal system: a free personal relationship between two persons based on mutual compatibility, entered into and dissolved without economic coercion. Engels believed this would lead to stable monogamous relationships, which corresponded to the "natural love instincts" of humanity, which had been distorted by the economic power of one partner and the subordination of the other into hypocrisy and infidelity.

Contrary to Marx's expectations, communist revolutions did not take place in advanced industrial countries but in countries engaged in overthrowing feudal and colonial regimes and just beginning to enter the industrial revolution. Women in prerevolutionary Russia or China had not yet experienced the expanded work roles for the enlarged education and civil rights of Western industrialism and liberalism. Their status was still that of chattel of fathers and husbands, who could be married, sold, and even killed at will.[16]

Communist revolutions have made good on the Marxist belief in the union between female emancipation and proletarian revolution by sweeping transformations of the status of women. Marriage codes established the complete civil equality of women, and comprehensive childcare and even communal kitchens, maternity leave and guaranteed re-employment, and campaigns to transform cultural consciousness strove to open the world of work to women on an equal basis. In China this policy of female emancipation demanded a literal uprooting and re-creating of the Chinese family.[17] Article 6 of the Constitution, adopted in September 1946, declared: "The People's Republic of China abolishes the feudal system which holds women in bondage. Women shall enjoy equal rights with men in political, economic, cultural, educational, and social life. Freedom of marriage for men and women shall be enforced."[18] All forms of concubinage and forced marriage were abolished, and divorce was to be granted by mutual consent. But the communalization of work conditions and the home, carried out much more radically in China than in Russia, has been the social basis for equality of women on the job and in their personal relations. The private work of women in the home has become communal work, freeing women on the job from the handicap of the double shift of home and work.

Women in the West can recognize the more systematic integration of women into work in Marxist countries. Communist regimes have been willing to recognize what liberal industrialism has always avoided, namely, that women cannot be equal on the job until there is a social reorganization of the economic relations of home and work and the unpaid roles of the home are no longer placed solely on the backs of women. As long as working women must solve this problem on an individual basis, paying out of their meager salaries for substitute homeworkers, only a small elite, often unmarried or childless, can hope for significant careers, while those who must work as heads of households are forced into desperate contradictions which often leave the vital roles of the home neglected. Women are made to feel guilty for this failure, instead of society taking responsibility for adjusting the relationship of work and home in just and rational ways.

However, women may well ask whether the social values created by the Marxist solution are a sufficient answer to the historic dependency of women. This is not only because women, especially in Eastern Europe and the Soviet Union, are often left with considerable residue of the unpaid second shift of the home, handicapped thereby on the job and still subject to sexual stereotyping of work. Even more, one might ask whether the Marxist solution does not make a male concept of alienated work the exclusive pattern for life and values. The Marxist solution envisions the integration of women into this type of alienated labor by drastically reducing the work of the home, collectivizing it in the public sector.

But what is called the "home" is nothing less than the original base of personal autonomy in the self-governing familial community, which has greatly shrunk in its economic, political, and cultural functions due to the alienation of these functions into public patterns of socialization. Marxism proposes to emancipate women by totalizing this process of alienation and collectivization, leaving the home little more than a bedroom and a nucleus of fleeting personal relations. As one function after another is collectivized outside the family, the family progressively loses its self-determination and becomes totally determined by social forces over which it has no power. The shrinking of the home, then, becomes the means of creating the totalitarian society where the self has lost its autonomous base. Socialists, as well as feminists, must rethink the social role of the home, if they are committed to a society of freedom as well as a society of equal work roles.

Socialism is based on a concept of women's rights that unquestionably assumes that this process is one of obliteration of the female sphere into the masculine sphere, that is to say, the alienation of local self-determination into macro-collectivization. The values cultivated in the home, the values identified with women, are thereby also obliterated for an exclusive definition of society through the values of conflict, work, and repression. Communism totalizes the society of alienated work and warfare, instead of, as Marx himself envisioned, abolishing this society. One must ask whether a society which seeks both freedom and equal work roles, both justice and humanization, must not envision the process of socialization the other way around, not by completing the historic process of alienating the functions of the home, but by resocializing the home by bringing access to work and political decision-making back into a more integral relationship to it. Communalization of home and work that puts the ownership and decision-making over these spheres in the hands of the local community represents a kind of socialization which restores rather than destroys the sphere of self-determination. The communal patterns of China, as the base for constructing the larger networks of society, or the kibbutz patterns embedded within the larger social system in Israel, are possible models for this development. Women are reintegrated into the larger world of work and decision-making, and society takes responsibility of communal childraising, not so much by abolishing the family and the home as by re-embedding it in a "tribe," or a network of relationships whose concrete functionings can be governed by the local group itself. Working and living complexes must still be integrated into larger structures of planning and exchange, but this does not mitigate against the possibility of a system where local communities make the concrete decisions that shape their own lives.

The bringing of work back to relationship to the base of autonomous community life also suggests the shaping of society by different values than those of alienated work and conflict, which have been, historically, shaped by the male roles. A humanized society must be one reintegrated into those values cultivated in the female sphere: co-operation, mutual support, leisure, celebration, free creativity, and exploration of feelings and personal relations. The priorities of human life must be re-examined. Work itself must be seen as a means to the end of self-expression, mutual help, and fulfillment of being, rather than all existence shaped by a program of alienated labor. We do not exist in order to work, but we work in order to be—not merely in the sense of minimal survival, but in the sense of that fulfillment of being when work is reunified with creative self-expression and takes place in the framework of a community of mutual affirmation. It is this vision of the recovery of the world of work for women, which is at the same time the dealienation of work and the rediscovery of community, that must be the distinctive value which women should bring to the question of work.

NOTES

1. David Hapgood, *Africa: From Independence to Tomorrow* (New York, 1970) pp. 35, 48.
2. Anne Firor Scott, *The Southern Lady: From Pedestal to Politics, 1830–1930* (Chicago, 1970) chap. 1.
3. Barbara Ehrenreich and Deirdre English, *Witches, Midwives and Nurses: A History of Women Healers* (Old Westbury, N.Y., 1972).
4. Quoted in August Bebel, *Women under Socialism* (New York, 1971; reprint of 1904) p. 205.
5. T. R. Forbes, *The Midwife and the Witch* (New Haven, 1969).
6. Alice Clark, *Working Life of Women in the 17th Century* (London, 1919; New York, 1968) *passim.*
7. Ann Gordon, Mari Jo Bhule, and Nancy Schrom, "Women in American Society: A Historical Contribution," *Radical America* 5, 4 (July–August 1971) 25–30.
8. Philip Aries, *Centuries of Childhood: A Social History of Family Life* (New York, 1962) esp. pp. 353–404.
9. Barbara Welter, "The Cult of True Womanhood, 1820–1860," *American Quarterly* 18 (1966) 152–74; also Dorothy Bass Fraser, "The Feminine Mystique, 1890–1910," *Union Seminary Quarterly Review* 27, 4 (Summer 1972), and *Suffer and Be Still: Women in the Victorian Age,* ed. Martha Vicinus (Bloomington, Ind., 1973).
10. *N.Y. Globe,* June 22, 1911 (Documents of the Catholic Bishops against Women's Suffrage, 1910–1920; Sophia Smith Collection, Smith College).
11. See Thorsten Veblen, *The Theory of the Leisure Class* (1899: New York, 1912) pp. 171–79, 338–57.
12. Bebel, *op. cit.,* pp. 146–66.
13. See K. Chesney, *The Victorian Underworld* (London, 1970) pp. 306–64; also E. M. Sigsworth and T. J. Wyke, "A Study of Victorian Prostitution and Venereal Disease," in *Suffer and Be Still,* pp. 77–99, and S. Marcus, *The Other Victorians: A Study of Sexuality and Pornography in Nineteenth Century England* (New York, 1974).
14. Eleanor Flexner, *Century of Struggle: The Women's Rights Movement in the United States* (New York, 1972) chaps. 9, 14, and 18.
15. Originally published in Zurich, 1884.
16. Sheila Rowbotham, *Women, Resistance and Revolution: A History of Women and Revolution in the Modern World* (New York, 1972) pp. 134–40, 170–83.
17. *Ibid.,* pp. 141–59, 184–99.
18. *Ibid.,* pp. 184–85.

28.

The Good-Provider Role: Its Rise and Fall

Jessie Bernard

Jessie Bernard's essay is a seminal piece of literature that traces the birth and demise of the good-provider role in U.S. society. The structure of the "traditional" American family, in which the husband-father is the provider and the wife-mother the housewife, lasted well over 150 years, from the 1830s to the present. The good-provider role had wide ramifications for all our thinking about families. It ushered in a new kind of marriage. The costs and rewards, for women and men alike, were copious.

There were always defectors from the role, and in recent years dissatisfaction with it increased. As more and more women entered the labor force and assumed a share of the provider role, the powers and prerogatives of the good-provider role became diluted. New demands were being made on it: (1) more intimacy, expressivity, and nurture; and, (2) more sharing of household responsibility and child care. This is leading to a reworking of family roles and to the emergence of a new social-psychological structure in marriage and family. A new day has dawned for role-playing!

Questions for reflection follow:

1. *If the good-provider role is on its way out, has a legitimate successor appeared? What would it look like?*
2. *Assess Bernard's statement: "everywhere the same image shows up: an image of women sharing the providing role and, at the same time, retaining responsibility for the household."*
3. *What would some of the possible effects be (on childcare, schooling, shopping) if men increased their participation in family affairs and household tasks?*
4. *What does the demise of the good-provider role do to male identity? To female identity? To marriage?*

From *American Psychologist* 36, 1, January, 1981: 1–12.

The Lord is my shepherd, I shall not want. He sets a table for me in the very sight of my enemies; my cup runs over (23rd Psalm). And when the Israelites were complaining about how hungry they were on their way from Egypt to Canaan, God told Moses to rest assured: There would be meat for dinner and bread for breakfast the next morning. And, indeed, there were quails that very night, enough to cover the camp, and in the morning the ground was covered with dew that proved to be bread (Exodus 16:12–13). In fact, in this role of good provider, God is sometimes almost synonymous with Providence. Many people, like Micawber, still wait for him, or Providence, to provide.

Granted, then, that the first great provider for the human species was God the Father, surely the second great provider for the human species was Mother, the gatherer, planter, and general factotum. Boulding (1976), citing Lee and deVore, tells us that in hunting and gathering societies, males contribute about one fifth of the food of the clan, females the other four fifths (p. 96). She also concludes that by 12,000 B.C. in the early agricultural villages, females provided four fifths of human subsistence (p. 97). Not until large trading towns arose did the female contribution to human subsistence decline to equality with that of the male. And with the beginning of true cities, the provisioning work of women tended to become invisible. Still in today's world it remains substantial.

Whatever the date of the virtuous woman described in the Old Testament (Proverbs 31:10–27), she was the very model of a good provider. She was, in fact, a highly productive conglomerate. She woke up in the middle of the night to tend to her business; she oversaw a multiple-industry household; *her* candles did not go out at night; there was a ready market for the high-quality linen girdles she made and sold to the merchants in town; and she kept track of the real estate market and bought good land when it became available, cultivating vineyards quite profitably. All this time her husband sat at the gates talking with his cronies.

A recent counterpart to the virtuous woman was the busy and industrious shtetl woman:

> The earning of a livelihood is sexless, and the large majority of women . . . participate in some gainful occupation if they do not carry the chief burden of support. The wife of a "perennial student" is very apt to be the sole support of the family. The problem of managing both a business and a home is so common that no one recognizes it as special. . . . To bustle about in search of a livelihood is merely another form of bustling about managing a home; both are aspects of . . . health and livelihood. (Zborowski & Herzog, 1952, p. 131)

In a subsistence economy in which husbands and wives ran farms, shops, or businesses together, a man might be a good, steady worker, but the idea that he was *the* provider would hardly ring true. Even the youth in the folk song who listed all the gifts he would bestow on his love if she would marry him—a golden comb, a paper of pins, and all the rest—was not necessarily promising to be a good provider.

I have not searched the literature to determine when the concept of the good provider entered our thinking. The term *provider* entered the English language in 1532, but was not yet male sex typed, as the older term *purveyor* already was in

1442. Webster's second edition defines the good provider as "one who provides, especially, colloq., one who provides food, clothing, etc. for his family; as, he is a good or an adequate provider." More simply, he could be defined as a man whose wife did not have to enter the labor force. The counterpart to the good provider was the housewife. However the term is defined, the role itself delineated relationships within a marriage and family in a way that added to the legal, religious, and other advantages men had over women.

Thus, under the common law, although the husband was legally head of the household and as such had the responsibility of providing for his wife and children, this provision was often made with help from the wife's personal property and earnings, to which he was entitled:

> He owned his wife's and children's services, and had the sole right to collect wages for their work outside the home. He owned his wife's personal property outright, and had the right to manage and control all of his wife's real property during marriage, which included the right to use or lease property, and to keep any rents and profits from it. (Babcock, Freedman, Norton, & Ross, 1975, p. 561)

So even when she was the actual provider, the legal recognition was granted the husband. Therefore, whatever the husband's legal responsibilities for support may have been, he was not necessarily a good provider in the way the term came to be understood. The wife may have been performing that role.

In our country in Colonial times women were still viewed as performing a providing role, and they pursued a variety of occupations. Abigail Adams managed the family estate, which provided the wherewithal for John to spend so much time in Philadelphia. In the 18th century "many women were active in business and professional pursuits. They ran inns and taverns; they managed a wide variety of stores and shops; and, at least occasionally, they worked in careers like publishing, journalism and medicine" (Demos, 1974, p. 430). Women sometimes even "joined the menfolk for work in the fields" (p. 430). Like the household of the proverbial virtuous woman, the Colonial household was a little factory that produced clothing, furniture, bedding, candles, and other accessories, and again, as in the case of the virtuous woman, the female role was central. It was taken for granted that women provided for the family along with men.

The good provider as a specialized male role seems to have arisen in the transition from subsistence to market—especially money—economies that accelerated with the industrial revolution. The good-provider role for males emerged in this country roughly, say, from the 1830s, when de Tocqueville was observing it, to the late 1970s, when the 1980 census declared that a male was not automatically to be assumed to be head of the household. This gives the role a life span of about a century and a half. Although relatively short-lived, while it lasted the role was a seemingly rocklike feature of the national landscape.

As a psychological and sociological phenomenon, the good-provider role had wide ramifications for all of our thinking about families. It marked a new kind of marriage. It did not have good effects on women: The role deprived them of many chips by placing them in a peculiarly vulnerable position. Because she was not reimbursed for her contribution to the family in either products or services, a wife was stripped to a considerable extent of her access to cash-mediated markets. By discouraging labor force participation, it deprived many women, especially affluent ones, of opportunities to achieve strength and competence. It deterred young women from acquiring productive skills. They dedicated themselves instead to winning a good provider who would "take care of" them. The wife of a more successful provider became for all intents and purposes a parasite, with little to do except indulge or pamper herself. The psychology of such dependence could become all but crippling. There were other concomitants of the good-provider role.

Expressivity and the Good-Provider Role

The new industrial order that produced the good provider changed not so much the division of labor between the sexes as it did the site of the work they engaged in. Only two of the concomitants of this change in work site are selected for comment here, namely, (a) the identification of gender with work site as well as with work itself and (b) the reduction of time for personal interaction and intimacy within the family.

It is not so much the specific kinds of work men and women do—they have always varied from time to time and place to place—but the simple fact that the sexes do different kinds of work, whatever it is, which is in and of itself important. The division of labor by sex means that the work group becomes also a sex group. The very nature of maleness and femaleness becomes embedded in the sexual division of labor. One's sex and one's work are part of one another. One's work defines one's gender.

Any division of labor implies that people doing different kinds of work will occupy different work sites. When the division is based on sex, men and women will necessarily have different work sites. Even within the home itself, men and women had different work spaces. The woman's spinning wheel occupied a different area from the man's anvil. When the factory took over much of the work formerly done in the house, the separation of work space became especially marked. Not only did the separation of the sexes become spatially extended, but it came to relate work and gender in a special way. The work site as well as the work itself became associated with gender; each sex had its own turf. This sexual "territoriality" has had complicating effects on efforts to change any sexual division of labor. The good provider worked primarily in the outside male world of business and industry. The homemaker worked primarily in the home.

Spatial separation of the sexes not only identifies gender with work site and work but also reduces the amount of time available for spontaneous emotional give-and-take between husbands and wives. When men and women work in an economy based in the home, there are frequent occasions for interaction. (Consider, for example, the suggestive allusions made today to the rise in the birth rate nine months after a blackout.) When men and women are in close proximity, there is always the possibility of reassuring glances, the comfort of simple physical presence. But when the division of labor removes the man from the family dwelling for most of the day, intimate relationships become less feasible. De Tocqueville was one of the first to call our attention to this. In 1840 he noted that

> almost all men in democracies are engaged in public or professional life; and . . . the limited extent of common income obliges a wife to confine herself to the house, in order to watch in person and very closely over the details of domestic economy. All these distinct and compulsory occupations are so many natural barriers, which, by keeping the two sexes asunder, render the solicitations of the one less frequent and less ardent—the resistance of the other more easy. (Tocqueville, 1840, p. 212)

Not directly related to the spatial constraints on emotional expression by men, but nevertheless a concomitant of the new industrial order with the same effect, was the enormous drive for achievement, for success, for "making it" that escalated the provider role into the good-provider role. De Tocqueville (1840) is again our source:

> The tumultuous and constantly harassed life which equality makes men lead [becoming good providers] not only distracts them from the passions of love, by denying them time to indulge in it, but it diverts them from it by another more secret but more certain road. All men who live in democratic ages more or less contract ways of thinking of the manufacturing and trading classes. (p. 221)

As a result of this male concentration of jobs and careers, much abnegation and "a constant sacrifice of her pleasures to her duties" (de Tocqueville, 1840, p. 212) were demanded of the American woman. The good-provider role, as it came to be shaped by this ambience, was thus restricted in what it was called upon to provide. Emotional expressivity was not included in the role. One of the things a parent might say about a man to persuade a daughter to marry him, or a daughter might say to explain to her parents why she wanted to, was not that he was a gentle, loving, or tender man but that he was a good provider. He might have many other qualities, good or bad, but if a man was a good provider, everything else was either gravy or the price one had to pay for a good provider.

Lack of expressivity did not imply neglect of the family. The good provider was a "family man." He set a good table, provided a decent home, paid the mortgage, bought the shoes, and kept his children warmly clothed. He might, with the help of the children's part-time jobs, have been able to finance their educations through high school and, sometimes, even college. There might even have been a little left over for an occasional celebration in most families. The good provider made a decent contribution to the church. His work might have been demanding, but he expected it to be. If in addition to being a good provider, a man was kind, gentle, generous, and not a heavy drinker or gambler, that was all frosting on the cake. Loving attention and emotional involvement in the family were not part of a woman's implicit bargain with the good provider.

By the time de Tocqueville published his observations in 1840, the general outlines of the good-provider role had taken shape. It called for a hardworking man who spent most of his time at his work. In the traditional conception of the role, a man's chief responsibility is his job, so that "by definition any family behaviors must be subordinate to it in terms of significance and [the job] has priority in the event of a clash'' (Scanzoni, 1975, p. 38). This was the classic form of the good-provider role, which remained a powerful component of our societal structure until well into the present century.

Costs and Rewards of the Good-Provider Role for Men

There were both costs and rewards for those men attached to the good-provider role. The most serious cost was perhaps the identification of maleness not only with the work site but especially with success in the role. ''The American male looks to his breadwinning role to confirm his manliness'' (Brenton, 1966, p. 194).[1] To be a man one had to be not only a provider but a *good* provider. Success in the good-provider role came in time to define masculinity itself. The good provider had to achieve, to win, to succeed, to dominate. He was a bread*winner*. He had to show ''strength, cunning, inventiveness, endurance—a whole range of traits henceforth defined as exclusively 'masculine''' (Demos, 1974, p. 436). Men were judged as men by the level of living they provided. They were judged by the myth ''that endows a money-making man with sexiness and virility, and is based on man's dominance, strength, and ability to provide for and care for 'his' woman'' (Gould, 1974, p. 97). The good provider became a player in the male competitive macho game. What one man provided for his family in the way of luxury and display had to be equaled or topped by what another could provide. Families became display cases for the success of the good provider.

The psychic costs could be high:

> By depending so heavily on his breadwinning role to validate his sense of himself as a man, instead of also letting his roles as husband, father, and citizen of the community count as validating sources, the American male treads on psychically dangerous ground. It's always dangerous to put all of one's psychic eggs into one basket. (Brenton, 1966, p. 194)

The good-provider role not only put all of a man's gender-identifying eggs into one psychic basket, but it also put all the family-providing eggs into one basket. One individual became responsible for the support of the whole family. Countless stories portrayed the humiliation families underwent to keep wives and especially mothers out of the labor force, a circumstance that would admit to the world the male head's failure in the good-provider role. If a married woman had to enter the labor force at all, that was bad enough. If she made a good salary, however, she was ''co-opting the man's passport to masculinity'' (Gould, 1974, p. 98) and he was effectively castrated. A wife's earning capacity diminished a man's position as head of the household (Gould, 1974, p. 99).

Failure in the role of good provider, which employment of wives evidenced, could produce deep frustration. As Komarovsky (1940, p. 20) explains, this is ''because in his own estimation he is failing to fulfill what is the central duty of his life, the very touchstone of his manhood—the role of family provider.''

But just as there was punishment for failure in the good-provider role, so also were there rewards for successful performance. A man ''derived strength from his role as provider'' (Komarovsky, 1940, p. 205). He achieved a good deal of satisfaction from his ability to support his family. It won kudos. Being a good provider led to status in both the family and the community. Within the family it gave him the power of the purse and the right to decide about expenditures, standards of living, and what constituted good providing. ''Every purchase of the family—the radio, his wife's new hat, the children's skates, the meals set before him—were all symbols of their dependence upon him'' (Komarovsky, 1940, p. 74–75). Such dependence gave him a ''profound sense of stability'' (p. 74). It was a strong counterpoise vis-à-vis a wife with a stronger personality. ''Whether he had considerable authority within the family and was recognized as its head, or whether the wife's stronger personality . . . dominated the family, he nevertheless derived strength from his role as a provider'' (Komarovsky, 1940, p. 75). As recently as 1975, in a sample of 3,100 husbands and wives in 10 cities, Scanzoni found that despite increasing egalitarian norms, the good provider still had ''considerable power in ultimate decision-making'' and as ''unique provider'' had the right ''to organize his life and the lives of other family members around his occupation'' (p. 38).

A man who was successful in the good-provider role might be freed from other obligations to the family. But the flip side of this dispensation was that he could not make up for poor performance by excellence in other family roles. Since everything depended on his success as provider, everything was at stake. The good provider played an all-or-nothing game.

Different Ways of Performing the Good-Provider Role

Although the legal specifications for the role were laid out in the common law, in legislation, in legal precedents, in court decisions, and, most importantly, in custom and convention, in real-life situations the social and social-psychological specifications were set by the husband or, perhaps more accurately, by the community, alias the Joneses, and there were many ways to perform it.

Some men resented the burdens the role forced them to bear. A man could easily vent such resentment toward his family by keeping complete control over all expenditures, dispensing the money for household maintenance, and complaining about bills as though it were his wife's fault that shoes cost so much. He could, in effect, punish his family for his having to perform the role. Since the money he earned belonged to him—was ''his''—he could do with it what he pleased. Through extreme parsimony he could dole out his money in a mean, humiliating way, forcing his wife to come begging for pennies. By his reluctance and resentment he could make his family pay emotionally for the provisioning he supplied.

At the other extreme were the highly competitive men who were so involved in outdoing the Joneses that the fur coat became more important than the affectionate hug. They ''bought off'' their families. They sometimes succeeded so well in their extravagance that they sacrificed the family they were presumably providing for to the achievements that made it possible (Keniston, 1965).[2]

The Depression of the 1930s revealed in harsh detail what the loss of the role could mean both to the good provider and to his family, not only in the loss of income itself—which could be supplied by welfare agencies or even by other family members, including wives—but also and especially in the loss of face.

The Great Depression did not mark the demise of the good-provider role. But it did teach us what a slender thread the family hung on. It stimulated a whole array of programs designed to strengthen that thread, to ensure that it would never again be similarly threatened. Unemployment insurance was incorporated into the Social Security Act of 1935, for example, and a Full Employment Act was passed in 1946. But there proved to be many other ways in which the good-provider role could be subverted.

Role Rejectors and Role Overperformers

Recent research in psychology, anthropology, and sociology has familiarized us with the tremendous power of roles. But we also know that one of the fundamental principles of role behavior is that conformity to role norms is not universal. Not everyone lives up to the specifications of roles, either in the psychological or in the sociological definition of the concept. Two extremes have attracted research attention: (a) the men who could not live up to the norms of the good-provider role or did not want to, at one extreme, and (b) the men who overperformed the role, at the other. For the wide range in between, from blue-collar workers to professionals, there was fairly consistent acceptance of the role, however well or poorly, however grumblingly or willingly, performed.

First the nonconformists. Even in Colonial times, desertion and divorce occurred:

> Women may have deserted because, say, their husbands beat them; husbands, on the other hand, may have deserted because they were unable or unwilling to provide for their usually large families in the face of the wives' demands to do so. These demands were, of course, backed by community norms making the husband's financial support a sacred duty. (Scanzoni, 1979, pp. 24–25)

Fiedler (1962) has traced the theme of male escape from domestic responsibilities in the American novel from the time of Rip Van Winkle to the present:

> The figure of Rip Van Winkle presides over the birth of the American imagination; and it is fitting that our first successful home-grown legend should memorialize, however playfully, the flight of the dreamer from the shrew—into the mountains and out of time, away from the drab duties of home . . . anywhere to avoid . . . marriage and responsibility. One of the factors that determine theme and form in our great books is this strategy of evasion, this retreat to nature and childhood which makes our literature (and life) so charmingly and infuriatingly "boyish." (pp. xx–xxi)

Among the men who pulled up stakes and departed for the West or went down to the sea in ships, there must have been a certain proportion who, like their mythic prototype, were simply fleeing the good-provider role.

The work of Demos (1974), a historian, offers considerable support for Fiedler's thesis. He tells us that the burdens thrust on men in the 19th century by the new patterns of work began to show their effects in the family. When "the [spatial] separation of the work lives of husbands and wives made communication so problematic," he asks, "what was the likelihood of meaningful communication?" (Demos, 1974, p. 438). The answer is, relatively little. Divorce and separation increased, either formally or by tacit consent—or simply by default, as in the case of a variety of defaulters—tramps, bums, hoboes—among them.

In this connection, "the development of the notorious 'tramp' phenonemon is worth noticing," Demos (1974, p. 438) tells us. The tramp was a man who just gave up, who dropped out of the role entirely. He preferred not to work, but he would do small chores or other small-scale work for a handout if he had to. He was not above begging the housewife for a meal, hoping she would not find work for him to do in repayment. Demos (1974) describes the type:

> Demoralized and destitute wanderers, their numbers mounting into the hundreds of thousands, tramps can be fairly characterized as men who had run away from their wives. . . . Their presence was mute testimony to the strains that tagged at the very core of American family life. . . . Many observers noted that the tramps had created a virtual society of their own [a kind of counterculture] based on a principle of single-sex companionship. (p. 438)

A considerable number of them came to be described as "homeless men" and, as the country became more urbanized, landed ultimately on skid row. A large part of the task of social workers for almost a century was the care of the "evaded" women they left behind.[3] When the tramp became wholly demoralized, a chronic alcoholic, almost unreachable, he fell into a category of his own—a bum.

Quite a different kettle of fish was the hobo, the migratory worker who spent several months harvesting wheat and other large crops and the rest of the year in cities. Many were the so-called Wobblies, or Industrial Workers of the World, who repudiated the good-provider role on principle. They had contempt for the men who accepted it and could be called conscientious objectors to the role. "In some IWW circles, wives were regarded as the 'ball and chain.' In the West, IWW literature proclaimed that the migratory worker, usually a young, unmarried male, was 'the finest specimen of American manhood . . . the leaven of the revolutionary labor movement'" (Foner, 1979, p. 400). Exemplars of the Wobblies were the nomadic workers of the West. They were free men. The migratory worker, "unlike the factory slave of the Atlantic seaboard and the central states, . . . was most emphatically 'not afraid of losing his job.' No wife and family cumbered him. The worker of the East, oppressed by the fear of want for wife and babies, dared not venture much" (Foner, 1979, p. 400). The reference to fear of loss of job was well taken; employers preferred married men, disciplined into the good-provider role, who had given hostages to fortune and were therefore more tractable.

Just on the verge between the area of conformity to the good-provider role—at whatever level—and the area of complete nonconformity to it was the non-good provider, the marginal group of workers usually made up of "the under-educated, the under-trained, the under-employed, or part-time employed, as well as the under-paid, and of course the unemployed" (Snyder, 1979, p. 597). These included men who wanted—sometimes desperately—to perform the good-provider role but who for one reason or another were unable to do so. Liebow (1966) has discussed the ramifications of failure among the black men of Tally's corner: The black man is

> under legal and social constraints to provide for them [their families], to be a husband to his wife and a father to his children. The chances are, however, that he is failing to provide for them, and failure in this primary function contaminates his performance as father in other respects as well. (p. 86)

In some cases, leaving the family entirely was the best substitute a man could supply. The community was left to take over.[4]

At the other extreme was the overperformer. Tocqueville, quoted earlier, was already describing him as he manifested in the 1830s. And as late as 1955 Warner and Ablegglen were adding to the considerable literature on industrial leaders and tycoons, referring to their "driving concentration" on their careers and their "intense focusing" of interests, energies, and skills on these careers, "even limiting their sexual activity" (pp. 48–49). They came to be known as work-aholics or work-intoxicated men. Their preoccupation with their work even at the expense of their families was, as I have already noted, quite acceptable in our society.

Poorly or well performed, the good-provider role lingered on. World War II initiated a challenge, this time in the form of attracting more and more married women into the labor force, but the challenge was papered over in the 1950s with an "age of togetherness" that all but apotheosized the good provider, his house in the suburbs, his homebody wife, and his third, fourth, even fifth, child. As late as the 1960s most housewives (87%) still saw breadwinning as their husband's primary role (Lopata, 1971, p. 91).[5]

Intrinsic Conflict in the Good-Provider Role

Since the good-provider role involved both family and work roles, most people believed that there was no incompatibility between them or at least that there should not be. But in the 1960s and 1970s evidence began to mount that maybe something was amiss.

Tocqueville had documented the implicit conflict in the American business-man's devotion to his work at the expense of his family in the early years of the 19th century; the Industrial Workers of the World had proclaimed that the good-provider role which tied a man to his family was an impediment to the great revolution at the beginning of the 20th century; Fiedler (1962) had noted that throughout our history, in the male fantasy world, there was freedom from the responsibilities of this role; about 50 years ago Freud (1930/1958) had analyzed the intrinsic conflict between the demands of women and the family on one side and the demands of men's work on the other:

> Women represent the interests of the family and sexual life; the work of civilization has become more and more men's business; it confronts them with ever harder tasks, compels them to sublimations of instinct which women are not easily able to achieve. Since man has not an unlimited amount of mental energy at his disposal, he must accomplish his tasks by distributing his libido to the best advantage. What he employs for cultural [occupational] purposes he withdraws to a great extent from women, and his sexual life; his constant association with men and his dependence on his relations with them even estrange him from his duties as husband and father. Woman finds herself thus forced into the background by the claims of culture [work] and she adapts an inimical attitude towards it. (pp. 50–51)

In the last two decades, researchers have been raising questions relevant to Freud's statement of the problem. They have been asking people about the relative satisfactions they derive from these conflicting values—family and work. Among the earliest studies comparing family–work values was a Gallup poll in 1940 in which both men and women chose a happy home over an interesting job or wealth as a major life value. Since then there have been a number of such polls, and a considerable body of results has now accumulated. Pleck and Lang (1979) and Hesselbart (Note 1) have summarized the findings of these surveys. All agree that there is a clear bias in the direction of the family. Pleck and Lang conclude that "men's family role is far more psychologically significant to them than is their work role" (p. 29), and Hesselbart—however critical she is of the studies she summarizes—believes they should not be dismissed lightly and concludes that they certainly "challenge the idea that family is a 'secondary' valued role" (p. 14).[6] Douvan (Note 2) also found in a 1976 replication of a 1957 survey that family values retained priority over work: "Family roles almost uniformly rate higher in value production than the job role does" (p. 16).[7]

The very fact that researchers have asked such questions is itself interesting. Somehow or other both the researchers and the informants seem to be saying that all this complaining about the male neglect of the family, about the lack of family involvement by men, just is not warranted. Neither de Tocqueville nor Freud was right. Men do value family life more than they value their work. They do derive their major life satisfactions from their families rather than from their work.

It may well be true that men derive the greatest satisfaction from their family roles, but this does not necessarily mean they are willing to pay for this benefit. In any event, great attitudinal changes took place in the 1960s and 1970s.

Douvan (Note 2), on the basis of surveys in 1957 and 1976, found, for example, a considerable increase in the proportion of both men and women who found marriage and parenthood burdensome and restrictive. Almost three fifths (57%) of both married men and married women in 1976 saw marriage as "all burdens and restrictions," as compared with only 42% and 47%, respectively, in 1957.

And almost half (45%) also viewed children as "all burdens and restrictions" in 1976, as compared with only 28% and 33% for married men and married women, respectively, in 1957. The proportion of working men with a positive attitude toward marriage dropped drastically over this period, from 68% to 39%. Working women, who made up a fairly small number of all married women in 1957, hardly changed attitudes at all, dropping only from 43% to 42%. The proportion of working men who found marriage and children burdensome and restrictive more than doubled, from 25% to 56% and from 25% to 58%, respectively. Although some of these changes reflected greater willingness in 1976 than in 1957 to admit negative attitudes toward marriage and parenthood—itself significant— profound changes were clearly in process. More and more men and women were experiencing disaffection with family life.[8]

"All Burdens and Restrictions"

Apparently, the benefits of the good-provider role were greater than the costs for most men. Despite the legend of the flight of the American male (Fiedler, 1962), despite the defectors and dropouts, despite the tavern habitué's "ball and chain" cliché, men seemed to know that the good-provider role, if they could succeed in it, was good for them. But Douvan's (Note 2) findings suggest that recently their complaints have become serious, bone-deep. The family they have been providing for is not the same family it was in the past.

Smith (1979) calls the great trek of married women into the labor force a subtle revolution—revolutionary not in the sense of one class overthrowing a status quo and substituting its own regime, but revolutionary in its impact on both the family and the work roles of men and women. It diluted the prerogatives of the good-provider role. It increased the demands made on the good provider, especially in the form of more emotional investment in the family, more sharing of household responsibilities. The role became even more burdensome.

However men may now feel about the burdens and restrictions imposed on them by the good-provider role, most have, at least ostensibly, accepted them. The tramp and the bum had "voted with their feet" against the role; the hobo or Wobbly had rejected it on the basis of a revolutionary ideology that saw it as enslaving men to the corporation; tavern humor had glossed the resentment habitués felt against its demands. Now the "burdens-and-restrictions" motif has surfaced both in research reports and, more blatantly, in the male liberation movement. From time to time it has also appeared in the clinicians' notes.

Sometimes the resentment of the good provider takes the form of simply wanting more appreciation for the life-style he provides. All he does for his family seems to be taken for granted. Thus, for example, Goldberg (1976), a psychiatrist, recounts the case of a successful businessman:

> He's feeling a deepening sense of bitterness and frustration about his wife and family. He doesn't feel appreciated. It angers him the way they seem to take the things his earnings purchase for granted. They've come to expect it as their due. It

> particularly enrages him when his children put him down for his "materialistic middle-class trip." He'd like to tell them to get someone else to support them but he holds himself back. (p. 124)

Brenton (1966) quotes a social worker who describes an upper-middle-class woman: She has "gotten hold of a man who'll drive himself like mad to get money, and [is] denigrating him for being too interested in money, and not interested in music, or the arts, or in spending time with the children. But at the same time she's subtly driving him—and doesn't know it" (p. 226). What seems significant about such cases is not that men feel resentful about the lack of appreciation but that they are willing to justify their resentment. They are no longer willing to grin and bear it.

Sometimes there is even more than expressed resentment; there is an actual repudiation of the role. In the past, only a few men like the hobo or Wobbly were likely to give up. Today, Goldberg (1976) believes, more are ready to renounce the role, not on theoretical revolutionary grounds, however, but on purely selfish ones:

> Male growth will stem from openly avowed, unashamed, self-oriented motivations. . . . Guilt-oriented "should" behavior will be rejected because it is always at the price of a hidden buildup of resentment and frustration and alienation from others and is, therefore, counterproductive. (p. 184)

The disaffection of the good provider is directed to both sides of his role. With respect to work, Lefkowitz (1979) has described men among whom the good-provider role is neither being completely rejected nor repudiated, but diluted. These men began their working lives in the conventional style, hopeful and ambitious. They found a job, married, raised a family, and "achieved a measure of economic security and earned the respect of . . . colleagues and neighbors" (Lefkowitz, 1979, p. 31). In brief, they successfully performed the good-provider role. But unlike their historical predecessors, they in time became disillusioned with their jobs—not jobs on assembly lines, not jobs usually characterized as alienating, but fairly prestigious jobs such as aeronautics engineer and government economist. They daydreamed about other interests. "The common theme which surfaced again and again in their histories, was the need to find a new social connection—to reassert control over their lives, to gain some sense of freedom" (Lefkowitz, 1979, p. 31). These men felt "entitled to freedom and independence." Middle-class, educated, self-assured, articulate, and for the most part white, they knew they could talk themselves into a job if they had to. Most of them did not want to desert their families. Indeed, most of them "wanted to rejoin the intimate circle they felt they had neglected in their years of work" (p. 31).

Though some of the men Lefkowitz studied sought closer ties with their families, in the case of those studied by Sarason (1977), a psychologist, career changes involved lower income and had a negative impact on families. Sarason's subjects were also men in high-level professions, the very men least likely to find marriage and parenthood burdensome and restrictive. Still, since career change often involved a reduction in pay, some wives were unwilling to accept it, with the result that the marriage deteriorated (p. 178). Sometimes it looked like a no-win game. The husband's earlier career brought him feelings of emptiness and alienation, but it also brought financial rewards for the family. Greater work satisfaction for him in lower paying work meant reduced satisfaction with life-style. These findings lead Sarason to raise a number of points with respect to the good-provider role. "How much," he asks, "does an individual or a family need in order to maintain a satisfactory existence? Is an individual being responsible to himself or his family if he provides them with little more than the bare essentials of living?" (p. 178). These are questions about the good-provider role that few men raised in the past.

Lefkowitz (1979) wonders how his downwardly mobile men lived when they left their jobs. "They put together a basic economic package which consisted of government assistance, contributions from family members who had not worked before and some bartering of goods and services" (p. 31). Especially interesting in this list of income sources are the "contributions from family members who had not worked before" (p. 31). Surely not mothers and sisters. Who, of course, but wives?

Women and the Provider Role

The present discussion began with the woman's part in the provider role. We saw how as more and more of the provisioning of the family came to be by way of monetary exchange, the woman's part shrank. A woman could still provide services, but could furnish little in the way of food, clothing, and shelter. But now that she is entering the labor force in large numbers, she can once more resume her ancient role, this time, like her male counterpart the provider, by way of a monetary contribution. More and more women are doing just this.

The assault on the good-provider role in the Depression was traumatic. But a modified version began to appear in the 1970s as a single income became inadequate for more and more families. Husbands have remained the major providers, but in an increasing number of cases the wife has begun to share this role. Thus, the proportion of married women aged 15 to 54 (living with their husbands) in the labor force more than doubled between 1950 and 1978, from 25.2% to 55.4%. The proportion for 1990 is estimated to reach 66.7% (Smith, 1979, p. 14). Fewer women are now full-time housewives.

For some men the relief from the strain of sole responsibility for the provider role has been welcome. But for others the feeling of degradation resembles the feelings reported 40 years earlier in the Great Depression. It is not that they are no longer providing for the family but that the role-sharing wife now feels justified in making demands on them. The good-provider role with all its prerogatives and perquisites has undergone profound changes. It will never be the same again.[9] Its death knell was sounded when, as noted above, the 1980 census no longer automatically assumed that the male member of the household was its head.

The Current Scene

Among the new demands being made on the good-provider role, two deserve special consideration, namely, (a) more intimacy, expressivity, and nurturance—specifications never included in it as it originally took shape—and (b) more sharing of household responsibility and child care.

As the pampered wife in an affluent household came often to be an economic parasite, so also the good provider was often, in a way, a kind of emotional parasite. Implicit in the definition of the role was that he provided goods and material things. Tender loving care was not one of the requirements. Emotional ministrations from the family were his right; providing them was not a corresponding obligation. Therefore, as de Tocqueville had already noted by 1840, women suffered a kind of emotional deprivation labeled by Robert Weiss ''relational deficit'' (cited in Bernard, 1976). Only recently has this male rejection of emotional expression come to be challenged. Today, even blue-collar women are imposing ''a host of new role expectations upon their husbands or lovers. . . . A new role set asks the blue-collar male to strive for . . . deep-coursing intimacy'' (Shostak, Note 4, p. 75). It was not only vis-à-vis his family that the good provider was lacking in expressivity. This lack was built into the whole male role script. Today not only women but also men are beginning to protest the repudiation of expressivity prescribed in male roles (David & Brannon, 1976; Farrell, 1974; Fasteau, 1974; Pleck & Sawyer, 1974).

Is there any relationship between the ''imposing'' on men of ''deep-coursing intimacy'' by women on one side and the increasing proportion of men who find marriage burdensome and restrictive on the other? Are men seeing the new emotional involvements being asked of them as ''all burdens and restrictions''? Are they responding to the new involvements under duress? Are they feeling oppressed by them? Fearful of them?

From the standpoint of high-level pure-science research there may be something bizarre, if not even slightly absurd, in the growing corpus of serious research on how much or how little husbands of employed wives contribute to household chores and child care. Yet it is serious enough that all over the industrialized world such research is going on. Time studies in a dozen countries—communist as well as capitalist—trace the slow and bungling process by which marriage accommodates to changing conditions and by which women struggle to mold the changing conditions in their behalf. For everywhere the same picture shows up in the research: an image of women sharing the provider role and at the same time retaining responsibility for the household. Until recently such a topic would have been judged unworthy of serious attention. It was a subject that might be worth a good laugh, for instance, as when an all-thumbs man in a cartoon burns the potatoes or finds himself bumbling awkwardly over a diaper, demonstrating his—proud—male ineptness at such female work. But it is no longer funny.

The ''politics of housework'' (Mainardi, 1970) proves to be more profound than originally believed. It has to do not only with tasks but also with gender—

and perhaps more with the site of the tasks than with their intrinsic nature. A man can cook magnificently if he does it on a hunting or fishing trip; he can wield a skillful needle if he does it mending a tent or a fishing net; he can even feed and clean a toddler on a camping trip. Few of the skills of the homemaker are beyond his reach so long as they are practiced in a suitably male environment. It is not only women's work in and of itself that is degrading but any work on the female turf. It may be true, as Brenton (1966) says, that "the secure man can wash a dish, diaper a baby, and throw the dirty clothes into the washing machine—or do anything else women used to do exclusively—without thinking twice about it" (p. 211), but not all men are that secure. To a great many men such chores are demasculinizing. The apron is shameful on a man in the kitchen; it is all right at the carpenter's bench.

The male world may look upon the man who shares household responsibilities as, in effect, a scab. One informant tells the interviewer about a conversation on the job: "What, are you crazy?" his hard-hat fellow workers ask him when he speaks of helping his wife. "The guys want to kill me. 'You son of a bitch! You are getting us in trouble.' . . . The men get really mad" (Lein, 1979, p. 492). Something more than persiflage is involved here. We are fairly familiar with the trauma associated with the invasion by women of the male work turf, the hazing women can be subjected to, and the male resentment of admitting them except into their own segregated areas. The corresponding entrance of men into the traditional turf of women—the kitchen or the nursery—has analogous but not identical concomitants.

Pleck and Lang (1979) tell us that men are now beginning to change in the direction of greater involvement in family life. "Men's family behavior is beginning to change, becoming increasingly congruent with the long-standing psychological significance of the family in their lives" (p. 1). They measure this greater involvement by way of the help they offer with homemaking chores. Scanzoni (1975), on the basis of a survey of over 3,000 husbands and wives, concludes that at least in households in which wives are in the labor force, there is the "possibility of a different pattern in which responsibility for households would unequivocally fall equally on husbands as well as wives" (p. 38). A brave new world indeed. Still, when we look at the reality around us, the pace seems intolerably slow. The responsibilities of the old good-provider role have attenuated far faster than have its prerogatives and privileges.

A considerable amount of thought has been devoted to studying the effects of the large influx of women into the work force. An equally interesting question is what the effect will be if a large number of men actually do increase their participation in the family and the household. Will men find the apron shameful? What if we were to ask fathers to alternate with mothers in being in the home when youngsters come home from school? Would fighting adolescent drug abuse be more successful if fathers and mothers were equally engaged in it? If the school could confer with fathers as often as with mothers? If the father accompanied children when they went shopping for clothes? If fathers spent as much time with children as do mothers?

Even as husbands, let alone as fathers, the new pattern is not without trauma. Hall and Hall (1979), in their study of two-career couples, report that the most serious fights among such couples occur not in the bedroom, but in the kitchen, between couples who profess a commitment to equality but who find actually implementing it difficult. A young professional reports that he is philosophically committed to egalitarianism in marriage and tries hard to practice it, but it does not work. He even feels guilty about this. The stresses involved in reworking roles may have an impact on health. A study of engineers and accountants finds poorer health among those with employed wives than among those with non-employed wives (Burke & Wier, 1976). The processes involved in role change have been compared with those involved in deprogramming a cult member. Are they part of the increasing sense of marriage and parenthood as ''all burdens and restrictions''?

The demise of the good-provider role also calls for consideration of other questions: What does the demotion of the good provider to the status of senior provider or even mere coprovider do to him? To marriage? To gender identity? What does expanding the role of housewife to that of junior provider or even coprovider do to her? To marriage? To gender identity? Much will of course depend on the social and psychological ambience in which changes take place.

A Parable

I began this essay with a proverbial woman. I close it with a modern parable by William H. Chafe (note 5), a historian who also keeps his eye on the current scene. Jack and Jill, both planning professional careers, he as doctor, she as lawyer, marry at age 24. She works to put him through medical school in the expectation that he will then finance her through law school. A child is born during the husband's internship, as planned. But in order for him to support her through professional training as planned, he will have to take time out from his career. After two years, they decide that both will continue their training on a part-time basis, sharing household responsibilities and using day-care services. Both find part-time positions and work out flexible schedules that leave both of them time for child care and companionship with one another. They live happily ever after.

That's the end? you ask incredulously. Well, not exactly. For, as Chafe (note 5) points out, as usual the personal is also political:

> Obviously such a scenario presumes a radical transformation of the personal values that today's young people bring to their relationships as well as a readiness on the part of social and economic institutions to encourage, or at least make possible, the development of equality between men and women. (p. 28)

The good-provider role may be on its way out, but its legitimate successor has not yet appeared on the scene.

NOTES

1. Rainwater and Yancey (1967), critiquing current welfare policies, note that they "have robbed men of their manhood, women of their husbands, and children of their fathers. To create a stable monogamous family we need to provide men with the opportunity to be men, and that involves enabling them to perform occupationally" (p. 235).

2. Several years ago I presented a critique of what I called "extreme sex role specialization," including "work-intoxicated fathers." I noted that making success in the provider role the only test for real manliness was putting a lot of eggs into one basket. At both the blue-collar and the managerial levels, it was dysfunctional for families. I referred to the several attempts being made even then to correct the excesses of extreme sex role specialization: rural and urban communes, leaving jobs to take up small-scale enterprises that allowed more contact with families, and a rebellion against overtime in industry (Bernard, 1975, pp. 217–239).

3. In one department of a South Carolina cotton mill early in the century, "every worker was a grass widow" (Smuts, 1959, p. 54). Many women worked "because their husbands refused to provide for their families. There is no reason to think that husbands abandoned their duties more often than today, but the woman who was burdened by an irresponsible husband in 1890 usually had no recourse save taking on his responsibilities herself. If he deserted, the law-enforcement agencies of the time afforded little chance of finding and compelling him to provide support" (Smuts, 1959, p. 54). The situation is not greatly improved today. In divorce child support is allotted in only a small number of cases and enforced in even fewer. "Roughly half of all families with an absent parent don't have awards at all. . . . Where awards do exist they are usually for small amounts, typically ranging from $7 to $18 per week per child" (Jones, 1976, abstract). A summary of all the studies available concludes that "approximately 20 percent of all divorced and separated mothers receive child support regularly, with an additional 7 percent receiving it 'sometimes'; 8 percent of all divorced and separated women receive alimony regularly or sometimes" (Jones, 1976, p. 23).

4. Even though the annals of social work agencies are filled with cases of runaway husbands, in 1976 only 12.6% of all women were in the status of divorce and separation, and at least some of them were still being "provided for." Most men were at least trying to fulfill the good-provider role.

5. Although all the women in Lopata's (1971) sample saw breadwinning as important, fewer employed women (54%) than either nonemployed urban (63%) or suburban (64%) women assigned it first place (p. 91).

6. Pleck and Lang (1979) found only one serious study contradicting their own conclusions: "Using data from the 1973 NORC [National Opinion Research Center] General Social Survey, Harry analyzed the bivariate relationship of job and family satisfaction to life happiness in men classified by family life cycle stage. In three of the five groups of husbands . . . job satisfaction had a stronger association than family satisfaction to life happiness" (pp. 5–6).

7. In 1978, a Yankelovich survey on "The New Work Psychology" suggested that leisure is now becoming a strict competitor for both family and work as a source of life satisfactions: "Family and work have grown less important than leisure; a majority of 60 percent say that although they enjoy their work, it is not their major source of satisfaction" (p. 46). A 1977 survey of Swedish men aged 18 to 35 found that the proportion saying the family was the main source of meaning in their lives declined from 45% in 1955 to 41% in 1977; the proportion indicating work as the main source of satisfaction dropped from 33% to 17%. The earlier tendency for men to identify themselves through their work is less

marked these days. In the new value system, the individual says, in effect, ''I am more than my role. I am myself'' (Yankelovich, 1978). Is the increasing concern with leisure a way to escape the dissatisfaction with both the alienating relations found on the work site and the demands for increased involvement with the family?

8. Men seem to be having problems with both work and family roles. Veroff (Note 3), for example, reports an increased ''sense of dissatisfaction with the social relations in the work setting'' and a ''dissatisfaction with the affiliative nature of work'' (p. 47). This dissatisfaction may be one of the factors that leads men to seek affiliative-need satisfaction in marriage, just as in the 19th century they looked to the home as shelter from the jungle of the outside world.

9. Among the indices of the waning of the good-provider role are the increasing number of married women in the labor force; the growth in the number of female-headed families; the growing trend toward egalitarian norms in marriage; the need for two earners in so many middle-class families; and the recognition of these trends in the abandonment of the identification of head of household as a male.

29.

Men Who Make Women Want to Scream

Connell Cowan and Melvyn Kinder

The state of men's lives is troubling and problematic. We seem confused, directionless, and stuck in patriarchal patterns. It has become fashionable in the current situation to engage in male bashing and blame men for all the ills of the world. This is neither therapeutic for men nor helpful to women who may be seeking a suitable life-partner.

The Cowan and Kinder essay offers a suggestive typology that acts as a critical examination of four types of men. Each of the men is initially charming, attractive, and intriguing—but all end up infuriating women: The Clam radiates a tough outer mystique; the Pseudo-Liberated male is disarmingly expressive; the Perpetual Adolescent shuns adult responsibilities: and the Walking Wounded wallows in self-pity and vulnerability.

Exploration of the various types should make for interesting, if not provocative, classroom discussion.

Questions to assist in processing the essay follow:

1. *What is initially attractive about men who display a tough, macho mystique? What is the dark side of this type of masculinity?*
2. *Emotional expressivity can be a form of narcissism. Why is this so? How can it be a false form of liberation for men?*
3. *Why do so many men resist adult responsibilities and commitments today?*
4. *Men carrying a mixture of hurt, bitterness, and rejection can make foolish choices. Why? Is a healing and healthy relationship possible for them?*

There are several types of men who very predictably end up infuriating women. Some are charming in the beginning and then change. Others are attractive because of the qualities women hope to find in them. All, sooner or later, make women want to scream in frustration.

The Clam

Some men radiate a tough mystique that grows out of a basically selfish, withholding, and guarded nature. This kind of man can be as dangerous as he is attractive and intriguing. A woman can be drawn to what she sees as strength in this man's insensitive toughness and may also feel potentially reassured by that "strength." We say "potentially," because she never quite feels part of such a man's strength, since the man doesn't really share or even truly open himself to the woman. He makes the woman do the emotional work for the two of them. He sets the stage and she dances around, attempting to read his mind. She knows she wants the security of feeling close to his strength. But he doesn't ever allow her to get too close. She loves it, she hates it. She knows she is drawn by the very characteristic she is bound and determined to change.

Arlene, 28, is a warm, gregarious bank loan officer. When she met Tom she knew this relationship was "it" for her. She described him as "a bit too emotionally guarded" for her tastes, but she thought all that would change once he realized he could trust her. She thought that she understood Tom's secretive tendencies, which she saw as reflecting his self-control or perhaps shielding an old hurt. He wasn't the least sensitive to her needs, but she talked herself into believing it was only because she hadn't communicated them to him clearly enough, and so it must be her fault.

They married eight months after they met. Arlene felt sure the kind of commitment they were making would open the door to at last feeling loved by Tom. She was absolutely convinced that if she loved him enough, with no holds barred, he would open himself to her. With love as the key, she would open his heart and finally reap the treasure that surely lay within. Tom's tough, controlled outer shell concealed a tough, controlled inner core. Tom claimed he loved Arlene, but she never felt it and he never showed the demonstrative affection she wanted and needed. She divorced him after one painful year.

Arlene made a mistake in the choice of her relationship with Tom. She interpreted his guarded, withholding nature as mystique. What she found was that instead of standing guard over some hidden treasure, he in fact was desperately trying to protect his insecurity from exposure. When Arlene realized this, Tom's strength was transformed in her eyes to brittle crumbling defenses. His wonderful mystique turned to fear.

The Clam either fears his dependency needs or has managed to convince himself that he doesn't have any. He is very attractive to many women, who mistake this trait for strength and self-containment. But problems soon emerge as the woman begins to want more. We all experience love, at least in part, through feeling needed by our partner, "needed" emotionally. The Clam can't allow himself to need anyone enough to form an intimate, satisfying bond. To do so would require confronting his fears of weakness and vulnerability. Ancient, scarred-over hurts may have destroyed his capacity to feel that deeply.

We all need, in a love relationship, to have our partners dependent upon us—not bloodsuckingly so, but needing us emotionally nevertheless. And this man will never allow himself to be dependent enough to be able to form a close, sharing relationship. He functions as a self-contained system. No matter how warm a woman's love, it will never melt his protective shield. It is too tough, too old.

Another necessary bonding agent in the man/woman relationship is trust. Trusting and being trusted. The Clam is recognized by his secretive qualities. The secretive person is protecting something he fears may be lost, betrayed, taken away. Women need to keep in mind when they meet a secretive man that his concealment is a result of his past and has nothing to do with them.

Trust develops through a process of give and take. It involves mutually disclosing deeper and more complex aspects of ourselves. The Clam cannot take a chance on important emotional exposure. He will not risk the danger of looking into old wounds stored away in the locked file of forgotten, painful memories. Most often, he doesn't even know just what it is he is protecting or even that he is, in fact, behaving in a self-protective and distrustful fashion. The Clam cannot trust and he does not open up.

He doesn't know how to love, for the process of giving and loving means exposing his needs and vulnerabilities. If he hasn't learned to love by the time he is an adult, a woman won't be able to teach him—no matter how patient she is. It's foolish to believe otherwise.

What is misguided in the pursuit of this man is the failure to correctly identify his real strengths and weaknesses. If you find yourself with this type, you may believe you possess the magic potion to change him, to release in him what you believe to be a capacity to love, but you don't. In fact, the more a woman loves and cares for this kind of man, the better the chances of driving him away. Intimacy is his enemy—it scares the hell out of him. If he doesn't run away first, you will become so frustrated with having to do all the emotional work, provide all the tenderness, that eventually you will end the relationship—if you're smart.

The Pseudo-Liberated Male

At the outset of a relationship, the Pseudo-Liberated Male can be disarmingly attractive to women. He is the living embodiment of the liberated man, the perfect and natural complement to today's woman. He accepts her changes, even encourages them. He seems gentle and sensitive, vulnerable, expressive, revealing—a real dream come true! But it's a dream that frequently turns into a nightmare.

This type of man interpreted the women's movement as an invitation to become more expressive emotionally. He distorts this new "freedom" as a license to whine and a rationalization to express endless fears and personal insecurities, often to the point of utter distraction. The Pseudo-Liberated Male is certainly quite different from the withholding man described earlier. Many women see him as a welcome change—someone who will share himself, be open with his feelings. That's great, but some of these men go overboard. Even when women begin to get a whiff of his excesses, they frequently don't trust their own instincts—they don't run.

In a way, women have been encouraged and made to feel as if they should like this man. After all, if they expect to be able to explore new and unfamiliar "masculine" parts of themselves, and if they expect men to accept and love them for it, then they, in turn, should be tolerant of men's becoming more expressive and vulnerable.

When Marv and Marlena came in for couples therapy, Marv said they were not having any specific problems living together—they wanted rather to make their relationship as dynamic and positive as possible, and they were both interested in the therapy process as a means of personal growth.

Marv, 32, is a free-lance carpenter and unpublished novelist. Marlena, 34, is an office administrator for an import-export company and the steady wage earner in their household. They're both active in antinuclear and liberal political causes, Marv more so than Marlena because he doesn't work steadily and has more time.

Marv and Marlena are both bright, attractive, and personable. But what became clear in the very first session, as Marv talked on and on, with occasional glances at Marlena for approval, was that Marv is a narcissistic Pseudo-Liberated Male. He wasn't interested in making his relationship with Marlena better. What he wanted was a fresh, larger audience for his seemingly inexhaustible insights about himself.

Marlena revealed that Marv preferred talking about their relationship and himself to just about any other activity. Marlena eventually confessed that she felt exhausted by the constant talk and by his incessant demands for attention and analysis of "where we're at with each other now."

This man hides the fact that he is an emotional drain, that he's a taker. He is so happy and relieved to have a chance to legitimize his insecurity and neediness that he doesn't realize that he is taking without giving. He sincerely believes that his emotional diarrhea is a gift. He hides his fears and passivity beneath a deceptive costume of gentleness and sensitivity—and hopes the woman won't see through his disguises.

During the early stages of the relationship, this man performs dazzlingly. He is a master with words—he may even be poetic. His verbal output is such that a woman thinks she should feel nourished. Instead, she feels drained. He wraps his need for reassurance in a pretty package, one that can make a woman feel privileged, needed. Eventually, she may become aware that all he ever seems interested in talking about is the relationship—or himself! She wants to like him. She thinks she should like him. After all, he is expressive, isn't he? "In touch" with his feelings? Why does he make her want to scream? Perhaps it's because she finally realizes that he would rather talk about a relationship than have one.

These men are sensitive, and that can be a refreshing experience. The problem is that as time goes by, it becomes increasingly apparent that their sensitivity is one-sided, directed consistently toward themselves.

We believe women do want to know how a man feels, but they don't want to hear about it all the time. A relationship with one of these overly emotional types can eventually make them feel crazy. Somewhere along the way, these women may sense they are drawing a curtain of insensitivity about themselves, much as they have accused men of doing in the past. They want to shout, "Will you just shut up and make love to me and stop this endless discussion about us?" "Where we're at" with this man is all too frequently talking about his feelings toward you, toward himself, and toward the relationship—"talking about" rather than letting it just happen.

The Clam is too contained, while the overly sensitive Pseudo-Liberated Male is too uncontained. He wears his insecurities like medals on his chest.

Trying to free this man from his emotional problem can make a woman feel powerful, but it's a trap. They are better left alone. You might even be doing them a favor, for then they would be forced to deal with their insecurities themselves, from the inside out, rather than attempting to foist the responsibility on some woman who will indulge them.

Some men who make women want to scream are fundamentally unredeemable. The smart woman passes on these men, regardless of how interesting or intriguing they may appear on the surface. The Clam and the Pseudo-Liberated Male are such men. Then there are two other types of men who are terribly frustrating to women, but who do have very redeemable features if a woman can tolerate the frustration and make her way through the obstacles they place in her way: the Perpetual Adolescent and the Walking Wounded.

The Perpetual Adolescent

The Perpetual Adolescent stopped developing in what is late adolescence for a man—around the mid-twenties. This man's unspoken and unconscious credo is "I'm going to be 25 forever." This stunted growth is not always easy to detect. It is reflected in his emotional construction and in his diminished capacity to participate fully in relationships rather than in the external surface features of his life.

Outwardly, he has many disarmingly attractive qualities. He may be boyish in a confident, brash way. This man often works with the public and is articulate, with an easy, charming manner. He makes people feel comfortable.

Greg, a handsome, athletic yacht broker, lives in an expensive condominium overlooking a marina. From his sundeck, he can see his sailboat bobbing in its slip as well as the pool and tennis courts crowded with tanned single men and women. At 36, Greg still considers himself young and needing to devote most of his time and energy to building his career. He feels no pressure to marry. In fact, he tells himself, as well as more than an occasional woman, that he needs more time before settling down—time for his work, time for travel, time to "have fun."

Greg describes his life-style as "fun." He jogs and works out daily. He looks youthful, tanned and toned. He dresses fashionably. He tells himself there's no hurry, plenty of time to find his "ideal woman."

Actually, these are excuses for Greg to live in a perpetual adolescence. He talks about responsibility and commitment but runs when a woman starts to demand it. He can be affectionate to a woman and mean it, but he is not willing to grow up and relate to her as an adult. When his relationships get to the stage where it is natural for them to move to a deeper level, Greg becomes frightened and pulls away. He typically dismisses the woman as "dependent, clingy, possessive, demanding," rather than facing his own fear and reluctance to enter adulthood. He is blind to his profound reluctance to mature, for his youthful posture serves as a shield and defense against intimacy.

The Perpetual Adolescent's greatest fear is entrapment, for he doesn't fully trust his own autonomy. "To have to give" and "to be able to receive" both detonate deep, underlying fears of dependency in him. This man hides his fears of intimacy from himself by coming very close to committing in a relationship. But he ultimately wards off those fears by always making sure that "very close" is only that, not marriage.

The Perpetual Adolescent has rather shallow views and interactions with women. For that matter, his friendships with men are equally shallow. He often perceives himself as an adventurer. But the greatest adventure of all—marriage—is an event he is never quite yet ready for.

Initially, he can be captivating, for he has fine-tuned many aspects of his external presence. He can trot out all the phrases that make him sound wonderful and make a woman feel wonderful. The trouble is, he's a deal opener, not a deal closer. The Perpetual Adolescent is extremely frustrating to women, for as they naturally want to deepen what seems like a nicely developing relationship, he slowly pulls away. If he only did something truly rotten, she could free herself and be glad to be rid of him. But he doesn't—maddeningly, his only real flaw is his unwillingness to grow up.

We have said that this man is redeemable, and he is. Given enough time and patience, most men eventually do grow up, marry, and have families. For this type of man, the critical age seems to be about 39. He begins to panic when he is unable to deny being middle-aged. Having learned to trust his own independence more solidly, he is less afraid of entrapment and connection. He has become acutely aware of his own mortality, and he doesn't want to become a lonely old man.

While we wouldn't recommend the younger version of this man, the older model isn't bad at all. Should you know someone like this and want to deepen the relationship, there are a couple of important factors to keep in mind.

This man, even though he fears it, is capable of becoming healthily dependent on a woman. The mistake most women make is in not understanding that he does need a woman and can make a connection. Typically, the woman pushes too quickly and succeeds only in pushing him away. It is not that the impulse to move forward is inappropriate on the woman's part, for it isn't, but the timing is critical. This man is most likely to connect deeply to the woman who has patience to let him develop a strong need for her first. Then, and only then, should she begin to make her healthy demands for commitment. By then, he is so involved that he wants to stay.

The Walking Wounded

After a separation or divorce, both men and women naturally feel a mixture of hurt, bitterness, and rejection. Fortunately for most of us, these wounds heal over time, and the best medicine is eventually to love again.

Men and women usually suffer equally, but there are wounds unique to men that merit understanding. The Walking Wounded man can drive women crazy for a time, but he does heal and definitely is redeemable. In fact, these men often make fine mates precisely because they are committed to long-term relationships.

There are two basic types of wounds. The most painful is, of course, the loss of one's mate and most likely the loss of family. The other is the loss of financial security resulting from the divorce. The loss of a family structure is devastating to most divorced men. Suddenly, he finds himself alone in an apartment or hotel room, feeling lost, disoriented, and forlorn. He envies his wife, who frequently continues living in the family home, in familiar and, at least in his mind, secure surroundings. For the first time, some men will sadly and poignantly realize how important it was to hear ''Daddy'' when they came home from work.

Contributing to this sense of isolation that divorced men experience is the constant apprehension that even in his grief, he must continue working hard to make money. There is a line from a western song that goes, ''I can't halve my half again.'' For many men, a divorce means money: the destruction of the financial security and comfortable life-style which they worked so long and hard to create. Women suffer equally from the financial fallout from divorce, but it is our purpose here to acquaint you with the male point of view.

Most men feel ''ripped off'' after a divorce. Regardless of the validity of this attitude, they are nevertheless embittered by the helplessness they felt during the process of marital dissolution. This helplessness is often in combination with the sense of futility they have regarding child custody. In addition, they have increased financial anxieties related to the demand of separate living expenses. In their anxious and dark moments, they're not sure they can make it.

Even though they may be freer to date than are their wives, they have a sense that it's all a dream. They tend to drink and to abuse drugs, which compounds their depression.

How do these men appear to the women who encounter them? If they are newly separated, they can actually appear quite attractive, because they haven't yet assumed the guarded mantel of men who have been single for a while. They may be vulnerable too, which can be appealing to women, especially those who like to nurture men.

The newly separated man is open, eager to talk and to reveal himself, though too often this evolves into a tedious self-pity which will eventually drive a woman crazy. Even so, his eagerness for contact and relationships is quite appealing to many women.

The recently separated man tends to talk about his ex-wife and bitch about any number of injustices he feels. This facet of him can become so boring that women quickly feel the urge to run. A word of advice: After a while, don't be such a good listener. It's bad for him to wallow in self-pity and definitely not romantic for any woman.

There is a common problem with the Walking Wounded that can break a woman's heart. A woman may be age 32 to 40 and childless and find herself involved with a divorced man of the same age or older who already has children. It is vitally important for that woman, if she wants children of her own, to make this desire known to the man early in the relationship. Many divorced men are well-meaning but frankly have no desire to start another family. Yet they will mutter vaguely, "Well, if it's really going well, I guess I might want to have another kid." That's not good enough. A woman needs a clear answer or else it's time to move on. To invest precious years in a relationship only to end up with a man who has very different dreams is tragic indeed.

Men who have been separated for a year or more are usually less appealing than the very vulnerable, freshly separated ones we've been exploring. But they often have another kind of attractiveness: They're ripe for the picking. This is true in spite of their seemingly hardened outer shell. Though wary and a bit suspicious about being hurt again, they will become involved. They can make good mates, and do wish genuine intimacy, but they are scared. The solution is simple: Don't push for commitment in the beginning, even in the first six months. Women who need reassurances right away will not do well with this type of man. He does need extra time, but not forever. After a period of exclusive involvement, it is appropriate for the relationship to deepen and become more involved. He will commit himself if the woman really means it. But in some cases it may take an ultimatum. The woman who acts as if she will wait forever is making a real mistake, because she will be taken for granted.

One final word on the Walking Wounded. There are women who advise friends and say to themselves, "Stay away from any man who has just come out of a relationship. They just want a nursemaid. As soon as they heal a little, they'll leave you to play the field." It is true that they may be overly dependent at first or need to date around a bit, but some of the best men are not out there very long. Men who have been in a marriage, even a bad marriage, want to be in a relationship again. The best men are not single for long, and shouldn't be dismissed foolishly.

30.

The New Man and Male Identity

Kieran Scott

Men are awakening today to a crisis of their gender identity. If patriarchy is spent and if we have seen enough of Rambo and the Marlboro man, where does that leave us? What does it mean to be a man? What is required of mature men? What is distinctive in our humanness? These questions are percolating today as an increasing number of restless men face ontological and ultimate questions in their lives.

Men are facing a profound vocational crisis, and many feel as if they are involved in a night battle in a jungle against an unseen foe. Exactly what we are supposed to become is not clear. At the root of the problem is a defective mythology of manhood, a kind of male mystique. This mystique, propagated during the industrial age, fostered an image of men as autonomous, rational, efficient, intensely self-interested, and disconnected from the earth. Men are slowly realizing that they cannot solve their problems within this current understanding of masculinity. It has led to a deep malaise in men's lives and a loss of their generative potential.

This essay points to the deep substance and structure of a reshaped male identity. It is a guide toward a recovery of the core of masculinity in the depths of the psyche and body. The attempt is to reappropriate an authentic male identity—a lived life of courage, responsibility, energy, and compassion.

Some questions for reflection on the essay follow:

1. *What are the causes of the current dislocation of mature masculinity?*
2. *How do you account for the near-universal demand made upon males to be "real men"?*
3. *What criteria or ideal measures manhood in the U.S. today? What ideals/ images of mature masculinity would you propose?*
4. *Is the contemporary men's movement reactionary (anti-feminist) or reformist (pro-maleness)?*
5. *What impact, if any, do you see the new men's movement having on marriage?*

Revised version of a paper presented at the annual convention of the American Professors and Researchers in Religious Education, New York, November 3–5, 1989.

Who is the American male? And what do men want? The simplest and truest answer is: we don't really know! This essay points to the current male predicament, links it to the misguided quest of finding male identity in the feminine, and redirects the search to the inner depths of men's lives.

The essay explores: 1) the historical evolution in male identity over the last quarter century, 2) the meaning of "the new man," with his gifts and liabilities, and 3) the distinctive character of male identity and the search for a new basis of its support.

Our normative cultural model of what it means "to be a man" has changed dramatically in the past thirty years. A brief historical sketch (while not inclusive nor exact) can disclose the sense of confusion there has been over what "a man is" in the United States in the latter part of the twentieth century.

The '50s Male

Before the rise of feminist consciousness, the image of the male in our society had massive inadequacies. During the '50s it was a fairly clear vision—aggressive, hardworking, emotionally unexpressive, athletic, and patriotic. He got to work early, labored responsibly, supported his wife and children, and admired discipline. He defined his masculinity, if not his identity, in these terms. This was not so much a matter of choice as of convention. Negligence in fulfilling these roles was considered a failure—a masculine failure. Ronald Reagan (or John Wayne) is a sort of mummified version of this dogged male: paternalistically patriarchal, boyishly patriotic, and aggressively antiplanet.

The '50s male didn't see women's souls well, but he looked at their bodies. He tended to treat them as objects. Men of this generation were encouraged to equate effeminacy with un-Americanism and to use their leisure to escape—into sports, hunting, or simply the basement—from women and all things feminine. The male persona was strong and positive, but underneath the charm and bluff remained much isolation, deprivation, and passivity. The vision was unbalanced. It lacked compassion. This man needed an enemy to feel alive—and he had it in communism. Vietnam was very much a '50s man war—it was a logical manifestation of errant masculinity. (Today this same man seems to have a sense of nostalgia for the Cold War.) Helen Mayer Hacker, in her classic 1957 essay, captures the '50s type succinctly: "The ideal American male personality has been described . . . as a 'red-blooded, gentlemanly, go-getter' and any confessions of doubt, uncertainties, or insecurities would tarnish this image, any sign of weakness might be taken for effeminacy. Perhaps this is the greatest burden of masculinity our culture imposes."

The '50s male had a clear vision of what a man is and what men were supposed to do, just as Oliver North had, but the vision involved massive inadequacies and flaws. It was terribly one-sided and disastrous for men, women, children, and the planet.

The '60s Male

The cultural upheaval of the '60s sparked a psychic upheaval in many men. The monolithic male model—bully, bluffer, breadwinner—began to crack and come apart. Another sort of man appeared. The Vietnam War made men question whether they knew what an adult male really was and the war helped discredit masculinity in its more lethal expressions. The "bell had tolled" on the '50s male. We now began to consider that the enemy was *within*.

The women's movement encouraged men to actually look at women—and made them sensitive to concerns and sufferings that the '50s male tended to avoid. As men began to look at women and women's sensibilities, some men began to notice their own feminine side and pay attention and be receptive to it. The process continues to this day. Most young men—at various levels of consciousness and to different degrees—are involved in it. This is a positive and significant development in our time.

The '70s–'80s Male

In the last twenty years we have seen the emergence of what is popularly known as "the new man." He is more thoughtful, more gentle, more receptive. He has journeyed from machismo to mutuality (Ruether and Bianchi). Mark Gerzon's *A Choice of Heroes* aptly captures this shift in his advocacy of five images of masculinity as an alternative to the destructive patterning of patriarchy. The polar images are outlined in the accompanying chart:

A Choice of Heroes—Mark Gerzon

The *Frontiersman* is the quick-fisted white male loner conquering the frontier—Daniel Boone, Kit Carson. His enemies were untamed nature, the outlaw, the savage redskin. Conquest, especially of the wilderness, was the key. The land was a virgin "she" and every real man wanted a piece.	In place of the frontiersman, Gerzon suggests a new image, the *Healer*. This is a man with a different view of himself and the land. He is aware of the need for healing the environment.
The *Soldier* is the defender image. Strong and courageous, he armors his body and emotions. He represses his feelings of vulnerability, his fears, and his sensitivities. He must be ready for violence. His sexuality, and his penis, become weapons of aggression.	An alternative model for the new man is *Mediator* . . . one who stands in the middle. Heroism here is not fighting but rather coexisting peacefully and cooperatively. Conflict resolution rather than battle is his directing imagery and energy.

The *Breadwinner* is head of the family and responsible for its economic support. Here is the patriarchal family, with a public man and a private woman, an absent father and a nurturing mother. His ethic is work and success. His manhood is established by the size of his paycheck.

An alternative image is that of the *Companion*. Companion, the word, is a composite of com (together) and panes (bread). The term suggests a shared life. The daily bread is not won by men and given to others, but rather made and eaten together. Mutuality is the hallmark.

The *Expert* is another traditional image. This man possesses knowledge and hence is in control. Knowledge here is power and a means of control. It is utilized, not to seek truth, but to serve his human interests of maintaining his position of authority.

A better image is the *Colleague*. The word literally means "to choose together." The Colleague respects competence and expertise but knows that its value lies in sharing. He resists hierarchies and champions shared reciprocal power.

Finally, *Lord* is the divine image of the masculine soul. God is male and male is god. The feminine in the divine is denied and repressed, and hence also the feminine in man. This man is authoritarian—lording it over all.

The alternative imagery, suggested by Gerzon, is the *Nurturer*. Authoritarianism or power over is out of place here. He is not burdened with the pressure of saving people but rather is receptive to a process of mutual empowerment allowing people to save themselves. This is man as midwife—delivering people to themselves.

Gerzon's analysis is helpful and constructive. He has taken seriously the feminist critique of a truncated masculinity. His alternative imagery is suggestive—reflecting awareness of the earth, of work and family, and of the human body, mind and soul. Furthermore, he claims, the human qualities symbolized by the images transcend sexual identity. Whereas the old archetypes were for men only, the emerging characteristics are for all. Our destiny is to be androgynous.

Barbara Ehrenreich (*The Hearts of Men* and "A Feminist's View of the New Man") agrees with the basic shifts and direction outlined by Gerzon. She is less benign, however, in her interpretation. Yes, the new man of the '80s has been, in a word, feminized but without necessarily becoming more feminist. In fact, feminists may not be eager to take credit for him at all. "In the 1970s," writes Ehrenreich, "it had become an article of liberal faith that a new man would eventually rise up to match the new feminist woman, that he would be more androgynous than any 'old' variety of man." This change, which was routinely described as the great evolutionary leap from John Wayne to Alan Alda, was uncritically assumed to be an unambiguous improvement. Ehrenreich is not so sure.

The new man emerging today, notes Ehrenreich, is not simply the old one minus the old prohibitions and anxieties. There *is* a new complex of traits and attitudes that has come to define manhood and a kind of new masculine gentility. We have witnessed the "feminization" of male tastes and sensibilities, and a transformation of the male psyche. But the transformation, she claims, has been superficial and self-indulgent. The old man was (and is) a tyrant and a bully. But the new man tends to be a fop. He is Narcissus, and lost in an androgynous drift. He shows few marks of ideological struggle, personal pain and arduous sacrifice, namely, the prerequisites for self-transformation. His is a state of "cheap grace" or pseudo-liberation.

The most striking characteristic of the new man, writes Ehrenreich, is that he no longer anchors his identity in his role as family breadwinner. He may *be* or *intend to be* the chief economic provider for his family, but his status comes from another source—himself, his own efforts and the self (persona) he presents to the world. The prototypical new male is likely to be from 25 to 40 years old, affluent, urban, and, more often than not, single. This is the man women are losing patience with today because of his "lack (or fear) of commitment" and his "refusal to grow up." Typically, he is focused on three major concerns: 1) consumerism; 2) physical well-being; and 3) presenting a sensitive persona before the world. Our new man, according to Ehrenreich, is an avid and self-conscious consumer, not only of clothes but of food, home furnishings, and visible displays of culture. He is highly class conscious and self-consciously elitist. In terms of physical well-being, one might say the new man is obsessed with his physical health and fitness. His devotion is now focused on sagging muscles and suspect arteries. The old man smoked, drank, and puttered at golf. The new man is a nonsmoker, drinks in moderation, and frequents gyms and spas. Death remains his mortal enemy and the only obscenity. Finally, in contrast to the old masculinity, the new man is concerned that people find him, not forbearing or strong, but genuine, open, and sensitive. Sensitivity has become the distinguishing mark of the educated, middle-class male. The old man defined himself *against* femininity. Currently, the new man defines himself *as* "feminine," proud of his sensitive feminine sensibilities.

"Is this the new man that we want?" asks Ehrenreich. While promising in some respects, he is not a model of authentic humanness. It is not enough, anymore, to ask that men become more like women; we should ask instead that they become like what both men and women *might be*—sensitive, socially just, and deeply committing. On this note, Ehrenreich rests her case.

Ehrenreich's observations are astute and painful. They call the bluff of the contemporary pseudo-liberated male. Yet her concluding and alternative proposal falls flat. It seems particularly unsatisfactory in terms of authentic masculinity. To be fully masculine means embracing something of gender foreignness, strange to our own male bodily experience. The key to nonpatriarchal masculinity lies in men turning to their sensitive side and appropriating what Jung called their feminine soul. The end result leaves male identity dependent on the feminine and tied psychologically to women.

James Nelson (*The Intimate Connection*), however, asks: Is there anything authentically male about men, independent of women's contribution, that is important to their male identity? Is there not something good, important, and *distinctive* about the experience of maleness itself? Something that can produce an energy that is not oppressive but rather creative and life-giving—and recognizably male? A "deep masculine" that men can find in themselves and justly celebrate? Nelson finds assistance on this question in the provocative work of the poet Robert Bly. Bly (*The Pillow and the Key* and *When a Hair Turns Gold*) is enthusiastic about men welcoming their own feminine consciousness and nurturing it—it is important—and yet he senses there is something wrong:

> The male in the past twenty years has become more thoughtful, more gentle. But by this process he has *not* become more free. . . . I see the phenomenon of what I would call the "soft male" all over the country today. . . . They're lovely, valuable people—I like them—and they're not interested in harming the earth or starting wars or working for corporations. There's something favorable toward life in their whole general mood and style of living. But something's wrong. Many of these men are unhappy: there's not much energy in them. They are life-preserving but not exactly *life-giving*. They are living provisional lives marked by a lack of passion.

Bly then turns to the Grimms' Fairy Tale "Iron John" ("Iron Hans" in the original). Here is the scenario: Once upon a time . . . something strange was happening in the forest near the king's castle. People are disappearing. One day a hunter shows up at the castle looking for work and volunteers to investigate the mystery. He goes into the forest with his dog and they come across a pond. As they are walking by the pond a large hand reaches out of the water, grabs the dog, and pulls it down. The hunter, not wanting to abandon his dog and being a sensitive man, returns to the castle for help. He recruits a team of men and together they go back with buckets to drain the pond. Lying at the bottom is a big man covered with reddish hair, the color of rusted iron. They capture him and take him back to the castle, where the king orders him to be placed in a cage in the courtyard.

Bly interrupts the story and interprets. When a male looks down into his psyche, through his feminine side, to the bottom of his deep pool he finds an ancient man covered with hair. This mythological and archetypical figure is symbolic of the instinctive, the sexual, and the primitive. Every modern male has this Wild Man (Iron John) at the bottom of his psyche. Making contact with his Wild Man is the process many men yearn for but have not explored. It is the step the '70s male or '80s male has not yet taken. It is the journey that lies ahead and the task to be undertaken in contemporary culture. The Wild Man will point them to forgotten depths and be their guide to male initiation.

The story continues. One day the king's eight-year-old son is playing in the courtyard with the golden ball he loves. The ball rolls into the cage, and the wild man grabs it. Iron John will return the ball only if the young prince will release him from the cage. This is going to be a problem. The Wild Man knows, however, that the key to the lock on the cage is under the queen's pillow. Since the boy's parents are away and since he wants the ball so badly, the young prince fetches the key and opens the cage. As Iron John begins to leave, the prince becomes terrified that his parents will be angry. He calls to Iron John for help. The Wild Man scoops him up, places him on his shoulders, and they go off into the forest together, where the prince will learn the secrets of manhood.

This is not the complete fairy tale, but it is all that Bly uses at this stage to make his point. Bly observes that the golden ball (a recurrent fairy-tale image) suggests wholeness, unity of personality, a sense of connectedness. This is the childhood stage of undifferentiated unity that we lost and spend the rest of our lives trying to get back. For some thirty years, men were told that the golden ball was in their feminine side, in receptivity, in cooperation, in nonaggressiveness. They entered but did not find the golden ball—because that's not where the ball rolled when it was lost. Bly asks us to consider the possibility, as the story suggests, that the golden ball lies within the magnetic field of the Wild Man. What he is suggesting here is: The deep, nourishing spiritual energy for the male lies in the *deep* masculine, not in the shallow, macho masculine, but in the deep masculine, the instinctive one who's underwater. It is something a woman cannot give a man. It has to be appropriated slowly and resolutely, bucket by bucket, with the help of other men. It is something like that which the Greeks called Zeus energy, which encompasses intelligence, robust health, compassion, decisiveness, goodwill, and positive power in the service of the community.

Bly is well aware that he can be misunderstood here. His proposal is not a patriarchal backlash, but may, in fact, be profound, if not provocative, in the next step toward male wholeness/liberation. To guard against misunderstanding, Bly insists that men and women both have to keep in mind two critical distinctions: between the Wild Man and the savage man; between fierceness and hostility. Male wholeness is toward engaging the Wild Man and reincorporating fierceness. Whereas the savage man and hostility are the embodiments of patriarchy. He explains. When a man gets in touch with the Wild Man, a true strength may be added. The kind of wildness the image implies is not the same as macho energy—which man already knows enough about. Rather, it is a form of energy, springing from the depths of masculinity, that leads to "forceful action undertaken, not without compassion, but with resolve." The savage, on the other hand, epitomizes what men have been trying to get away from: the destructive, chauvinistic, unrounded, uncultivated.

The New Age '80s men and some feminists tend also to confuse fierceness with hostility. Noting the distinction made by anthropologist Ashley Montague, Bly explains. Men need fierceness in their lives, and it is important that we stop slandering it by identifying it with hostility. The instinct for aggression is in the genes; but hostility is learned in families. The ability to defend our community is present in our DNA; in hostility, we follow the modeling given by our parents. The instinct and heat of fierceness we possess at birth; the copying and coldness of hostility we learn in the family. The ability to be fierce does not imply the habit of treating people as if they were objects or ravaging the land as if it were a utility. Violence and brutality toward women, children, and the earth are not the function of fierceness but evidence of the absence of it. Fierceness implies passion, positive energy, response, alertness to boundaries, defense of one's desires and interests. The soft man of the '80s mistakenly wanted to root out all these traits. He only wanted to be receptive.

But men (and women) will be fierce at different times or in different situations. In every human relationship something fierce is needed once in a while. This may show itself as Eros, in love for each other. Parents today need some daily fierceness in order to resist the endless demands of their children. Law-enforcement officials need fierceness to guard the welfare of their communities. Children need the heat of fierceness if they are to develop steadfastness, endurance, and vigilance. And we all need fierceness to protect the planet.

Bly's use of the fairy tale to examine the meanings of gender images is original and suggestive. It opens new avenues of possibilities for men. It directs them to explore their deepest and most distinctive selves. And it is at this point where a *religious* dimension of their maleness will be revealed to them. Bly enables us to tap into images of the Wild Man and visualize him. In doing so, he puts men in touch with their own natural male energy. This enables men to reclaim their identity and re-image their masculinity.

However, the integration of the Wild Man may not be the final chapter in the male's journey to maturity. Additional work may need to be done. With a purpose similar to Bly's, but with an alternative image and archetype, Eugene Monick (*Phallos: Sacred Image of the Masculine*) inquires: What is the place, if any, of that age-old distinctive emblem of manhood, the male genitals? Monick explores the psychic and religious dimensions of the male experience of his phallus, his erect penis. He starts with the assumption that men need to discover another basis of support for their maleness. Patriarchy is spent. Psychologically, it is on the rocks. Its substance can no longer support male identity or male bonding. The new basis of that support, he suggests, is phallus. His proposal is provocative. It tends to make us uneasy. And, again, it may lend itself to fears and misunderstandings.

In Monick's interpretative scheme, phallus is the archetypical image of the masculine. It is a fundamental characteristic and universal attribute of maleness—its standard, stamp, and subjective authority. As a symbol, the erect penis embodies the mysteries of male masculinity. It opens the door to masculine depths. From time immemorial it has been the source of *mysterium tremendum* and functioned as a divine image. Men have to search no further than phallus to ground their distinctive identities. There is no other source to return to, no other support structure, no other spring of energy to return us to our original wholeness. Yet, males feel highly ambivalent about the phallus.

Furthermore, Monick notes, there is a double-sidedness to the phallic experience. One dimension is the *earthy* phallus. This is the phallic experience as hot, throbbing raw sexuality. In some measure it is Bly's Iron John maleness. We may be tempted to hide from this side of our sexuality and judge it severely. But if we do, we will lack life-giving energy, lose the possibility of ecstasy, and run the serious risk of becoming a shell of masculinity.

Earthy phallus is distrusted. And there is reason for distrust—it has a shadow side. There can be an ugly, brutal side to the earthy phallus that uses others for gratification. It can be characterized by insatiable desires, possessiveness, domination, ruthless competition and violence. Life is replete with examples of its stupid and devastating behavior. Yet without the positive presence of earthy energy a man is bland. He is gentle without strength, peaceful without vitality, tranquil without vibrancy.

Men also experience *solar* phallus. Solar (from the sun) represents higher things. It means enlightenment. It puts a man in touch with the excitement of achievement. It is the pride a man takes in his social reputation. It is what he would like to see noted in his obituary. Solar phallus is a man's profession. It is how he speaks, thinks, intellectualizes. A solar man wants the facts, loves institutionalization, and strains to go further intellectually, physically, and socially. Carl Jung believed this to be the substance of masculinity.

As with earthy phallus, there is a shadow side to the experience of the solar phallus. It is motivated by conquest, ideological principles, and can tyrannize whatever is considered to be in error. It is patriarchal oppression, proving one's worth through institutional accomplishment, and the use of technical knowledge to dominate. This is the arrogant academic, the political ideologue, the social snob. Feminists and environmentalists attack patriarchal phallus. Yet without the integrative positive energy of solar phallus, a man lacks direction and movement. He remains mediocre and is blocked from transcendence. What we are witnessing in our culture, Monick maintains, is a perversion of phallus as patriarchy. The journey to the center of masculine identity, however, lies in placing phallus—earthy and solar—at the root of male consciousness.

Monick's explorations are interesting, if not suggestive. James Nelson substantially agrees with him as far as he goes. Monick's proposal, according to Nelson, falls short. For Monick, phallus, the erect penis, is *the* archetype and sacred image of the masculine. That seems to be enough. But it is not. Yes, it is a vital part of the male's experience of his sexual organs. But it is only a part. A man's (flaccid) *penis* is as genuinely his reality as is his phallus, and just as important to his male humanity. We tend to overvalue phallus (erection) and undervalue penis. That is the road to patriarchy. On the other hand, the affirmative experience of penis gives men permission to embrace their own corporality and sexuality *as it is*. Acceptance of their corporal givenness allows men to receive and respond to the web of relations in which their lives are immersed. We named that style androgynous. It is time to move beyond the usual (literal and conventional) meanings of androgyny. The term itself is not free from difficulty and misunderstanding. As a transitional concept, it is helpful in pointing toward the transformation of sex-role stereotypes and human integration. However, the notion of androgyny, argues James Nelson, operates with a "combinationist assumption." It begins with a fixed notion of masculine traits and a fixed notion of feminine traits. Then it moves to the contention that these fundamental different qualities can and should be combined in men and women.

The androgynous perspective is grounded in an underlying dualism. From its assumption it follows that, in developing "the feminine" in himself, a man will add on a different (or foreign) set of gender traits to that which is essentially himself. But, as Nelson indicates, in a basic sense, we do not have to *become* androgynous, for we essentially *are* . . . unique individuals, female and male, each with the capacity to be both firm and tender, receiving and giving, rational and intuitive. We have been given ''bilingual bodies.'' Even if one language has been developed more than the other, the second language is not foreign to us. It is not something we need to add on. It is just as originally part of us as the language with which we have been most familiar.

To be fully male, then, does not mean embracing something of gender foreignness. The vision for men is not to develop ''feminine'' energies (or for women to develop ''masculine'' energies). Rather the vision for men is the *fullest* development of our *masculine* energies. These are the only human energies we have, and the invitation is to develop them more richly.

VII

Divorce/Annulment

31.

Children After Divorce: Wounds That Don't Heal

Judith S. Wallerstein

In many discussions of divorce, the question of a breakup's effect on children does not arise. The discussion centers on what divorce means for the divorcing couple. In this section on divorce and annulment, we have put the question of children in divorce first, for two reasons. First of all, some students using this book will themselves be children of divorce or may be dating or engaged to a child of divorce. They may find in the following essay by Judith Wallerstein insight into their own reactions or those of their beloved. Secondly, it seems important to name all those wounded by divorce. Doing so raises the stakes involved in careful preparation for marriage.

Wallerstein reports research among the sons and daughters of divorce, where she finds far more emotional pain than many expected. Apparently, healing the wounds of divorce in children can take many years. Some questions follow:

1. *Which of the findings of the follow-up studies of children of divorce were for you the most surprising?*
2. *What is the difference between seeing divorce as a single circumscribed event and seeing it as a process?*
3. *What does Wallerstein mean by the "sleeper effect" of divorce?*
4. *What is the diminished parenting consequence and how is it related to the overburdened child?*
5. *What interpretation do you give to the incident of the boy who piled all the furniture on top of the baby dolls?*

As recently as the 1970s, when the American divorce rate began to soar, divorce was thought to be a brief crisis that soon resolved itself. Young children might have difficulty falling asleep and older children might have trouble at school.

From *The New York Times Magazine*, 22 January 1989, pp. 19–21, 41–44.

Men and women might become depressed or frenetic, throwing themselves into sexual affairs or immersing themselves in work.

But after a year or two, it was expected, most would get their lives back on track, at least outwardly. Parents and children would get on with new routines, new friends and new schools, taking full opportunity of the second chances that divorce brings in its wake.

These views, I have come to realize, were wishful thinking. In 1971, working with a small group of colleagues and with funding from San Francisco's Zellerbach Family Fund, I began a study of the effects of divorce on middle-class people who continue to function despite the stress of a marriage breakup.

That is, we chose families in which, despite the failing marriage, the children were doing well at school and the parents were not in clinical treatment for psychiatric disorders. Half of the families attended church or synagogue. Most of the parents were college educated. This was, in other words, divorce under the best of circumstances.

Our study, which would become the first ever made over an extended period of time, eventually tracked 60 families, most of them white, with a total of 131 children, for 10, and in some cases 15, years after divorce. We found that although some divorces work well—some adults are happier in the long run, and some children do better than they would have been expected to in an unhappy intact family—more often than not divorce is a wrenching, long-lasting experience for at least one of the former partners. Perhaps most important, we found that for virtually all the children, it exerts powerful and wholly unanticipated effects.

Our study began with modest aspirations. With a colleague, Joan Berlin Kelly—who headed a community mental-health program in the San Francisco area—I planned to examine the short-term effects of divorce on these middle-class families.

We spent many hours with each member of each of our 60 families—hearing their firsthand reports from the battleground of divorce. At the core of our research was the case study, which has been the main source of the fundamental insights of clinical psychology and of psychoanalysis. Many important changes, especially in the long run, would be neither directly observable nor easily measured. They would become accessible only through case studies: by examining the way each of these people processed, responded to and integrated the events and relationships that divorce brings in its wake.

We planned to interview families at the time of decisive separation and filing for divorce, and again 12 to 18 months later, expecting to chart recoveries among men and women and to look at how the children were mastering troubling family events.

We were stunned when, at the second series of visits, we found family after family still in crisis, their wounds wide open. Turmoil and distress had not noticeably subsided. Many adults were angry, and felt humiliated and rejected, and most had not gotten their lives back together. An unexpectedly large number of children were on a downward course. Their symptoms were worse than they had been immediately after the divorce. Our findings were absolutely contradictory to our expectations.

Dismayed, we asked the Zellerbach Fund to support a follow-up study in the fifth year after divorce. To our surprise, interviewing 56 of the 60 families in our original study, we found that although half the men and two thirds of the women (even many of those suffering economically) said they were more content with their lives, only 34 percent of the children were clearly doing well.

Another 37 percent were depressed, could not concentrate in school, had trouble making friends and suffered a wide range of other behavior problems. While able to function on a daily basis, these children were not recovering, as everyone thought they would. Indeed most of them were on a downward course. This is a powerful statistic, considering that these were children who were functioning well five years before. It would be hard to find any other group of children—execpt, perhaps, the victims of a natural disaster—who suffered such a rate of sudden serious psychological problems.

The remaining children showed a mixed picture of good achievement in some areas and faltering achievement in others; it was hard to know which way they would eventually tilt.

The psychological condition of these children and adolescents, we found, was related in large part to the overall quality of life in the post-divorce family, to what the adults had been able to build in place of the failed marriage. Children tended to do well if their mothers and fathers, whether or not they remarried, resumed their parenting roles, managed to put their differences aside, and allowed the children a continuing relationship with both parents. Only a handful of kids had all these advantages.

We went back to these families again in 1980 and 1981 to conduct a 10-year follow-up. Many of those we had first interviewed as children were now adults. Overall, 45 percent were doing well; they had emerged as competent, compassionate and courageous people. But 41 percent were doing poorly; they were entering adulthood as worried, underachieving, self-deprecating and sometimes angry young men and women. The rest were strikingly uneven in how they adjusted to the world; it is too soon to say how they will turn out.

At around this time, I founded the Center for the Family in Transition, in Marin County, near San Francisco, which provides counseling to people who are separating, divorcing or remarrying. Over the years, my colleagues and I have seen more than 2,000 families—an experience that has amplified my concern about divorce. Through our work at the center and in the study, we have come to see divorce not as a single circumscribed event but as a continuum of changing family relationships—as a process that begins during the failing marriage and extends over many years. Things are not getting better, and divorce is not getting easier. It's too soon to call our conclusions definitive, but they point to an urgent need to learn more.

It was only at the 10-year point that two of our most unexpected findings became apparent. The first of these is something we call the sleeper effect.

The first youngster in our study to be interviewed at the 10-year mark was one who had always been a favorite of mine. As I waited for her to arrive for this interview, I remembered her innocence at age 16, when we had last met. It was she who alerted us to the fact that many young women experience a delayed effect of divorce.

As she entered my office, she greeted me warmly. With a flourishing sweep of one arm, she said. "You called me at just the right time. I just turned 21!" Then she startled me by turning immediately serious. She was in pain, she said.

She was the one child in our study who we all thought was a prime candidate for full recovery. She had denied some of her feelings at the time of divorce, I felt, but she had much going for her, including high intelligence, many friends, supportive parents, plenty of money.

As she told her story, I found myself drawn into unexpected intricacies of her life. Her trouble began, typically, in her late teens. After graduating from high school with honors, she was admitted to a respected university and did very well her freshman year. Then she fell apart. As she told it, "I met my first true love."

The young man, her age, so captivated her that she decided it was time to have a fully committed love affair. But on her way to spend summer vacation with him, her courage failed. "I went to New York instead. I hitchhiked across the country. I didn't know what I was looking for. I thought I was just passing time. I didn't stop and ponder. I just kept going, recklessly, all the time waiting for some word from my parents. I guess I was testing them. But no one—not my dad, not my mom—ever asked me what I was doing there on the road alone."

She also revealed that her weight dropped to 94 pounds from 128 and that she had not menstruated for a year and a half.

"I began to get angry," she said. "I'm angry at my parents for not facing up to the emotions, to the feelings in their lives, and for not helping me face up to the feelings in mine. I have a hard time forgiving them."

I asked if I should have pushed her to express her anger earlier.

She smiled patiently and said, "I don't think so. That was exactly the point. All those years I denied feelings. I thought I could live without love, without sorrow, without anger, without pain. That's how I coped with the unhappiness in my parents' marriage. Only when I met my boyfriend did I become aware of how much feeling I was sitting on all those years. I'm afraid I'll lose him."

It was no coincidence that her acute depression and anorexia occurred just as she was on her way to consummate her first love affair, as she was entering the kind of relationship in which her parents failed. For the first time, she confronted the fears, anxieties, guilt and concerns that she had suppressed over the years.

Sometimes with the sleeper effect the fear is of betrayal rather than commitment. I was shocked when another young woman—at the age of 24, sophisticated, warm and friendly—told me she worried if her boyfriend was even 30 minutes late, wondering who he was with and if he was having an affair with another woman. This fear of betrayal occurs at a frequency that far exceeds what one might expect from a group of people randomly selected from the population. They suffer minute to minute, even though their partners may be faithful.

In these two girls we saw a pattern that we documented in 66 percent of the young women in our study between the ages of 19 and 23; half of them were seriously derailed by it. The sleeper effect occurs at a time when these young women are making decisions with long-term implications for their lives. Faced with issues of commitment, love and sex in an adult context, they are aware that the game is serious. If they tie in with the wrong man, have children too soon, or choose harmful life styles, the effects can be tragic. Overcome by fears and anxieties, they begin to make connections between these feelings and their parents' divorce:

"I'm so afraid I'll marry someone like my dad."

"How can you believe in commitment when anyone can change his mind anytime?"

"I am in awe of people who stay together."

We can no longer say—as most experts have held in recent years—that girls are generally less troubled by the divorce experience than boys. Our study strongly indicates, for the first time, that girls experience serious effects of divorce at the time they are entering young adulthood. Perhaps the risk for girls and boys is equalized over the long term.

When a marriage breaks down, men and women alike often experience a diminished capacity to parent. They may give less time, provide less discipline and be less sensitive to their children, since they are themselves caught up in the maelstrom of divorce and its aftermath. Many researchers and clinicians find that parents are temporarily unable to separate their children's needs from their own.

In a second major unexpected finding of our 10-year study, we found that fully a quarter of the mothers and a fifth of the fathers had not gotten their lives back on track a decade after divorce. The diminished parenting continued, permanently disrupting the child-rearing functions of the family. These parents were chronically disorganized and, unable to meet the challenges of being a parent, often leaned heavily on their children. The child's role became one of warding off the serious depression that threatened the parents' psychological functioning. The divorce itself may not be solely to blame but, rather, may aggravate emotional difficulties that had been masked in the marriage. Some studies have found that emotionally disturbed parents within a marriage produce similar kinds of problems in children.

These new roles played by the children of divorce are complex and unfamiliar. They are not simple role reversals, as some have claimed, because the child's role becomes one of holding the parent together psychologically. It is more than a caretaking role. This phenomenon merits our careful attention, for it affected 15 percent of the children in our study, which means many youngsters in our society. I propose that we identify as a distinct psychological syndrome the "overburdened child," in the hope that people will begin to recognize the problems and take steps to help these children, just as they help battered and abused children.

One of our subjects, in whom we saw this syndrome, was a sweet 5-year-old girl who clearly felt that she was her father's favorite. Indeed, she was the only person in the family he never hit. Preoccupied with being good and helping to calm both parents, she opposed the divorce because she knew it would take her father away from her. As it turned out, she also lost her mother who, soon after the divorce, turned to liquor and sex, a combination that left little time for mothering.

A year after the divorce, at the age of 6, she was getting herself dressed, making her own meals and putting herself to bed. A teacher noticed the dark circles under her eyes, and asked why she looked so tired. "We have a new baby at home," the girl explained. The teacher, worried, visited the house and discovered there was no baby. The girl's story was designed to explain her fatigue but also enabled her to fantasize endlessly about a caring, loving mother.

Shortly after this episode, her father moved to another state. He wrote to her once or twice a year, and when we saw her at the five-year follow-up she pulled out a packet of letters from him. She explained how worried she was that he might get into trouble, as if she were the parent and he the child who had left home.

"I always knew he was O.K. if he drew pictures on the letters," she said. "The last two really worried me because he stopped drawing."

Now 15, she has taken care of her mother for the past 10 years. "I felt it was my responsibility to make sure that Mom was O.K.," she says. "I stayed home with her instead of playing or going to school. When she got mad, I'd let her take it out on me."

I asked what her mother would do when she was angry.

"She'd hit me or scream. It scared me more when she screamed. I'd rather be hit. She always seemed so much bigger when she screamed. Once Mom got drunk and passed out on the street. I called my brothers, but they hung up. So I did it. I've done a lot of things I've never told anyone. There were many times she was so upset I was sure she would take her own life. Sometimes I held both her hands and talked to her for hours I was so afraid."

In truth, few children can rescue a troubled parent. Many become angry at being trapped by the parent's demands, at being robbed of their separate identity and denied their childhood. And they are saddened, sometimes beyond repair, at seeing so few of their own needs gratified.

Since this is a newly identified condition that is just being described, we cannot know its true incidence. I suspect that the number of overburdened children runs much higher than the 15 percent we saw in our study and that we will begin to see rising reports in the next few years—just as the reported incidence of child abuse has risen since it was first identified as a syndrome in 1962.

The sleeper effect and the overburdened-child syndrome were but two of many findings in our study. Perhaps most important, overall, was our finding that divorce has a lasting psychological effect on many children, one that, in fact, may turn out to be permanent.

Children of divorce have vivid memories about their parents' separation. The details are etched firmly in their minds, more so than those of any other experiences in their lives. They refer to themselves as children of divorce, as if they share an experience that sets them apart from all others. Although many have come to agree that their parents were wise to part company, they nevertheless feel that they suffered from their parents' mistakes. In many instances, conditions in the post-divorce family were more stressful and less supportive to the child than conditions in the failing marriage.

If the finding that 66 percent of the 19- to 23-year-old young women experienced the sleeper effect was most unexpected, others were no less dramatic. Boys, too, were found to suffer unforeseen long-lasting effects. Forty percent of the 19- to 23-year-old young men in our study 10 years after divorce, still had no set goals, a limited education and a sense of having little control over their lives.

In comparing the post-divorce lives of former husbands and wives, we saw that 50 percent of the women and 30 percent of the men were still intensely angry at their former spouses a decade after divorce. For women over 40 at divorce, life was lonely throughout the decade; not one in our study remarried or sustained a loving relationship. Half the men over 40 had the same problem.

In the decade after divorce, three in five children felt rejected by one of their parents, usually the father—whether or not it was true. The frequency and duration of visiting made no difference. Children longed for their fathers, and the need increased during adolescence. Thirty-four percent of the youngsters went to live with their fathers during adolescence for at least a year. Half returned to the mother's home disappointed with what they had found. Only one in seven saw both mother and father happily remarried after 10 years. One in two saw their mother or their father undergo a second divorce. One in four suffered a severe and enduring drop in the family's standard of living and went on to observe a lasting discrepancy between their parents' standards of living.

We found that the children who were best adjusted 10 years later were those who showed the most distress at the time of the divorce—the youngest. In general, preschoolers are the most frightened and show the most dramatic symptoms when marriages break up. Many are afraid that they will be abandoned by both parents and they have trouble sleeping or staying by themselves. It is therefore surprising to find that the same children 10 years later seem better adjusted than their older siblings. Now in early and mid-adolescence, they were rated better on a wide range of psychological dimensions than the older children. Sixty-eight percent were doing well, compared with less than 40 percent of older children. But whether having been young at the time of divorce will continue to protect them as they enter young adulthood is an open question.

Our study shows that adolescence is a period of particularly grave risk for children in divorced families. Through rigorous analysis, statistical and otherwise, we were able to see clearly that we weren't dealing simply with the routine angst of young people going through transition but rather that, for most of them, divorce was the single most important cause of enduring pain and anomie in their lives. The young people told us time and again how much they needed a family structure, how much they wanted to be protected, and how much they yearned for clear guidelines for moral behavior. An alarming number of teen-agers felt abandoned, physically and emotionally.

For children, divorce occurs during the formative years. What they see and experience becomes a part of their inner world, influencing their own relationships 10 and 15 years later, especially when they have witnessed violence between the parents. It is then, as these young men and women face the developmental task of establishing love and intimacy, that they most feel the lack of a template for a loving relationship between a man and a woman. It is here that their anxiety threatens their ability to create new, enduring families of their own.

As these anxieties peak in the children of divorce throughout our society, the full legacy of the rising divorce rate is beginning to hit home. The new families being formed today by these children as they reach adulthood appear particularly vulnerable.

Because our study was such an early inquiry, we did not set out to compare children of divorce with children from intact families. Lacking fundamental knowledge about life after the breakup of a marriage, we could not know on what basis to build a comparison or control group. Was the central issue one of economics, age, sex, a happy intact marriage—or would any intact marriage do? We began, therefore, with a question—What is the nature of the divorce experience?—and in answering it we would generate hypotheses that could be tested in subsequent studies.

This has indeed been the case. Numerous studies have been conducted in different regions of the country, using control groups, that have further explored and validated our findings as they have emerged over the years. For example, one national study of 699 elementary school children carefully compared children six years after their parents' divorce with children from intact families. It found—as we did—that elementary-age boys from divorced families show marked discrepancies in peer relationships, school achievements and social adjustment. Girls in this group, as expected, were hardly distinguishable based on the experience of divorce, but, as we later found out, this would not always hold up. Moreover, our findings are supported by a litany of modern-day statistics. Although one in three children are from divorced families, they account for an inordinately high proportion of children in mental-health treatment, in special-education classes, or referred by teachers to school psychologists. Children of divorce make up an estimated 60 percent of child patients in clinical treatment and 80 percent—in some cases, 100 percent—of adolescents in inpatient mental hospital settings. While no one would claim that a cause and effect relationship has been established in all of these cases, no one would deny that the role of divorce is so persuasively suggested that it is time to sound the alarm.

All studies have limitations in what they can accomplish. Longitudinal studies, designed to establish the impact of a major event or series of events on the course of a subsequent life, must always allow for the influence of many interrelated factors. They must deal with chance and the uncontrolled factors that so often modify the sequences being followed. This is particularly true of children, whose lives are influenced by developmental changes, only some of which are predictable, and by the problem of individual differences, about which we know so little.

Our sample, besides being quite small, was also drawn from a particular population slice—predominantly white, middle class and relatively privileged suburbanites.

Despite these limitations, our data have generated working hypotheses about the effects of divorce that can now be tested with more precise methods, including appropriate control groups. Future research should be aimed at testing, correcting or modifying our initial findings, with larger and more diverse segments of the population. For example, we found that children—especially boys and young men—continued to need their fathers after divorce and suffered feelings of rejection even when they were visited regularly. I would like to see a study comparing boys and girls in sole and joint custody, spanning different developmental stages, to see if greater access to both parents counteracts these feelings of rejection. Or, does joint custody lead to a different sense of rejection—of feeling peripheral in both homes?

It is time to take a long, hard look at divorce in America. Divorce is not an event that stands alone in children's or adults' experience. It is a continuum that begins in the unhappy marriage and extends through the separation, divorce and any remarriages and second divorces. Divorce is not necessarily the sole culprit. It may be no more than one of the many experiences that occur in this broad continuum.

Profound changes in the family can only mean profound changes in society as a whole. All children in today's world feel less protected. They sense that the institution of the family is weaker than it has ever been before. Even those children raised in happy, intact families worry that their families may come undone. The task for society in its true and proper perspective is to strengthen the family—all families.

A biblical phrase I have not thought of for many years has recently kept running through my head: "Watchman, what of the night?" We are not, I'm afraid, doing very well on our watch—at least for our children. We are allowing them to bear the psychological, economic and moral brunt of divorce.

And they recognize the burdens. When one 6-year-old boy came to our center shortly after his parents' divorce, he would not answer questions; he played games instead. First he hunted all over the playroom for the sturdy Swedish-designed dolls that we use in therapy. When he found a good number of them, he stood the baby dolls firmly on their feet and placed the miniature tables, chairs, beds and, eventually, all the playhouse furniture on top of them. He looked at me, satisfied. The babies were supporting a great deal. Then wordlessly, he placed all the mother and father dolls in precarious positions on the steep roof of the doll house. As a father doll slid off the roof, the boy caught him and, looking up at me, said, "He might die," Soon, all the mother and father dolls began sliding off the roof. He caught them gently, one by one. "The babies are holding up the world," he said.

Although our overall findings are troubling and serious, we should not point the finger of blame at divorce per se. Indeed, divorce is often the only rational solution to a bad marriage. When people ask whether they should stay married for the sake of the children, I have to say, "Of course not." All our evidence shows that children exposed to open conflict, where parents terrorize or strike one another, turn out less well-adjusted than do children from divorced families. And although we lack systematic studies comparing children in divorced families with those in unhappy intact families, I am convinced that it is not useful to provide children with a model of adult behavior that avoids problem-solving and that stresses martyrdom, violence or apathy. A divorce undertaken thoughtfully and realistically can teach children how to confront serious life problems with compassion, wisdom and appropriate action.

Our findings do not support those who would turn back the clock. As family issues are flung to the center of our political arena, nostalgic voices from the right argue for a return to a time when divorce was more difficult to obtain. But they do not offer solutions to the wretchedness and humiliation within many marriages.

Still, we need to understand that divorce has consequences—we need to go into the experience with our eyes open. We need to know that many children will suffer for many years. As a society, we need to take steps to preserve for the children as much as possible of the social, economic and emotional security that existed while their parents' marriage was intact.

Like it or not, we are witnessing family changes which are an integral part of the wider changes in our society. We are on a wholly new course, one that gives us unprecedented opportunities for creating better relationships and stronger families—but one that also brings unprecedented dangers for society, especially for our children.

32.

Remarried Catholics—Searching for Church Belonging

James J. Young

Paulist Father James Young explains here why so many divorced and remarried Catholics are still living as active Roman Catholics practicing their faith in Catholic parishes. Some of these persons—but far from all—have remarried after receiving annulments declaring their former marriages invalid. Young cites Pope John Paul II's letter, On the Family, *encouraging divorced and remarried Catholics to remain members of the Church, though officially not allowed to receive communion.*

Young points out a way such Catholics may still claim their right to the Eucharist. His discussion of this question deserves careful study. Some questions for reflection follow:

1. *What are the deep conflicts Young claims beset Catholics who remarry without the Church's approval?*
2. *What are the reasons Catholic parishes are reaching out to divorced Catholics?*
3. *If the Pope says divorced and remarried Catholics may not receive Communion, what is the path of good conscience by which they may receive?*
4. *Explain the theological thinking that permits communion even to those clearly in invalid second marriages?*
5. *What is your overall reaction to Young's essay?*

Note: *Jim Young died suddenly in 1987 at a relatively young age. Most of his priestly life, he had worked with tireless creativity in ministry to divorced Catholics.*

Sometimes they come to the parish house with a child to be baptized, or appear at parent classes for First Communion preparation. They may volunteer to help with

---

From Steven Preister and James J. Young, *Catholic Remarriage: Pastoral Issues and Preparation Models* (New York: Paulist Press, 1986), pp. 39–42.

the parish feeding program or sign up to visit the elderly. A husband or wife may be met on a hospital call or at a prayer meeting. At first, they may seem awkward and ill-at-ease, even evasive. They're remarried Catholics, and more of them are surfacing every day in American parishes.

Understandably, they often make other Catholics or those in positions of leadership somewhat nervous. They carry with them the suggestion of marriages abandoned, vows violated, and Church discipline ignored. Some Catholics are anxious that being too friendly or too accommodating to the remarried may undermine the Church's teaching on marital permanence and even encourage divorce. As predictable as these concerns may be, our pastoral experience is painting a far more complex and challenging portrait of remarried Catholics.

We are learning that they typically are men and women who exhausted every resource available to save a failing marriage, and only decided to divorce after prolonged consultation with counsellors and pastors. None divorced easily; the guilt, stress and upset that follows every broken marriage testifies to that. It may well be that because they were so Catholic, shaped by the Church's high valuation of lasting marriage, the pain was even more intense. Even though all divorce recovery programs hold up establishing an autonomous single existence as the major goal for the separated person, it is easy to understand why remarriage emerges so early as the obvious solution to this painful transition. For most people, a new marriage provides the only imaginable way of finding love again, of being happy again, or having a place in society again. Further, somewhat paradoxically, the best parts of a bad marriage may provide the appetite for a better marriage. Daily we meet divorced men and women who have put together a satisfying single life after divorce, but most admit that they would readily marry again if a suitable partner appeared.

The limited surveys we have suggest that, by and large, Catholics are much like the population at large. This means that almost three-fourths of divorced Catholics are dating seriously within a year, and half of them are remarried within three years of the civil divorce. Eventually as many as five out of six of the men and three out of four of the women will remarry.

The benefits of remarriage seem obvious to most. A new marriage brings a new partner with the companionship, sexual intimacy and support marriage provides. For most women, only remarriage helps them return to the financial security they knew in their former marriage. Most single parents are convinced that their children will be better off with a stepfather or stepmother, which is why fully 60% of remarriages bring children into the new household. Even though remarried living and stepparenting are unfamiliar situations with few accepted models of behavior, most will risk the unknown when the opportunity appears rather than continue raising children alone or living alone.

There may be as many as a million Catholics who have remarried over the past fifteen years with the Church's blessing. The increased availability of expeditious annulment procedures has allowed some 800,000 Catholics to receive annulments; many more, who never married as required in a Church ceremony the first time, have been able to marry again with the Church's approval. The Catholic community has worked very hard in recent years, using all the remedies available, to help Catholics remarry and remain in good standing in the community. Yet our best estimates are that well over 75% of remarried Catholics live in presumably invalid second marriages. They are the men and women who are presenting such difficult pastoral problems in Catholic parishes today.

Catholics who remarry without the Church's approval are usually caught in deep conflict. Many of them of lifelong devout Catholics, who attend Mass regularly, pray regularly, and live good Christian lives. When asked how they could go against the Church's discipline which does not provide for such second marriages, they often answer, "I hated to get married outside the Church, but I knew God would understand. I knew he didn't want me to be so lonely, and I knew he wanted my children to have a mother." Others say that though they loved the Church and being Catholic was in their bones, they felt the Church was too strict on divorce and remarriage and didn't appreciate the hardships people endure. Most say they agonized for months, sought spiritual counsel, and prayed at length before deciding to go ahead with a prohibited second marriage. Afterwards many were tormented by fears of "living in sin" and mistaken notions of being excommunicated from the Church. For some such guilt became a burdensome factor working against the success of the second marriage.

Those who have ministered extensively to divorced Catholics insist that the remarried are not men and women who have rejected the Church's teaching on the permanence and indissolubility of marriage. They are not people who promote divorce. Almost unanimously they profess a high regard for lifelong marriage, and insist they would never wish a divorce on anyone. "At times I still can't believe I'm divorced and remarried," a woman told me, "I'm sure if I was still caught in that first marriage, I'd be in a mental hospital now."

Further, surveys indicate that widespread divorce and remarriage among Catholics does not reflect a lessening of traditional Catholic family values among remarried Catholics. Recently, the Notre Dame Study of Parish Life found remarrying Catholics reapproaching and reidentifying with Catholic parishes. The remarried are "normal" again and want to live like ordinary families again. For traditional Catholics that means being part of a parish and raising children in the Church. Social critic Michael Novak believes that widespread divorce among Catholics stems more from the increased pressure on the family rather than lack of commitment to family values. Many commentators cite such contemporary factors as emotional problems, joblessness, addiction, mobility and loss of supportive family relationship, poverty, crime, and effects of Viet Nam—all of which tear marriages apart. To come close to divorced people is to look through a painful window at the dark underside of American life and the many forces that make lasting marriage difficult. For most, remarriage is a second chance to live and love again; another chance to salvage a broken life.

For this reason including these remarried couples in parish life may be an important way of helping the parish ground its life in the realities of Christian living today. The Catholic community has always been close to its people and their pain, and that basic pastoral instinct may be dramatized no more clearly today than in the ever-widening process of reconciliation of the remarried. And as this reconciliation has grown, rather than causing scandal and promoting more divorce and remarriage, the opposite actually has been the case. Since 1981 the U.S. divorce rate has been declining and the remarriage rate slowing. Could it be that understanding divorce better and the difficulties of remarriage has challenged more persons in troubled marriages to work harder at making them last?

The most important reason, however, for reaching out to remarried Catholics is the fact that they remain baptized members of the Church and deserve our pastoral care. Pope John Paul II clearly made this point in his 1980 letter *On the Family*. "I earnestly call upon pastors and the whole community of the faithful to help the divorced (and remarried) and with solicitous care to make sure that they do not consider themselves as separated from the Church, for as baptized persons they can and indeed must share in her life." He calls upon pastors to be especially sensitive to those "who have sincerely tried to save their first marriages and have been unjustly abandoned" and "those who have entered into a second union for the sake of the children's upbringing and who are sometimes subjectively certain in conscience that their previous and irreparably destroyed marriage had never been valid." (*On the Family*, #84) He goes on to say that remarried Catholics should be encouraged to attend Mass, listen to the word of God, persevere in prayer, contribute to works of charity and to community efforts in favor of justice, and to bring up their children in the faith.

In 1977 the American Bishops removed the American Church law which had attached a penalty of automatic excommunication to second marriage for Catholics who had previously been married in a Catholic ceremony and had not obtained an annulment of their first marriage. They wrote about their action, "It welcomes back to the community of believers in Christ all who may have been separated by excommunication. It offers them a share in all the public prayers of the Church community. It restores their right to take part in church services. It removes certain canonical restrictions upon their participation in church life. It is a promise of help and support in the resolution of the burden of family life. Perhaps above all, it is a gesture of love and reconciliation from the other members of the Church."

That love and reconciliation, of which the bishops wrote, is surfacing daily in American Catholic parishes. The papal and episcopal statements indicate the clear pastoral responsibility on the diocesan and parish level to search out and find such alienated remarried Catholics. Sadly some may have no interest in being an active Catholic again, but in recent years diocesan and parish efforts have turned up thousands of married Catholics most interested in being part of the Catholic community again. Pulpit appeals for reconciliation which clarify the place of remarried Catholics in the Church today continue to be needed; some parishes have deputized lay visitors to call on remarried couples who may be alienated and invite them back. There may be several million alienated remarried Catholics in the United States.

In his same 1980 letter, the Pope reaffirmed the Church's general practice of not admitting the remarried to Eucharistic communion. "They are unable to be admitted thereto from the fact that their state and condition of life objectively contradict that union of love between Christ and the Church which is signified and effected by the Eucharist.'' He added a second reason for the traditional exclusion. ''If these people were admitted to the Eucharist the faithful would be led into error and confusion regarding the Church's teaching about the indissolubility of marriage.'' It must be noted that the Pope has already affirmed the place of remarried Catholics in the Church community and stressed the goodness of many of their lives with a most approving statement of their position in the Church. Yet he feels that the continuing existence of a prior marriage assumed to be binding until death bars them from the Eucharist. Since Eucharistic reception is a sign of accepting the teaching of the Church and living up to that teaching, those who have married a second time without Church approval are not properly disposed to take Communion.

There are two paths which make Eucharistic reception for the remarried possible. The first is an annulment of a prior marriage. Fortunately, this healing remedy is readily available in all of our American dioceses. Where properly explained to remarried Catholics, most choose to pursue an annulment since they have a strong desire to have their new marriage accepted by the Church community and restored to Communion. Catholic belonging always seems incomplete without Eucharist.

The second path is the ''good conscience'' solution by which a pastoral minister helps the remarried make a judgment about the appropriateness of their taking Communion when an annulment has not been possible. If the Catholics involved have a moral certitude that their first marriage was invalid, i.e., not a true Catholic marriage, then they are not bound by that prior marriage. This means that the second marriage can be considered a true Catholic marriage even though it cannot be publicly celebrated in the Church. This practice was urged on pastors by Cardinal Seper, then Prefect of the Congregation for the Doctrine of the Faith in Rome in 1973.

Where this solution has been applied and where couples in marriages not blessed by the Church have been encouraged by their pastors to take Communion, a compassionate readiness on the part of the Catholic people to welcome them to the Lord's table seems quite common. Even though there is always the danger of scandal in such cases, there seems to be little evidence of such scandal. It may be that given the mobility of our society and the largeness of our congregations people are not well-known enough for their personal marital circumstances to be public knowledge. Or it may be that now that most Catholics know someone who has struggled through divorce and remarriage, often their own family members, there is an understandable desire to see such remarried welcomed and accepted by the Church. Many pastors report a charitable openness on the part of most parishioners in supporting the Church's outreach to the remarried.

All Catholics in second marriages are not covered by the ''good conscience'' solution. A distinction is made between those who are morally certain that their first marriage was invalid and those who are certain that their first marriage was valid. There are many persons who insist that their first marriage was never a marriage; it was undermined by emotional illness or serious personality defects from the start. Yet others insist that they had a very good Catholic marriage for many years, but it died. Dramatic personality changes in mid-life, some personal tragedy which seemed to destroy the husband-wife relationship, or another person who wins away a spouse—all destroy marriages and lead to divorce. Many sincere Catholics, after the breakup of such marriages, refuse to apply for annulments, convinced they had a good marriage and would still be married, if only. . . . The traditional position is that those who are sure their marriage was not a good one from the start, not ever valid, may receive Communion even if they do not have an annulment. Whereas those who are convinced that their first marriage was a good one for a long time, may not. Those in the second category, it is proposed, are bound by the first marriage, and so the second marriage is certainly invalid. Those in invalid second marriages, as we have seen, may not receive Communion.

As might be expected that latter position is being questioned by theologians and canonists today. Must those who have been unable to live up to the Church's teaching and laws on marriage always be excluded from Eucharistic sharing? Those open to the reception of Communion by such remarried persons suggest that where a first marriage is irretrievably lost, and where one or both parties have entered into a stable new marriage where he or she is faithful to obligations which remain from that first marriage—such as raising children in the faith or financial support—they should be offered the Eucharist as a spiritual resource to help them handle the demands of a new marriage. The Pope and bishops have insisted that the remarried are to live up to all the obligations of Christian life. How can they be asked to bear such burdens and not be offered the ordinary food of Christians? There is growing evidence that some pastors are supporting such remarrieds in taking the Eucharist, convinced that they are good people and need the Eucharist. Further, they ask, is not the Eucharist a meal of reconciliation for the flawed and imperfect? Did not the Lord share meals in the Gospels with the outcast and the suffering?

This is the frontier area of ministry to the remarried in the Church today. An enormous amount of progress has been made in cleansing the Catholic community of negative, condemnatory attitudes towards the remarried and a process of reconciliation is underway. How many of these people, and for what reason, can be offered the Eucharist needs much further reflection and lived experience.

Reconciling the remarried has many pastoral benefits for the parish community. It has also saved whole families for the Church; a decade ago many of these families would have been cut off with their children. Further, there are numerous reports of Catholics remarried to persons of no prior religious affiliation who are now coming forward and requesting to be baptized or received into the Catholic Church. Most of all, it brings into the life of the parish the rich Christian experience of men and women who have endured the heartbreak of broken marriage in faith and dared to love again.

33.

What God Has Joined Together . . .

Bernard Cooke

One of the most important essays in this collection is the following one by Bernard Cooke. It deserves and probably will require more than one reading. In addition, most students will probably need to be introduced to some of its technical language and to the process thought on which it is partly based.

Cooke looks at marriage from the points of view of biology, social institution, distinctive personal relationship, and biblically based covenant. He also brings to his discussion the three theological shifts outlined at the start of his essay. Basically, he proposes that marriage as the paradigm form of human friendship should at its best mature into an increasingly indissoluble bond between persons. It is his argument for this position that calls for study. Two sets of questions will aid in examining it, those Cooke himself asks at the very end of his essay and these following ones:

1. *Cooke takes very seriously the deep sacramental value of marriage. Which of his statements were for you the clearest expression of this value?*
2. *What does Cooke mean by saying "The source of whatever indissolubility attaches to a particular marriage must be the character of the marriage itself" and its symbolic import as a Christian sacrament?*
3. *What do you say to Cooke's point that a marriage becomes increasingly indissoluble as it becomes increasingly Christian?*
4. *Is Cooke contradicting himself when he says that a couple can be truly married but at the same time still in the process of becoming married to each other?*
5. *How would you summarize Cooke's position on indissolubility?*
6. *How would you summarize his position on sexual intercourse?*

Among the pastoral problems to which Catholic theology should address attention, few have as widespread impact as the question of the indissolubility of

From *Commonweal* (27 March 1987): 178–182.

Christian marriages. That we are seriously reexamining this element of Catholic teaching reflects pastoral anxiety for the well-being of the millions of women and men in situations that have separated them from their Catholic roots. But it reflects also the broadened context of doing theology today, and it is to this aspect of reflection on indissolubility that I wish to direct my remarks.

Today's developments in theology constitute a multifaceted phenomenon; within this complex change, it seems to me that three shifts are of special relevance to the topic of our discussion. (1) Today we are using the life experience of believing Christians, as individuals and as communities, as the starting point of our theological reflection. While other sources of insight—Scripture, traditional teaching, liturgy, etc.—enter in as principles of interpretation, it is the providential action of God in people's lives that provides the immediate "word" of revelation with which we must deal as theologians. (2) We are gradually absorbing into our theological process the historical consciousness, the awareness of *process,* and the general acceptance of evolution that are hallmarks of modern Western thought. In doing so, we have rediscovered the eschatological perspective that characterizes biblical thought. (3) We are beginning to theologize ecumenically, realizing that we cannot ignore other Christian traditions—for that matter, religious traditions other than Christian—in our attempts to understand more deeply and accurately the workings of the divine with humans.

Let us, then, draw upon the first of these methodological shifts, namely the use of Christian experience as a basis for reflection. Here we are faced with the concrete and unavoidable reality: according to every ordinary observable measure, large numbers of Catholic marriages do, in fact, dissolve. Can we in the face of this widespread experience justifiably say that these marriages continue to exist?

Any response to that question must distinguish among several meanings of "marriage." For example, at the most elemental biological level, where marriage involves two people mating for the continuation of the race, it is undeniable that in many cases such a strictly biological relationship does not and need not continue beyond a certain point. As a social institution providing stability for the process of begetting and raising children, marriage can take various forms, including, in modern societies, persons being involved in a sequence of marriage-divorce-remarriage. As a distinctive personal relationship involving a unique sexual commitment, marriage does suggest some aspect of indissolubility—at least many people do believe and hope as they marry that this special self-giving is "forever." Nonetheless, the large number of people who have given up this attitude for one of remaining together "as long as things work out" suggests that there is no self-evident and adequate grounding for indissolubility in some promise intrinsic to marital sexual self-giving.

Finally, as a paradigm form of human friendship, marriage at its best should certainly mature into an increasingly indissoluble bond between persons; but human experience teaches us the bitter lesson that friendships, even long-standing and treasured ones, do not always stand the test of time. While it may always be "eternally true" that two married persons *were* close friends, if the friendship does cease, one simply cannot assert that it continues and constitutes indissolubility.

The reproductive drive of the species, society's concern for successful child-bearing, marital sexual intimacy, human friendship—all these dimensions of marriage certainly point to some degree of permanence, but not to sufficient grounds for universally attributing indissolubility to all marriages, including Catholic marriages.

We enter a somewhat different realm, however, when we regard Catholic marriage in the light of the bibilical/theological category of covenant. In this context, the contractual aspect of the pledge between woman and man in marriage takes on added dimensions: the couple commit themselves to one another, but they also commit themselves *as a couple* to participate sacramentally and ministerially in the life of the Christian community; they commit themselves to shared discipleship and a life together of working for the establishment of the Kingdom of God. Not that all Catholic couples as they begin their married life are conscious of and open to this broader meaning of their marital contract, but this is the intrinsic reality of Christian marriage which we can hope will become understood and appreciated by people.

Certainly, we are closer to a grounding for indissolubility when we regard Catholic marriage as Christian covenant, for the promise involved has a clearly eschatological orientation; it reaches in its significance to the divine. But what are we to say when the contract has been broken by one or both parties? We might in some cases say that there has been infidelity that extends beyond the two persons to the Christian community and to God, that there has been sinful negligence or malice, that some responsibilities may still remain from the earlier covenant commitment. But can we say, for example, that an innocent and betrayed person in a marriage, a person who has, clearly been irrevocably deserted, is still involved in a one-sided contract? Can a person remain committed to the Christian community to live out a sacramental relationship that is existentially impossible?

One can, of course, give an essentially legal response to this question: we have a law, a law that gives expression to a view of Catholic marriage which we are not free to abandon. Much as it pains us, the overall common good requires that exceptions not be made, so that the indissoluble character of Christian marriage will be safeguarded. But does the preservation of this ideal demand the absolutely universal implementation of this rule? Perhaps this law itself is meant to be the statement of an ideal toward which Catholics should strive with varying degrees of success or failure. Having raised that question, let us bracket it for the moment and come back to it after we have treated some other elements of sacramental theology.

A final possibility for grounding the indissolubility of Christian marriage lies in the sacramentality of the two Christian persons as they live in relationship to one another. They are the sacrament, not simply because they are recognizable in the community as the two who publicly bound themselves by marital contract, but because and *to the extent* that they can be recognized as translating Christian faith into their married and family life. For Christians the parameters, of personal destiny, of personal responsibility and commitment, of personal development and achievement, in brief of human life, are broadened by the revelation contained in the life and death and resurrection of Jesus of Nazareth. This is true of individual human existence; it is true of the shared existence that is marriage.

When two Christians are married they commit not only their growth as persons to one another; they commit their faith, their relation to God in Christ to one another—obviously, not totally, but to a very considerable degree. The concrete interaction with one another in their daily life will unavoidably serve as "word of God" in the light of which they will develop their self-image, their freedom, their values, their faith and hope and love.

But God's word, no matter what the medium of its transmission, has always been a promise of unconditioned divine fidelity. No characteristic is more emphasized in the biblical literature; Israel's God is a faithful God. When we come to the New Testament, the raising of Jesus from the dead is seen as the culminating fulfillment of God's promises, the supreme proof of divine fidelity. And the question comes then: Can a Christian marriage truly sacramentalize, i.e., both speak of and make present, this divine fidelity unless it itself bears the mark of unfailing, irrevocable endurance? Can a marriage speak experientially about a divine love that never fails, unless it itself is lived as a relationship that is indissoluble? Or—to change the question slightly, but perhaps importantly—if it is not lived this way can one speak of it as sacramental?

In this context, we can return to the questions raised earlier about the commitment implicit in marital intercourse. That there is some special personal commitment signified by this action is hard to deny, but it is also hard to deny that it is signified only to the extent that this act is one of genuine personal love, expressive of each person's selfhood and honest respect of the other's selfhood. The extent to which an actual situation of sexual interchange symbolizes an irrevocable, i.e., indissoluble, commitment of each to the other seems, then, to be commensurate with the attitudes, understandings, etc., of the two people engaged in marital intercourse. Apparently we must ask, in a somewhat more restricted form, the question we just raised about the broader reality of Catholic marriage: When are we justified in applying the term "sacramental"?

Without suggesting any final answer to these questions, it does seem that we can associate the indissolubility of Christian marriage more satisfactorily with the sacramentality of marriage than with any other aspect. Historical studies have pointed out how the meaning of "sacrament" as applied to marriage has shifted from the emphasis on "binding promise" which it had in Augustine's explanation of Christian marriage to greater stress in medieval and subsequent centuries on the meaning of "Christian symbol." On the other hand, comtemporary sacramental theology has increasingly broadened the scope of sacrament beyond simply the liturgical ritual; and it has moved away from the "automatic effect" mentality that characterized so much post-Tridentine explanation of sacraments and has instead re-emphasized the extent to which the sanctifying effectiveness of sacraments depends on the awareness and decisions of the Christian people involved in one or other sacramental context.

Inadequate as our understanding of the sacramentality of Christian marriage is, it does seem to provide some focus for the practical pastoral judgments about indissolubility that we face at this moment in Christian history. Perhaps we can sharpen the focus a bit by raising the question: If indissolubility is in some way and to some degree ''intrinsic'' to Christian marriage, what is the source of this indissolubility *in a particular case?*

Is God the source—or, to put it more bluntly, is God doing something extra to make a particular Christian marriage indissoluble? Unless I misread present theological developments, it seems that we are presently moving toward a reinterpretation of ''providence'' in terms of the divine *presence* in the lives of humans. But if this is so, and if we then apply this to marriage, it would accentuate the importance of awareness and free decision in the sacramentality of any given marriage, for God's presence to humans is conditioned by their conscious and free acceptance of the divine saving love.

Is the church the source? Does the Christian community, more specifically do the bearers of authority in the church, have the power to make Catholic marriages dissoluble? And if they do have such power, is their exercise of this power the cause of Catholic marriages being indissoluble? I know of no theological voice that would clearly respond ''yes,'' that would go beyond claiming for the church the power to proclaim and defend and socially implement (within the church's own internal life) an indissolubility that already exists in Christian marriage prior to any church action or regulation.

But has not the official church, at least as far back as Trent, claimed the power to govern the *existence* of Catholic marriages by its legal activity? Despite the most Christian self-giving on the part of two devoted Catholics, the absence of the legally established form or of proper delegation on the part of the witnessing cleric rendered their marriage invalid.

For example, years ago, when I was studying the canon law of marriage, the teacher highlighted the importance of ''proper form'' by repeating a canonical ''horror story''—whether factual or not, the story quite clearly made its point. According to the account, a socially prominent young couple, wishing to avoid all the fuss of a big public wedding celebration, went for advice to the chancellor of a large U.S. diocese, since he was a close friend of the woman's family. Sympathetic to the young people's desire, he offered to marry them privately in his office; so, he requested his secretary to join them as witness to the marriage, the marriage was performed, and the young couple on their honeymoon informed their respective families of the fait accompli. However, the next day the chancellor—obviously with great embarrassment—realized the lack of due form because there had been only the one witness to the marriage. Clearly, it would have been catastrophic to contact the newly married in the midst of their honeymoon and ask them to return so that they could be married. Legalism was able, however, to triumph: the chancellor obtained a ''sanatio in radice'' and the young couple never had to know that they began their married life in a state of material sin.

Common sense seems to say that there is something wrong here. Let us suppose that the diocesan chancellor had never realized his error, and that without any legal "sanation" the two people had lived a life together that reflected to their children and to all who knew them the transforming presence of God's love. Could one truly say that there did not exist a deeply sacramental Christian marriage? My purpose in citing this example is not to ridicule canonical arrangements in the church; rather, it is to raise some basic questions about ecclesiastical claims to make things be or not be. More precisely, it is to question ecclesiastical power to condition the indissolubility of marriages.

We seem to be left, then, with no other clear alternative than the one we have already discovered; the source of whatever indissolubility attaches to a particular marriage must be the character of the marriage itself, more specifically its symbolic import as a Christian sacrament.

Up to this point our reflection together could quite justifiably be faulted for the static way in which it has treated marriage, so let us examine the indissolubility of Christian marriage from the perspective of *marriage as process*. Marriages come into existence over a considerable length of time, conditioned by any number of occurrences and experiences and choices, progressing—if they do progress—through stages of change that find their Christian explanation in terms of the mystery of death and resurrection. Men and women are gradually initiated into marriage as a human relationship and a Christian sacrament; the initiation is never completed in this life—no more than is a person's lifelong initiation into Christianity, for becoming married is for most Christians a major element in the broader initiation into the Christ.

It would seem, then, that one should not talk about a marriage as being completely or absolutely indissoluble but as becoming increasingly indissoluble as it becomes increasingly Christian; the more profoundly Christian a marriage relationship becomes, the more inseparable are the two persons as loving human beings, and the more does their relationship sacramentalize the absolute indissolubility of the divine-human relationship as it finds expression in the crucified and risen Christ. Exactly how all this will occur in a given instance is as diverse and distinctive as are the people involved and the overall social situation of a given culture or historical period.

To put it in biblical terms, a Christian marriage, like any other created realities, does not exist absolutely; like anything in creation, particularly anything in human history, a marriage exists eschatologically; it is tending toward its fulfillment beyond this world. However, the fact that it does not yet have in full fashion the modalities—such as indissolubility—that should characterize it does not mean that it is devoid of them. A Christian marriage is indissoluble, but short of the eschaton it is *incompletely indissoluble*. Perhaps we could profitably borrow a notion from recent New Testament scholarship, namely "realized eschatology." Christian marriage already realizes to some degree the indissolubility which can mirror the divine fidelity to humans, but it cannot yet lay claim to the absoluteness which will come with the fullness of the Kingdom. Similarly, two Christians can be very genuinely and sacramentally married, but they are still being married to one another; their union can become yet richer and stronger.

One wonders if the understanding of Christian marriage has not for centuries suffered the fate of being overly structured and frozen by the use of Greek categories of thought with their presumptions of universality and absoluteness. Since "absolute" is a characteristic reserved to divinity, one cannot strictly speaking apply it to any created reality or to any bit of human knowledge. On the other hand, the view of all creation as eschatological accords with the first of all biblical commands, "I alone am the Lord, your God."

Indissolubility is an aspect of the intrinsic finality of any marriage, more so of a Christian marriage because of its amplified significance. As such, it shares in the responsibility to fulfill that finality which a woman and a man undertake when they enter upon a marriage. Indissolubility is something they should strive to intensify in their shared life. But that does not say that it is impossible for them to fail at this task, impossible for the actual indissolubility of a marriage to gradually weaken and ultimately disappear.

Perhaps we can and must say that the *promise* not to engage in marital intimacy with any other person, the promise that each party made at the time of beginning their marriage, remains in force no matter what happens. Perhaps we can and must say the *responsibility* for the other rests permanently on each of them. But how can we say that a relationship that in its human and existential aspects, and therefore in its sacramentality, has dissolved is indissoluble?

The contemporary church is rapidly regaining its sense of Christian existence as a process, a lifelong initiation into relationship with the Christian community and with the risen Lord. This is the clear import of the post-Vatican II revision of the rite for the initiation of adults. As in the past, liturgical action points the way for our theological reflection and our doctrinal clarification: *lex orandi, lex credendi.* "Being Christian" is something a person only gradually and incompletely achieves.

For Christians, married life is meant to share in this initiation into Christ. The clear conclusion is that an individual Christian marriage does not from its first moments completely reflect the Christ-mystery, completely reflect the indissoluble bond of saving love that links Christ with his spouse, the church, any more than a person is completely Christian with baptism. One *becomes* Christian; one *becomes* married.

By way of corollary, it might be well to extend these remarks to the notion of marital consummation. There is a long history of the role of first sexual intercourse between a couple as establishing a societal bond, and along with this a long history of Christianity considering first marital intercourse as somehow intrinsic to the marriage contract and therefore to the very existence of the marriage. I have no intention of summarizing, even briefly, that history. Suffice it to recall the operative church law that regards a marriage soluble if it is only *ratum* and not *consummatum*.

What I do wish to do is suggest the impropriety of such an abstract understanding of sexual intercourse, especially of marital intercourse. It is true that for two people deeply in love, there is often profound meaning in their first full sexual intimacy, but theirs will be a sad married life if they do not progress in their self-giving far beyond this first experience. Too much of the discussion of sexual intercourse among moral theologians and canonists has forgotten that it is a *human* activity, even though they have verbally nodded in that direction. Precisely because it is so human—distinctive with each couple, fragilely linked with all the other elements of a couple's relationship to one another, symbolically expressive of so much that cannot find explicit verbalization yet is itself in need of communication between persons to make its meaning clear—truly human sexual intercourse needs to be learned over a long period of time. And when one introduces Christian significance into this action so that it can become the heart of the marriage's sacramentality, the need for lifelong learning becomes only too apparent. Sexual intercourse does consummate Christian marriage, but only in this context of ongoing personal intimacy, for it can only authentically say what the two Christians honestly are for one another.

Tragically, very many marriages are scarcely consummated as personal relationships; they do not grow. Among these are many that begin in a Catholic wedding ceremony. If consummation is intrinsic to the establishment of a Christian marriage, one can only wonder how many marriages qualify as "Christian," and therefore how much claim they can lay to indissolubility.

What can one say by way of conclusion? A list of questions:

- To what extent does modern process view of reality affect the way in which we consider a particular Christian marriage as indissoluble?
- If Christian couples themselves are the sacrament of Christian marriage, and couples obviously differ greatly in the extent to which they are genuinely Christian, to what extent is a particular marriage truly sacramental, to what extent does it actually symbolize the love between Christ and the church?
- And if the special indissolubility of *Christian* marriage is tied to sacramentality, in what way does indissolubility pertain to marriages that seem to have lost all operative sacramentality?
- Or are we to say that the covenant pledge, with one's partner and with the Christian community, which one took at the wedding ceremony remains a promise to the community even if the actual human marriage relationship dissolves? In this case the indissolubility attaches to the overall ecclesial sacramentality of the institution of Christian marriage rather than to the sacramentality of this or that particular marriage union.
- But, to return to our emphasis on doing theology out of experience, is not the experience of "getting married" and the significance (sacramentality) attached to it one of promise to the other person rather than to the community?
- Finally, it seems that we need a somewhat new though tradition-respecting look at indissolubility to discover whether we are justified in applying it as absolutely as we Catholics have done in more recent centuries. It strikes me that a more flexible and individualized approach will still continue to honor the teaching that Christian marriage is of its nature indissoluble.

34.

Canonical and Theological Perspectives on Divorce and Remarriage

Charles Guarino

Though the following review of Church teaching and law on marriage may at first seem very different from Bernard Cooke's essay, they are related in various ways. While Cooke works out a theological reconsideration of indissolubility, Charles Guarino lays out for us the current Church law and the thinking behind it.

This information is crucial for understanding current church teaching and church law on annulment and indissolubility. Like every other society, the Church has laws governing marriage. Guarino's presentation is notable for the clarity with which he presents current official teaching and the law flowing from it.

1. *What is Guarino's rationale for saying that annulment is not the Catholic Church's form of divorce?*
2. *What are the three requirements for a valid marriage in the Church?*
3. *Why is consent such a key issue in the annulment process?*
4. *How do the "subtle pressures that can interfere with . . . valid consent" explain the special concern for marriage preparation in the Catholic church?*
5. *How are the criteria for marriage preparation programs helpful or unhelpful?*
6. *In your own words, explain the "internal forum" solution.*

From Steven Preister and James J. Young, *Catholic Remarriage*, Pastoral Issues and Preparation Models (New York: Paulist Press, 1986), pp. 47–55.

Juridical Requirements for Marriage in the Church After a Divorce

In the Church's theology of marriage the "covenant or community of life and love," is meant to be a stable, permanent and faithful bond. Thus, the Church cannot simply allow marriage after divorce. In order for marriage to be celebrated in the Church after a divorce, the previous, presumably valid, sacramental bond must be sufficiently examined and a decision made about the nullity of that sacramental bond, determining whether it is truly a binding union.

This is *not* the Catholic Church's form of divorce, as some mistakenly view the annulment process. Rather, it is a logical progression from our theological belief that the marital relationship is a permanent and faithful union and, therefore, not to be tampered with or interfered with or "put asunder" by any authority, civil or ecclesiastical. In other words, no power on earth can dissolve a truly sacramental union of two spouses.

In *The Sacraments in Theology and Canon Law,* Volume 38 of the Concilium series, we are given the following, very concise explanation of the indissolubility of marriage:

> In principle, every marriage is indissoluble; the sacramental character of the marriage between two baptized partners reinforces this indissolubility. It is totally impossible for the partners to break the bond themselves. This is called the intrinsic indissolubility which allows of no exceptions. The sacramental marriage bond between two baptized partners, when consummated (*matrimonium ratum et consummatum*) cannot be set aside by ecclesiastical authority. This is called the extrinsic indissolubility. Any marriage which is not sacramental, or does not exist between two baptized partners, and/or is not consummated, can be dissolved by the Church in principle and under certain conditions (Edelby, et al.: 45).

While upholding its belief in the permanence of the sacramental marriage bond, the Church is also responsible for providing justice for those whose marriages have failed and who now seek to enter what they often see as the true sacramental marriage, one which bears the marks of a relationship that is more mature and developed than a previous impulsive or ill-defined relationship. If a previous marriage was truly lacking from the very beginning at least one essential element for a true and valid sacramental bond (i.e., for "a community of life and love," trust, respect and all of those essential elements that specify the community of life) then the Church rightfully declares such a marriage null and thereby recognizes those spouses to be free to prepare to enter a true sacramental marriage for the first time.

The question might be raised: What about the condemnation of divorce and remarriage found in the teaching of the Synoptic Gospels and St. Paul? All agree that what is contained there is the authentic teaching of Jesus Himself, who clearly repudiated the kind of divorce which existed in His day, and requires that "man may not separate what God has united."

When the Church allows divorced people who have been declared free to marry by juridical procedure to celebrate marriage canonically (i.e., before a priest and two witnesses [canon 1108]), does it, in effect, contradict Christ's condemnation of divorce and His teaching on the permanence of the marital bond? Clearly not. When the Church declares the *nullity of a marriage,* it is stating that a particular element required for valid matrimonial consent or the ability to fulfill the marital commitment was lacking at the time of the wedding, and so the marriage was not valid. (This is explained in greater detail in the section on *The Annulment Process and Its Effects*). When the Church grants a *dissolution of a marriage,* it is not declaring a marriage invalid but rather setting aside the presumably valid marriage in favor of another marriage permitted in certain, specified circumstances (this, too, is explained later in this chapter).

Whether declaring the nullity of a marriage or granting a dissolution of a marriage, the Church is not sanctioning divorce or denying the indissolubility of marriage. Divorce is the result of civil court proceedings whereby married persons become legally separated and the marriage is declared null and void. In declaring the nullity of a marriage, the Church is stating that it was invalid from the very beginning. In granting a dissolution of a marriage, the Church is setting aside a nonsacramental marriage so that a person who has become divorced may marry in the Church. It is important to understand this distinction between divorce and the nullity and dissolution of a marriage to appreciate that in its ministry to the separated and divorced, the Church continues to uphold Christ's teaching.

The Church recognizes that divorce is a tragic human failure. But when these things happen irretrievably, the apodictic condemnation of their possibility does not help us deal with their reality. The Church must always be in a position of reconciling Christ's teaching on marriage with God's call to life in peace within the context of a society where nearly one half of all marriages are ending in divorce.

The Church is always called to be the great healer, the source and sacrament of God's mercy and reconciliation. Those who can be declared free from a previous failed marriage through acceptable juridical procedures of the Church Tribunal, should be freed and given every possibility to marry in the Church and be nourished by its rich sacramental life.

The Church's ministry, therefore, to the separated and divorced, is nothing less than the fulfillment of her role as Christ present in the needs and pain of His wounded people.

Canonical Requirements for Marriage in the Church

Recognizing the Church's need to minister to the separated and divorced, let us consider what is required for marriage in the Church. What is stated here and specified in the canons of the Code cited in the following pages applies to all those preparing to enter marriage in the Church, including those whose previous marriages have ended in divorce.

Marriage in the Church presupposes the following:

1. The understanding that marriage is a covenant or a community of life and love that is permanent, faithful and fulfills itself in the procreation and education of children.
2. The freedom to marry.
3. The ability to enter marriage and fulfill its essential obligations.

These requirements for valid marriage are clearly enunciated in the following canons of the Revised Code of Canon Law:

> *Canon 1096:* For matrimonial consent to be valid it is necessary that the contracting parties at least not be ignorant that marriage is a permanent consortium between a man and a woman which is ordered toward the procreation of offspring by means of some sexual cooperation. Such ignorance is not presumed after puberty (*Code:* 399).
>
> *Canon 1055:* The matrimonial covenant, by which a man and a woman establish between themselves a partnership of the whole of life, is by its nature ordered toward the good of the spouses and the procreation and education of offspring; this covenant between baptized persons has been raised by Christ the Lord to the dignity of a sacrament. For this reason a matrimonial contract cannot validly exist between baptized persons unless it is also a sacrament by that fact (*Code:* 387).
>
> *Canon 1056:* The essential properties of marriage are unity and indissolubility, which in Christian marriage obtains a special firmness in virtue of the sacrament (*Code:* 387).
>
> *Canon 1057:* Marriage is brought about through the consent of the parties, legitimately manifested between persons who are capable according to law of giving consent; no human power can replace this consent.
>
> Matrimonial consent is an act of the will by which a man and a woman, through an irrevocable covenant, mutually give and accept each other to establish marriage (*Code:* 387).
>
> *Canon 1134:* From a valid marriage arises a bond between the spouses which by its very nature is perpetual and exclusive; furthermore, in a Christian marriage, the spouses are strengthened and, as it were, consecrated for the duties and dignity of their state by a special sacrament (*Code:* 411).

These canons deal with the required understanding of what marriage is and of what marital consent consists.

Other canons in the Revised Code further specify the basic requirement found in canon 1066, which states: "Before marriage is celebrated, it must be evident that nothing stands in the way of its valid and licit celebration" (*Code:* 391). These further requirements are described in the following canons:

> *Canon 1067:* The conference of bishops is to issue norms concerning the examination of the parties, and the marriage banns or other appropriate means for carrying out the necessary inquiries which are to precede marriage. The pastor can proceed to assist at marriage after such norms have been diligently observed (*Code:* 391).
>
> *Canon 1069:* All the faithful are obliged to reveal any impediments they are aware of to the pastor or to the local ordinary before the celebration of marriage. (*Code:* 391).
>
> (Canons 1073–1082 explain the diriment impediments as those that make a

person incapable of validly contracting a marriage, and canons 1083–1094 list those impediments.)

Canon 1108: Only those marriages are valid which are contracted in the presence of the local ordinary or the pastor or a priest or deacon delegated by either of them, who assist, and in the presence of two witnesses, according to the rules expressed in the following canons, with due regard for the exceptions mentioned in canons 144, 1112 1, #116, and 1127 #s 2 and 3.

The one assisting at a marriage is understood to be only that person who, present at the ceremony, asks for the contractants' manifestation of consent and receives it in the name of the Church (*Code:* 403).

Canon 1063: Pastors of souls are obliged to see to it that their own ecclesial community furnishes the Christian faithful assistance so that the matrimonial state is maintained in a Christian spirit and makes progress toward perfection. This assistance is especially to be furnished through:

1. Preaching, catechesis adapted to minors, youths and adults, and even the use of media of social communications so that through these means the Christian faithful may be instructed concerning the meaning of Christian marriage and the duty of Christian spouses and parents;
2. Personal preparation for entering marriage so that through such preparation the parties may be predisposed toward the holiness and duties of their new state;
3. A fruitful liturgical celebration of marriage clarifying that the spouses signify and share in that mystery of unity and of fruitful love that exists between Christ and the Church;
4. Assistance furnished to those already married so that, while faithfully maintaining and protecting the conjugal covenant, they may day by day come to lead holier and fuller lives in their families (*Code:* 389).

(Canons 1064–1072 further describe the pastoral care needed to prepare couples for marriage and what must precede the celebration of marriage in the Church.)

Freedom to Marry

Besides these canonical requirements for marriage in the Church, those whose marriages have ended in divorce have the further obligation, before attempting to enter a canonical union, to be declared free to marry. It must be determined by juridical procedure that one of the following situations is true:

1. The previous marriage was never valid because of "a lack of form," i.e., the Catholic spouse, who is required to be married before a priest and two witnesses (canonical form—canon 1108), married without observing this requirement and without any dispensation from canonical form.
2. The previous marriage was not valid because of the existence of at least one diriment impediment at the time of the marriage (canons 1073–1094).
3. The previous marriage was not valid because of the presence of certain nullifying factors, i.e., canonically acceptable and provable norms that prevent a valid union (canons 1095–1107).

Beyond these more common grounds for declaring a marriage null because of invalidating causes, the Church has historically allowed for the dissolution of a marriage in certain cases:

1. The oldest case, finding its origin and scriptural basis in the writings of St. Paul, is that of a marriage between two partners who are not baptized. This is referred to in 1 Corinthians 7: 12–16, and is designated by the name of "Pauline Privilege." The conditions for this are found in canon 1143 which states: "A marriage entered by two non-baptized persons is dissolved by means of the pauline privilege in favor of the faith of a party who has received baptism by the very fact that a new marriage has been contracted by the party who has been baptized, provided the non-baptized party departs.

"The non-baptized party is considered to have departed if he or she does not wish to cohabit with the baptized party or does not wish to cohabit in peace without insult to the Creator unless, after receiving baptism, the baptized party gave the other party a just cause for departure" (*Code,* canon 1143: 413).

Also in canon 1147 reference is made to the pauline privilege: "For a serious cause the local ordinary can permit the baptized party who employs the pauline privilege to contract marriage with a non-Catholic party, whether baptized or not, while observing the prescriptions of the canons on mixed marriages" (*Code:* 413).

2. The second case dates from the Middle Ages. It concerns a marriage that is not consummated. The conditions in this case are that at least one of the partners wants to terminate the marriage and the lack of consummation has been established with moral certitude. Such a dissolution must come from the Pope. Canon 1142 of the Revised Code states: "A non-consummated marriage between baptized persons or between a baptized party and a non-baptized party can be dissolved by the Roman Pontiff for a just cause, at the request of both parties or of one of the parties, even if the other party is unwilling" (*Code:* 413).
3. The third case, known as "the privilege of faith" is the dissolution of a marriage between two non-Catholics, of whom at least one is not baptized either before the wedding or during the entire common life of the marriage. The conditions are that one partner joins the Catholic Church or wishes to marry a Roman Catholic and marital life with the former spouse can no longer be continued. This impossibility of continuing marital life with the former spouse must not be attributable to the person seeking the privilege of faith.

   This dissolution is granted by Papal authority through the Sacred Congregation for the Doctrine of the Faith.

   It is also possible to apply this privilege in a situation where the petitioner does not wish to convert but desires to marry a Catholic who wishes to continue to live his or her baptismal commitment.
4. The fourth case, still more recent, is that of a dissolution of a marriage between a Catholic and a non-baptized partner after a dispensation from the impediment of disparity of worship. This non-sacramental marriage has irremediably broken down. Now the Catholic wishes to enter into a sacramental marriage, i.e., with a validly baptized Christian. This case also is an

extension of the Pope's role as Vicar of Christ and part of his pastoral concern for his flock.

It should be noted that this dissolution will not be granted to permit the Catholic to enter into another disparity of cult marriage (i.e., marriage with an unbaptized person).

Having briefly described above the various solutions historically employed by the Church in its ministry to the divorced, we now concentrate our attention on the formal juridical procedure known as annulment.

The Annulment Process and Its Effects

An annulment is an official declaration by the Catholic Church's external forum, the tribunal, that a particular marriage *de facto* and *de jure* was lacking, from the very beginning, some essential element for a valid, permanent, sacramental union.[1]

A declaration of nullity is granted only when it can be shown, through the facts of a particular marital history and the substantiating testimony of witnesses, that some juridical defect rendered a particular marriage not valid, despite all outward appearances, despite even the good faith of the partners and despite the procreation of children subsequent to the wedding. (It should be noted here that an annulment in no way affects the legitimacy of children born of such a marriage. In fact, canon 1137 specifically states that any children born of a marriage later declared null are legitimate.)

Since marriage occurs by consent, freely given and with full knowledge (canon 1057), the question often placed before a tribunal concerns this notion of consent under one of these headings (or in some cases, under all three):

1. Did both partners clearly understand the nature of marriage as a "community of life" and what such marriage would require of them?
2. Did both partners freely accept marriage as a lifelong commitment?
3. Did both partners have the personal capacity to carry out that to which they consented, i.e., to form a community of life with the chosen partner?

Let us take a closer look at the quality of this marital consent that is, by nature of the vocation of marriage, far more exacting than the consent given in ordinary decisions.

Consent to marry is the most momentous decision a person can make because its effects endure beyond the here and now; it is a lifelong choice with far-reaching effects.

Consent must be free and discerning. External or internal pressure, which can significantly reduce freedom and undermine critical judgment, could impair consent to such a degree that essential requirements for such a binding decision as marriage are not fulfilled.

For centuries, theologians have recognized that strong emotion and external pressure could weaken free choice and diminish culpability as far as sin was concerned. While the Church has been more cautious in applying these same principles to marital consent, it accepts the findings of modern psychology that show how unconscious motives and situational pressures can get in the way of freedom and judgment in decisionmaking. Such findings have greatly helped Church tribunals assess the adequacy of marital consent in particular cases.

Consider the following examples of subtle pressures that can interfere with freedom and discernment necessary for valid consent (Keefe):

1. Take the couple who have been sexually active and now the woman is pregnant. She rightfully refuses abortion. She does not want to give up the baby for adoption. The man responsible for the pregnancy feels trapped. He may have fine intentions, feeling honor bound to do "the right thing." One or both may see marriage as the only way out. Is this decision a free, mature choice of a lifetime marital partner, or is it a pressured solution to a problem?
2. What about the consent of the teenager, overwhelmed by infatuation with the only person ever dated, in love more with love than the person he or she consents to marry? Or the youngster with no critical appraisal of the character of the intended partner, and with meager appreciation of the financial responsibilities or marriage or the burdens of parenthood? Add to the picture, perhaps, the desperate need to escape an unhappy home life, marred by alcoholism or quarrels.
3. How would we assess the widower, still grieving for his deceased wife? He has a demanding job and is anxious for his young children. So he hastily remarries. Is he giving prudent, thoughtful marital consent or enlisting a housekeeper and stepmother for his children?
4. What sort of marital consent is given by a person with lukewarm, nominal faith, who has absorbed the divorce mentality which pervades United States culture, and the philosophy of casual sex which is daily T.V. fare?

While there is no automatic answer about the quality of marital consent in each of these examples, the average person would question the wisdom of such marriages and have serious misgivings about the freedom or discretion of the immature or agonized person taking marriage vows under such circumstances. In each case, if the tribunal thoroughly investigated the premarital history, it might well conclude that one or both of the partners could not freely and maturely choose to marry at that time, so marital consent was invalid.

Marital consent should not be free from undue pressures, but should also imply the capacity to carry out that to which the partners consent, namely a community of conjugal life that is perpetual and exclusive. Both partners must have the maturity to establish and sustain a mutually supportive communal relationship with one another.

Saying "yes" without the capacity to carry it out is invalid, even though a person takes marriage vows in good faith and with the best of intentions. St. Thomas Aquinas provided the basic principle that guides us here: "No one can oblige himself to what he can neither give nor do."

Prior to our more sophisticated understanding of human behavior, people presumed that everyone had what it takes to make a marriage work except, of course, the most overtly disturbed individuals. Before Vatican II, the Church considered the marriage contract principally in terms of procreative rights and obligations. The wider issue of a mutually supportive human relationship with all that is involved in creating and sustaining such a relationship was, while not ignored, given inadequate attention.

Given the understanding of marriage as a community of life and love, and the deeper appreciation of what goes into a successful interpersonal relationship, the Church now recognizes that for some persons, psychological problems are the consuming, motivating force of life, and so prevent a person from having the capacity to establish and maintain the close, empathic, cherishing relationship with a spouse that marriage requires. In other words, such a person does not have what it takes to develop the community of life that is the substance of the marital pledge.

In the past, the Church recognized that psychoses—the disintegrative mental illnesses such as schizophrenia, and the pathological condition of manic depression—could so impair mental and emotional stability that one's consent to marriage lacked the necessary discernment or capacity. More recently, with the contributions made by the behavioral sciences, the Church acknowledges that other dysfunctions of personality may render a particular marriage covenant impossible. For example, a homosexual orientation may prevent a person from fulfilling the demands of heterosexual intimacy required in marriage. Psychological impairment caused by alcoholism often undermines the capacity for a permanent community of life and love. Another group of emotional disturbances carries the label "personality disorders." Although these do not include acute episodes or bizarre features of psychoses or the disabling anxiety or symptoms of neuroses, they are, nevertheless, marked by deeply ingrained maladaptive patterns of behavior, usually recognizable by adolescence or earlier, which continue throughout most of adult life. While such persons may function well enough in certain areas, they are often psychologically unable to meet one essential criterion of marriage, the close and intimate personal relationship of mutual support and affection.

In general, when a diagnosis is given of a personality disorder, recent jurisprudence in the Church shows these disorders can affect the validity of marriage in one or all of three ways:

1. By depriving a person of the due discretion necessary for true marital consent.
2. By depriving a person of the internal freedom required to give that consent.
3. By rendering a person incapable of fulfilling the essential obligations of marriage.

Most annulments granted today in the United States are based upon what is called a "psychological incapacity for marriage," which is often attributable to the presence of these personality disorders in one or both of the spouses at the time of the marriage. The precise clinical labels of these disorders are not important in the annulment process. What is important is the realization that such psychopathology can make a particular marriage a morally impossible venture. The tribunals of the Church do not seek to assign blame for the marital breakup, but rather they seek only to understand the causes for the failed marriage and determine whether either or both partners lacked proper marital consent or the ability to carry out that consent.

Requirements of the Annulment Process

While procedures for an annulment may differ from one diocese to another, these are some basic steps that are fairly uniform throughout the Church that can serve as a guide to those seeking an annulment as well as those ministering to the separated and divorced who want to marry in the Church. The following outline, while not pretending to be a complete and detailed exposition of the annulment process, provides an understanding of the general procedures that must be pursued in the petition for an annulment:

1. Initial contact is usually made with the parish priest or other parish minister who should be equipped to evaluate the marital situation and make appropriate recommendations.
   The parish minister is the resource person who introduces a person to the ministry of the Church's tribunal and then assists that person in whatever may be required by the tribunal.
2. When the person makes contact with the tribunal, he or she will be required to provide a summary of the principal facts concerning family background, the courtship, the marriage and what led to the marital breakup. The salient facts of the premarital history are especially significant in the annulment process, since an annulment must be based upon the condition of both parties at the time of the marriage.
   Church law allows an annulment process only for those marriages that have in fact ended in divorce. Therefore, it is best for the parties concerned that the divorce be finalized before the annulment process is begun.
3. Church law requires that the other partner be informed of the petition for annulment and given the opportunity to present testimony. In addition, witnesses are needed to corroborate the facts presented by both partners and provide additional information where it is available. It is best if the witnesses are family members or friends who knew both partners before the marriage and would be able to testify about that important period in the marital history. It should be noted here that all persons are contacted individually and efforts are made in order to maintain the strictest confidentiality. However, both parties have the right to see the testimony presented unless the tribunal exercises its right to withhold that testimony for the good and protection of all parties concerned.
4. The formal hearing, normally held at the diocesan tribunal office, can include a tribunal judge, an advocate assigned to each of the parties to protect their rights, and a defender of the bond who monitors the proceed-

ings and sees to it that Church law is observed properly. Sometimes a psychiatrist or psychologist may be consulted as an expert witness and be requested to be present at the hearing.

5. When the investigation has been completed and a decision made about the nullity of the marriage, either partner has the right to appeal the decision if he or she feels aggrieved by it. If a formal appeal is made, it is appended to the case and forwarded to the next higher Church tribunal. The Defender of the Bond also has the right to appeal any affirmative decision.
6. The Code of Canon Law requires a mandatory review of every case in first instance which has declared the nullity of a marriage. This review, consisting of three judges and a defender of the bond, evaluates the merits of the case in first instance and either ratifies the decision of the presiding judge or calls for a new trial.

   While the review process has added to the amount of time needed to complete an annulment investigation, it also guarantees that greater care will be given to the total evaluative effort of discerning grounds of nullity.

The entire annulment process, unlike a divorce proceeding, is designed to give the partners in a marriage the opportunity to learn from these procedures the real reasons for the breakdown of the marital relationship, be healed of the wounds of that broken relationship, and grow and mature as a result of their healing and new understanding.

While many persons complain that it is wrenching for them to recall and sort out painful memories, they also agree it was a catharsis that helped them to discover some meaning in the tragedy of a broken marriage. They appreciate their new insights about themselves which help to deepen their sense of values. As a result, the annulment process can foster both psychological and spiritual growth.

Some people may argue that a decision within one's own conscience is sufficient for a person to marry and be reconciled to God. While some Catholic clergy and laity may subscribe to this conscience solution, most people whose marriages have ended need external confirmation that what they sincerely intended to be a permanent marriage lacked from the very beginning all that was required for a valid marriage. It is important to understand that the requirement for this external juridical process stems from the need to deal with an external reality, the marriage entered in a public forum. Since marriage is a public event, with both religious and civil effects, many believers feel the need for an external, independent, religious judgment that their marriage was not valid and that, therefore, they are free to marry.

But the greatest benefit of the annulment process for many who have already established a happy and stable union within another marriage is their ability to return to the sacraments, especially the reception of the Lord in Holy Eucharist, and the deepening of religious practice as a family celebration.

Such an effect of the annulment process establishes the real healing which the ministry of the tribunal seeks to provide for those whose previous marriages have ended in divorce and who, therefore, find themselves in a vulnerable position. Often it is this healing aspect of the tribunal process that convinces people of the value of pursuing an annulment even when there has been no remarriage or any present intention to marry in the Church.

In its pastoral concern, it is not uncommon for the tribunal to place a caution *vetitum* on certain people. A *vetitum* often requires counseling before marriage in the Church.

The *vetitum,* sometimes known as a prohibition, is a caution expressed before allowing marriage in the Church. In the strict sense, it is the responsibility of the proper local Ordinary to review situations involving a *vetitum* before giving his permission for the licit celebration of marriage in the Church. The *vetitum* safeguards the integrity of the sacrament of marriage and its essential elements, while requiring needed counselling of future spouses.

This prohibition becomes sensible especially in annulment cases involving psychological incapacity. If an annulment has been granted on such grounds, then before marriage can be celebrated in the Church, it must be determined that such psychological incapacity has been resolved.

The canonical basis for making such a requirement for counselling is found in canon 1066 which states:

> Before a marriage is celebrated, it must be evident that nothing stands in the way of its valid and licit celebration (*Code:* 391).

This is further strengthened by canon 1077, which allows:

> In a particular case the local Ordinary can prohibit the marriage of his own subjects wherever they are staying and of all persons actually present in his own territory, but only for a time, for a serious cause and as long as that cause exists (*Code:* 393).

And canon 1684 adds:

> After the sentence which first declared the nullity of marriage has been confirmed at the appelate level either by decree or by another sentence, those persons whose marriage was declared null can contract new marriages immediately after the decree or the second sentence has been made known to them *unless a prohibition is attached to this sentence or decree, or it is prohibited by a determination of the local ordinary* (*Code:* 605) (emphasis added).

Canon 1685 emphasizes this prohibition even more by requiring:

> Immediately after the sentence has been executed, the judicial vicar must notify the ordinary of the place in which the marriage was celebrated about this. He must take care that notation be made quickly in the matrimonial and baptismal registers concerning the nullity of the marriage and any *prohibitions which may have been determined* (*Code:* 605) (emphasis added).

Criteria for Evaluating Post-Annulment, Pre-Marital Requirements and Preparation Programs

In order to evaluate the effectiveness of all premarital requirements from counseling to catechesis, we suggest the following theological and canonical criteria:

THEOLOGICAL CRITERIA

1. A clear understanding of Christian marriage as a sacrament, i.e., as a covenant relationship by which a man and a woman establish between themselves a partnership of the whole of life, which by its nature is ordered towards the good of the spouses and the procreation and education of offspring (canon 1055), and acceptance of marriage as a permanent and faithful union (canon 1056).
2. Development of a spirituality of marriage based on an active practice of the faith which should include participation in the sacramental life of the Church and a willingness to be involved, according to their capacity, in their own parish community, especially in those programs geared to maturation in the faith and the development of marriage enrichment (canon 1063).
3. Examination of the motives for marriage and an appreciation of interrelated love-giving and life-giving aspects of the sexual relationship within marriage (canon 1096).
4. Facing up to the responsibilities of married life and parenthood, including the need to alter individual objectives into mutual goals; also the need to adjust to the responsibilities incurred from a previous marriage and which continue to be a responsibility in the present marriage (e.g., children from a previous marriage) (canon 1071 #1, 3).

CANONICAL CRITERIA

1. Before marriage in the Church, it must be determined that the parties are free to marry (canons 1058, 1060, 1066, 1077, 1707).
2. If an annulment has been received, were recommendations made during the annulment process for pre-marital counselling (usually to include the prospective spouse, especially if he/she has also been previously married), and if so, were these counselling requirements completed to the satisfaction of the Chancery office of the diocese in which the new marriage is to be celebrated (canons 1066, 1077)?
3. In the annulment process, was a *vetitum* placed on either party before marriage in the Church (e.g., was the requirement for counselling made or some other caution given)? If so, did the local Ordinary grant permission to celebrate marriage licitly (canons 1066, 1077, 1684)?
4. Is the spouse from a previous marriage fulfilling any and all responsibilities that are part of a separation and divorce decree (e.g., human and financial obligations regarding spouse and children of a former marriage, visitation rights, spiritual and educational concerns, etc.) (canon 1071 #1, 3)?
5. Those on the pastoral level charged with the responsibility of preparing couples for marriage in the Church must see to it that all other canonical requirements have been fulfilled for the valid and licit celebration of the marriage (e.g., valid baptismal certificates; freedom from diriment impediments; proper canonical form, etc.) (canons 1063, 1065, 1067, 1068, 1072).

The motivation of the Church carefully to prepare couples for marriage stems from its desire to be of service to God's people and its deep sense of responsibility to provide for our society and our world a strong model of stable Christian marriage, permanent and faithful to the end. It is not an ideal for the few who share our faith and theology of marriage, but rather should be a goal for all who seek marriage in the Church and a lifetime commitment to a community of love. Preparation for marriage in the Church, therefore, should include not only the juridical requirements for the reception of a valid sacrament, but also an evaluation of the faith dimension and understanding of those seeking to marry, so that the ministers of the Church can be assured that what the persons are seeking is, in fact, what the Church teaches marriage to be.

Extraordinary Procedures

In addition to the various canonical procedures that can free a person from a previous marital bond, there are also some individuals who, for one reason or another, are not able to pursue these ordinary juridical procedures, but are nevertheless anxious to participate fully in the Church's sacramental life, having already remarried or wanting to remarry. These extraordinary procedures, which we will briefly describe here, are called "the internal forum solution" and "the brother/sister relationship," applicable in very few and specific situations.

Internal Forum Solution

The tribunal is referred to as the "external forum" of the Church, because it is the public domain and it protects the rights of those persons who seek justice within the Church's judicial system, as it also protects the sacrament of marriage and the good of the Church. An official annulment granted in the external forum has legal standing in the Church and enables one to marry with the full public solemnity of Catholic rites, as well as participate in the sacramental life of the Church.

The "internal forum" is the forum of conscience; it is private, personal and not, in any way, public. It usually involves guidance from a priest or other ministerial person whose role is to discern the quality of the informed conscience and test the promptings of the Spirit in the person or persons whose previous marriage is believed to have been invalid. The previous marriage cannot, however, be annuled in the external forum because of one of several possibilities:

1. There is a lack of demonstrable evidence and witnesses.
2. The parties are unable to endure the annulment process or any other juridical procedure since it would cause excessive emotional strain and could seriously jeopardize the well-being of the parties.
3. There was lack of access to a tribunal, etc.

The internal forum solution is usually not programmed publicly, but rather, is discovered while one is living out the faith dimension of life, and discerns from the promptings of the Spirit that one is freed from any sin from a previous bond and desires strongly to receive the sacraments of the Church.

The internal forum solution is never used as a "shortcut" approach in place of the external forum procedures. One is always required to attempt the external forum first, and only if that becomes an impossible recourse, can one seek this extraordinary procedure. Obviously, great caution must be exercised in making use of this procedure since an excessive use of the solution could greatly damage the good order of the Church and the integrity of the Church's witness regarding marriage.

Brother/Sister Relationship

The "brother/sister solution" refers to those situations where either the external or internal forum solutions cannot be applied. The brother/sister solution allows the parties to receive the sacraments of the Church if they refrain from sexual activity.

Considering our present understanding of marriage as a community of life and love, wherein the sexual relationships is an expression of the covenant love shared, this "solution" seems excessively harsh and unrealistic.

It should be noted, nonetheless, that historically this "solution" has been part of Catholic practice in dealing with seemingly unsolvable marriage cases.

It might be helpful to seek this brother/sister solution in certain cases where sexual activity is not of paramount importance in the marital relationship. Since the marital exchange includes more than rights over another's body, it is possible that, for some who live together in a mutually caring and supportive relationship that is fulfilling and satisfying in itself and who see the brother/sister approach as the only solution to their desire to participate in the Church's sacramental life, this might be recommended with pastoral prudence and discretion.

Conclusion

It has been the objective of this chapter to show that the Church's ministry to the separated and divorced is not an exercise in legalism nor a rejection of those whose marriages have ended in civil divorce. Rather, it is the result of the Church's desire to minister to the needs of those who feel isolated from their own particular community, family, and Church because they are divorced. The Church, through its theology of marriage and its consequent legal system, constantly seeks to hold before its people and the world the ideal of stable marriage, faithful and permanent in its conjugal love. At the same time, recognizing the weakness and the sin inherent in human nature, the Church strives to heal those whose marriages have not survived. Through its judicial system, in one of the several procedures discussed here, the Church allows marriage whenever a previous bond can be annulled or some other acceptable procedure can enable a previously married person to be married in the Church and/or participate in the Church's sacramental life.

But always, the first obligation of the Church is to be true to itself and its firm belief in the permanence of the marriage bond, a belief clearly expressed in the magisterial teaching and tradition that the Church constantly seeks to uphold. In granting annulments and in allowing any other procedure, whether of the internal or external forum, the Church never sets aside its belief in the ideal of stable marriage. Rather, the Church is paving the way for what may become the true sacramental union in the lives of those separated and divorced people who seek marriage in the Church as the fulfillment of their faith commitment and the spiritual consummation of the love they share.

NOTE

1. The Catholic Church respects and presumes the validity of all marriages (canon 1060: "Marriage enjoys the favor of the law; consequently, when a doubt exists, the validity of a marriage is to be upheld until the contrary is proven"). This is true also of those marriages of non-Catholic and non-Christian people where the celebration of the marriage took place in other churches, temples, mosques, etc., as well as those purely civil ceremonies of marriage, such as that before a justice of the peace, judge or other civil magistrate.

VIII

Interreligious Perspectives on Marriage

35.

Marriage in the Jewish Tradition

Blu Greenberg

To this point, the readings in this volume have been an exploration of marriage (and its set of related issues) mainly from a Christian perspective. This is apt, in light of our primary reading audience. However, a rich examination of marriage today cannot remain sealed within a narrow denominational context. There are two reasons for this: (1) interreligious marriages are flourishing across all boundaries, and (2) viewing marriage from diverse perspectives can foster tolerance, understanding, and appreciation. This section teaches us that comparison and contrast is the spice of life!

Blu Greenberg begins with a detailed sketch of marriage in the Jewish tradition. Marriage is the Jewish way. It is considered an ideal state, a primary community norm, the optimal way to live. It is good, very good. In fact, it is the ultimate paradigm for the relationship between God and the Jewish people.

Greenberg wisely requests a temporary suspension of a feminist critique of the tradition so as to examine the essence of the (biblical and rabbinic) sources—i.e., the centrality of marriage in the literature and life of the people. The voyage is intriguing and illuminating. It is all the more interesting when one compares the Jewish sources, ceremonies, and laws with Christian marital forms, rituals, and codes. Likewise, the impact of the women's movement and intermarriage on Jewish marriage can be seen to bear striking resemblances to the reshaping of marital forms in Christian circles today.

Questions to assist in focusing the discussion follow:

1. *What constitutes the core of the traditional Jewish ideal of marriage?*
2. *What impact has the contemporary women's movement had on Jewish marriages?*
3. *Name some of the factors involved in the rapid rise in the intermarriage rate for Jews in the U.S.*
4. *If marriage is the primary communal model and norm in Judaism (and the only legitimate adult union), what implications does this hold for other sexual forms that fall outside the marital paradigm?*

From *Journal of Ecumenical Studies* 22:1 (Winter 1985): 3–20. Reprinted with permission.

In Judaism, and from the very moment of origins of the Jewish people, marriage was considered to be the ideal state. Although the institution of marriage has been newly called into question by a marginal few whose values are shaped first and foremost by contemporary culture,[1] marriage nevertheless continues to be a fundamental socio-religious principle of Judaism and a primary community norm. Almost from birth, all things work toward marriage. The greeting recited at a boy's circumcision ceremony, on the eighth day of life, is: "As this child has been entered into the covenant, so may he be entered into a life of Torah study, the wedding canopy, and good deeds."[2]

The marital relationship takes precedence over every other human connection. For example, the Fifth Commandment—filial piety—is of ultimate value and reaches into many areas of life. There is much in the ethical and halachic literature that urges that a child not marry against the parents' will or without consent, but the law is unequivocal: "If the father objects to his son's marriage or woman of his choice, the son is not obliged to listen to his father . . . for the belovedness of the partners is of paramount value."[3]

Marriage is not only the optimal way to live, but it is also central to the theology of Judaism. The entire success of the covenant rests on the marriage premise and its procreative impulse. The biological family, born of marriage, is the unit that carries the promises and the covenant, one generation at a time, toward their full completion and realization.[4]

In this regard—the centrality of marriage—Judaism can be said to be an earthy religion, dealing with intimate relationships, sex, procreation, and the powerful though often untidy bonds of family life. In this regard, too, Judaism differs from more spiritually oriented religions. Or, to put it another way, the spirit comes as much through marriage and family life in Judaism as it does through the individual's experience of *mysterium tremendum*.

Having said all these lovely things about marriage and family in the Jewish tradition, I find myself in an unenviable position, for much of the literature, law, and language surrounding marriage and divorce reflects hierarchy and sexism. As a feminist writing to my own faith cohorts, I feel perfectly uninhibited to point up and critique inherent sexism in the tradition. As a Jew writing to Christians, Muslims, and members of Far Eastern religions—persons who would not naturally feel the love and appreciation an insider feels and who would, therefore, come away with a one-sided and uncorrected view—I feel somewhat constrained.

So I ask the reader to grant me two requests: First, I ask that you suspend for the moment a feminist critique, that you set aside questions of sexism, inequity, and noninclusive language. Let us assume these givens: that imbalance did exist, that patriarchy was the social and psychic mode, that role distinctiveness lent itself too easily to hierarchy. To all this we shall soon return. Second, I ask that you bear in mind these general truths: that the tradition was sexist more in theory than in practice, more in certain cultures than in others, more in the past than in the present, more in legal formulation than in actual relationships, more in ancient law than in scriptural narrative.[5]

Now, having temporarily set aside this issue, let us proceed to examine the essence of the sources—the centrality of marriage in the literature and life of the people as it moved through 4,000 years of human history.

The Sources

Oddly enough, although marriage is undeniably the only legitimate adult union, we find throughout the Scriptures not a single explicit command to marry. Procreate? Yes. Marry? Not one commandment. Rather, the information comes to us in the form of description and recommendation.[6] In what is surely among the most romantic verses in the Bible, we read: "It is not good that man should live alone . . . therefore, shall a man leave his mother and father and cleave unto his wife, and they shall become as one flesh" (Gen. 2:18, 24). The use of the word "cleave" (*davok*) signifies sensuality, intimacy, interdependency, and a long-term relationship—staples, I would say, of a good marriage.[7]

In contrast to law, biblical narrative is very rich in details of marriage. More than the sacred literature of any other people, the Torah is the story of family, of marriages, and not prettied-up versions, either, but the stuff of real marriages—love, romance, anger, deceit, honor, faithfulness, distrust, infidelity, companionship, intimacy. There is not a single idealized version of marriage in the entire Bible—not one marriage without some flaw or weakness. Perhaps that explains why marriage becomes the ultimate paradigm for the relationship between God and the Jewish people.[8] Just as a good marriage has its high and low points, yet the partners do not sever the bonds, so does a covenantal relationship. This theme—of the lasting marriage between God and the Jewish people despite the latter's backsliding—is sounded repeatedly by the Prophets.

Jewish tradition is often referred to as halacha—the Jewish way. It is the sum of several parts: (a) divine revelation, (b) the Jewish historical experience, and (c) the exegetical enterprise of connecting (a) to (b). What is the process whereby this remarkable feat—of connecting revelation to people's lives through the sweep of generations and in a diversity of cultures—is achieved? That process is none other than rabbinic interpretation, explication, and expansion of the law. Rabbinic literature (Talmud and Midrash), in contrast to the Bible, is much more explicit on all matters concerning marriage and divorce. In fact, six of the sixty-three tractates of the Talmud deal in whole or in large part with such matters.[9]

The Obligation to Marry

Responsibility for marriage rests on everyone—parents, community, and individual:[10]

> What are the essential duties of father to son? . . . to circumcise, redeem, teach him Torah, take a wife for him, and teach him a craft. (Kid. 29A)

The tradition goes so far as to say:

> If an orphan wishes to marry [and has no means] the community should purchase for him a dwelling, a bed, and all necessary household utensils and then marry him off. (Ket. 67B)

Choice of Mate

The marriage had to be characterized by *shalom bayit* (a peaceful and harmonious household); consequently, great care had to be taken in selecting a mate:

> A man should be matched to a woman according to the measure of his deeds. (Sot. 2A)

All measures of compatibility were to be considered: character, background, values, the extended family, even genetic makeup.[11] Wealth, however, was not to be a consideration,[12] but mutual desire was a requisite:

> A father is forbidden to betroth his daughter to another while she is a minor. He must wait until she grows up and says, "I want to marry So-and-so." (Kid. 41A)[13]

Predestination

Still, mere mortals cannot succeed in mate selection without a measure of divine assistance:

> . . . What has God been occupied with since the six days of creation? . . . with the task of finding appropriate life mates for his earthly creatures . . . Though it looks easy to make a match, even for God the task is as difficult as splitting the Red Sea. (Gen. Rab. 67.3)[14]

Duties in Marriage

Many of the obligations of husband to wife were spelled out in the Talmud; some of these were detailed in the *ketubah,* the marriage contract. Though equality did not fully exist, and role distinctions were clearcut, there was nevertheless a kind of complementarianism to the Jewish marriage:

> These are the tasks that a wife must carry out for her husband: She must grind corn, bake, cook, suckle her child, make his bed, and work in wool. (Ket. 5:5)

> Her husband is liable for her support, her ransom, and her burial, Rabbi Judah says even the poorest in Israel must not furnish less than two flutes and one woman wailer [at his wife's funeral]. (Ket. 4:4)

> One must always observe the honor due to his wife, because blessings rest on a man's home only on account of his wife. (B.M. 59A)

Sexuality

Sex, in the context of marriage, was of positive value. In contrast to Christianity (Mt. 19:10; 1 Cor. 7), celibates were frowned upon, even if they were considered to be among the greatest scholars (Yev. 63B). The sexual urge was considered a basic and normal need that required satisfaction:

> Should a man marry first, or devote himself to Torah and then marry . . . Marry first . . . for one who is not married will be possessed the day long with sexual thoughts . . . and be unable to concentrate on his studies. (Kid. 29B)[15]

Procreation

The very first biblical commandment, "Be fruitful and multiply" (Gen. 1:28), was a fundamental obligation of the marriage partners.[16] The institutions of polygyny and levirate marriage (a man's obligation to marry his deceased brother's widow if the brother died childless) can be understood only in the context of procreation.[17] But procreation had to be balanced with sexual passion in marriage, and mutual desire at that. This explains why (a) polygyny rarely was practiced and eventually was forbidden; (b) levirate marriages could be circumvented by a release ceremony of *halitza;* and (c) marital rape was explicitly forbidden.[18] It also explained why divorce, though not a popular or esteemed option, nevertheless was permitted, for divorce was deemed to be better than a marriage of unhappy or ill-suited partners.[19] In sum, then, the ideal was a relationship characterized by romance, sexuality, compatibility, harmony, fidelity, mutual care, and the business of raising children. The midrashist summed it up well:

> He who has no wife lives without peace, without help, without joy, without forgiveness, without life itself. He is not a whole person; he diminishes the image of God in the world. (Gen. Rab. 17:3)

The Act of Marriage

In its most technical sense, marriage in Judaism is a change in personal status. Neither sacrament nor mere legal transaction, it enjoys the trappings of both: an aura of sacredness, the language of sanctification, the richness of ceremony and rite, the sanction of religious leaders. It also involves a contract, a formal declaration, witnesses, signatures, and an exchange of monetary value. To understand how this conglomerate nature came to be, one must examine the Jewish wedding ceremony from its very beginnings.

The Bible provides relatively few clues to the act of creating a marriage. About all we know is that it is a man's initiative. A man "takes" a woman in marriage (Gen. 29:27; Ex. 2:1; Dt. 22:13). There is hint of preparation, feasting, and celebration (Gen. 24; Gen. 29; Jg. 14:12), but no description whatsoever of an actual ceremony, and no explanation of the verb "take."

The Talmud,[20] reflecting inherited tradition and building upon it, explains the procedures. The Mishnah teaches: "A woman is acquired [in marriage] in three ways and acquires herself [her independence] in two. She is acquired by means of money, deed [document] or cohabitation; she acquires herself by divorce or death of her husband" (Kid. 1:1). In a lengthy discussion of this Mishnah, the Gemara explains that the biblical "take" refers to money; however, this money does not signify purchase but rather is the symbol of a legal transaction.[21] What, then, does acquisition mean? For that, we turn to the next chapter in the Mishnah: "A man betrothes a woman . . ." (Kid. 2:1). The Hebrew for betrothal is *kiddushin,* derived from the root word *kadosh*—holy, set apart, off limits to all others. A woman is set aside for her husband and her husband alone. It is not merely a social commitment such as an engagement is today; rather, it is a halachic procedure. The transfer of an item of monetary value, however small, symbolizes this "setting aside."

A marriage, as mentioned above, can be constituted in two other ways: first, through a deed, a written document drawn to the satisfaction of all parties concerned; second, through cohabitation. What these three methods have in common is a certain privatism for the marriage partners: a man initiates, and a woman consents.[22] Technically, in each instance marriage can be effected in the absence of any sort of control or sanction by community. The sages of Gemara, however, point to the inadmissability of such methods. An exchange of coin of insignificant value was considered inappropriate and hasty for something as serious as marriage; intercourse as a means of betrothal carried the taint of licentious behavior. Thus, the rabbis forbade these procedures as means of effecting a marriage, not by striking the laws from the books, but by interpreting and circumscribing them to bring them more in line with community norms. Those who disregarded community norms could be subject to fines, flogging, and even annulment of the marriage (Yev. 110A). In tracing development of the law, then, we see that control over procedures of marriage was legally shifted from parties of the first part onto the community—probably as it had always been in ancient Jewish life.

Part of this ongoing process was the introduction (circa third century B.C.E.) of the *ketubah*—the marriage contract that stipulated obligations of husband to wife during the course of the marriage as well as financial protections in the event of divorce. The *ketubah* was drawn between the parties prior to the betrothal and was witnessed as any other legal document.

To further prevent marriages of whim, the wedding ceremony was divided by rabbinic fiat into two parts: *erusin* (or *kiddushin*), the betrothal; and *nisuin,* the completion of the marriage procedures. The two parts were separated by a period of time, of up to a year. In *erusin* (or *kiddushin*), the groom gave the bride an object of value, in the presence of witnesses, then made a formal declaration to her: "Behold, thou art consecrated unto me, with this ring, according to the law of Moses [inherited tradition] and of Israel [sanction of community]."[23] After the set period of time had elapsed, the couple was joined in *nisuin* under the wedding canopy, and a series of blessings was recited in the presence of a quorum of ten men, the unit that technically constitutes a Jewish congregation. In time, it was observed that the arbitrary separation between betrothal and marriage could lead to problems, so it became standard procedure, and then law (twelfth century), to perform the two ceremonies at the same time. In order to maintain the distinctiveness of each, the *ketubah* was now read aloud between *erusin* and *nisuin.*

Increasingly, then, marriage came under religious control and communal sanction and celebration. Still, it retained its original bi-party transactional nature as described in the Mishnah: a ring as symbol of transaction, the marriage declaration in the presence of witnesses as oral deed, and the bridal canopy as suggestive of intimate relations.

The Wedding Ceremony

A traditional wedding ceremony performed today would likely consist of the following order of events:

1. *Kabbalat kinyan* (acquisition): Prior to the ceremony, the groom formally undertakes the obligations specified in the *ketubah.* He signifies this by holding aloft for a moment a handkerchief held out to him by the officiating rabbi, who stands in for the bride. This action is witnessed by two men who sign their names to the *ketubah,* certifying that the conditions therein were accepted by the parties involved.

2. *Bedeken* (veiling the bride): The groom, accompanied by male family and friends, is escorted to the room where the bride is seated with female members of her party standing about her. The groom and his entourage approach in song. He gently draws the bridal veil over her face. Often, at this moment, the fathers of the couple will each in turn bless the bride, laying their hands on her head as they give blessing. Family and friends look on. It is an indescribably sweet moment, and many eyes glisten in joy.

3. *Chuppah* (bridal canopy): The Jewish wedding has taken on the full coloration of Western culture—including the processional with bridesmaids, ushers, ringbearer, etc. But there are some differences: First, the ceremony must be performed under a *chuppah,* which symbolizes their intimate household. Second, bride and groom do not stand alone during the ceremony; they are accompanied down the aisle by their parents, a symbol that bride and groom do not marry in isolation, nor will they construct their future home in absence of familial or communal support. Third, the last part of the processional at a traditional Jewish wedding is a most unusual one—the bride's encirclement of the groom. When she reaches the *chuppah,* she walks around the groom seven times, with her mother and mother-in-law following her. Various interpretations have grown around this custom: by drawing a circle with her own body, she creates an invisible wall and then steps inside—a symbol of togetherness and distinctiveness from the rest of society.[24] This tradition is also based on a messianic reference in Jeremiah: "a woman shall encircle a man" (31:22). (Though some consider it a sexist ritual, to me it has always seemed a wonderfully sexual one, as if she is wrapping him up to take him home with her!)

4. *Erusin:* The two betrothal blessings, one over wine and one over the act of betrothal that will follow, are recited by a rabbi. The rabbi does not drink the wine, but gives the goblet to the groom and then to the bride, for each to take a sip. The rabbi is known as the *mesader kiddushin,* one who "arranges" the betrothal rather than performs it. Unlike civil law, where a Justice of the Peace pronounces them husband and wife, Jewish marriage is an act between two partners. The core element of a Jewish marriage remains as it was from the very beginning: man taking woman in marriage. The groom places a ring on the

bride's right index finger and recites the ancient declaration, "Behold, thou art consecrated unto me, according to the laws of Moses and of Israel." Her acceptance of the ring is considered assent.

5. *The ketubah:* The marriage contract is read aloud in its original Aramaic. Often, it is read also in translation. The person who reads the *ketubah*—an honored guest—hands it to the groom, who hands it to his bride. It is hers for safekeeping, but for the moment she will likely pass it to a parent or friend during the ceremony. At this juncture, the officiating rabbi often will address the bride and groom.

6. *Nisuin* (marriage): The final part of the ceremony consists of seven different blessings of celebration and hope. Customarily, seven different guests are given the honor, each reciting one blessing. The seventh blessing is as follows:

> Blessed art Thou, Lord our God, creator of the universe Who has created joy and gladness, bridegroom and bride, laughter and exultation, pleasure and delight, love, brotherhood, peace, and fellowship. May it be soon, O Lord our God, that there be heard in the cities of Judah and in the streets of Jerusalem the voice of joy and gladness, the voice of bridegroom and of bride, the jubilant voice of bridegrooms from their canopies and of youths from their feasts of song. Blessed art Thou, O Lord, Who enables bridegroom to rejoice with the bride.

Upon completion of the blessings, of which the first is the benediction over wine, the couple drinks from the wine goblet. Again, in order to distinguish between *erusin* and *nisuin,* a new goblet of wine is used.

Before the couple leaves the *chuppah,* one additional ritual is performed. A glass is wrapped in a napkin and placed on the floor, where the groom shatters it with a well-placed stomp. The breaking of the glass reminds Jews of the destruction of Jerusalem and its ancient holy Temple in the year 70 C.E. The act also suggests to all present that the world is not yet redeemed, and, therefore, our great joy on this day is incomplete—but only for a moment, for now everyone in the audience calls out, "Mazel tov! Congratulations, good luck."

7. *Yichud:* The public ceremony is completed. After the couple return down the aisle, they are escorted to a private room for *yichud,* a few moments of precious privacy. Two witnesses are posted at the door to ensure that no one disturbs them. *Yichud* has both historical and halachic referents. In biblical times, the bride and groom would have their first intercourse immediately after marriage, and the bride would bring forth tokens of her virginity—the sheet with bloodstains—which her family would then display as a badge of honor (Dt. 22: 13–17).[25] The reader will remember that according to the Mishnah there were three ways to effect a marriage. Intercourse, the third, was rejected by the rabbis as not being a legitimate mode of marrying and could only consummate a properly performed marriage. Today, of course, no couple consummates marriage at this time, but the privacy of *yichud* symbolically represents consummation and thus finalizes the halachic requirements of a Jewish marriage.

8. *Seudah:* A wedding feast is required. It is a characteristic of Judaism that ritual and rite are formally celebrated with feasting.

9. *Shevah berachot:* The wedding feast closes with a special grace, which

includes a repetition of the seven *nisuin* marriage blessings. Here, too, the honors are distributed among the guests.

So what have we in a Jewish wedding? As in much else in Judaism, a dialectic: ancient, yet overlaid with custom that grew in stages; a private transaction, yet one that requires full participation of others and is subject to broad communal norms; a joyous occasion, yet allowing the intrusion of reality and the memory of sorrowful times; straightforward legal procedures, yet replete with rituals of sanctification and tones of sacredness; sexist in structure, yet according honor and deference to the bride. Perhaps it could be no other way in a rite that was shaped over the course of several millennia.

The ceremony I have described is a traditional one. Conservative, Reconstructionist, and Reform Jews observe these procedures in varying degrees. Some assimilated Jews marry in a civil ceremony, but any person who chooses to be married in a Jewish ceremony will follow the core rituals of *erusin* and *nisuin*.

Marriage and Divorce in Relation to the State

Marriage and divorce come wholly under the jurisdiction of the religious courts. In a Jewish state, such as existed prior to the Hurban in 70 B.C.E., and again after 1948 with the rebirth of modern-day Israel, the religious courts were/are an integral part of state machinery. As such, they enjoy a natural autonomy and control in matters of family law.

What about the broad and extensive Diaspora experience, Jews living among the nations of the world as part of, yet distinct from, the host culture? Varied though that experience was, whether under the Zoroastrians of Persia, the Visigoths of Spain, the Muslims of Turkey, or the Roman Catholics of Italy, Jews performed and were married in Jewish marriages.

Under Julius Caesar, Jews were granted status as a *religio licita,* to live according to their own laws.[26] Gradually, as Christianity became the established church throughout Europe, the Jews were relegated increasingly to subordinate status. Life became more difficult and often more oppressive for Jews, but in many areas they were left largely to their own internal devices. Even where their rights and freedoms, including religious ones, were circumscribed, the privilege of regulating marriage was never taken from them.[27]

While internal autonomy was the rule regarding performance of marriage, there were instances where hostile governments imposed their will. In Germany, Austria, and Russia, at various intervals during the eighteenth and nineteenth centuries, the Jews feared the dreaded *familiaten,* government edicts which controlled the number of Jewish marriages that could be contracted in any given locale. Not to be able to marry or to marry off one's children was considered a great tragedy and deprivation. In some communities, child marriages were arranged, for who knew what the morrow would bring?[28]

With emancipation, Jews became citizens of the state and no longer members of a sub-national group with chartered rights, "a state within a state." While emancipation brought with it an end to diabilities, it also brought an end to rights and privileges of self-government and the internal control that this implies. Modern nation-states have continued to respect religious freedoms, and marriage and divorce have continued to be the prerogative of religious corporations. But now there was an alternative—civil procedures. For some Western Jews, emancipation provided the sanction "to reject Jewish civilization in its wider ethnic and cultural implications."[29] Emancipation also spawned the denominations—Reform, Conservative, and Orthodox Judaism—each taking a position regarding the binding nature of religious codes versus civil law. By the 1870s, the liberal branch of Reform Judaism placed marriage in the hands of the civil authorities.[30] In matters of divorce, too, the debate raged loudly between the liberal and traditional wings, the former accepting civil divorce as adequate and legitimate, the Orthodox and Conservative totally rejecting such a view.[31]

Marriage as a Changing Institution

Many of the halachic dictates regarding marriage come to us as ethical guidelines rather than commandments or prohibitions. Not every Jewish male married, nor did every woman bring joy and light into the home, nor did all couples marry compatibly or forever. Each person is an individual and therefore cannot be pressed into a mold; each relationship has its own dynamic and chemistry. With all its structure, Jewish law accommodated this reality. That is why there is no specific commandment to marry, and why the nondogmatic regulations and prescriptions could be subject to different interpretations in later generations.

In addition to personal variations, cultural factors affected the nature of Jewish marriage in any given generation. For example, there was greater sexism in Moslem than in Christian cultures. It was Maimonides of Spanish-Arabic culture, and not Rabbi Moses Isserles of the North European Christian one, who enacted more restrictive legislation regarding wives.[32] Polygyny persisted among Yemenite Jews until the twentieth century, for they lived among a people for whom several wives was a sign of status and success. Ashkenazic Jews, living in Christian Europe, never practiced polygamy, and it was formally outlawed in the Middle Ages.[33] Similarly, levirate marriages continued to be enacted in Oriental cultures long after they ceased elsewhere.

What about today? The primary cultural influence on marriage has been the women's movement. Jewish marriages are no less subject to redefinition in light of its far-reaching impact. I will mention five differences in Jewish marriage today.

First, except for the most traditional sectors of a community, Jewish women are now marrying later than ever before, often acquiring professional degrees and securing themselves in a career before they enter into marriage. No more biding one's time, no more waiting around for "Mr. Right" to come along and chart the future. On the one hand, this is a healthy sign of independence; on the other, there is some cause for concern in the Jewish community, because later marriages and the growing number of singles affect population growth. Second, marriages—whether early or late—are producing fewer children. The Jewish community suffers not from zero population growth but from negative population growth, with approximately 1.6 children per couple. This is far below replacement levels, producing a crisis in the community. Birth control is permitted in Jewish law. There are certain restrictions as to method and family requirements, but there is a great deal of room for choice,[34] and many couples make it in favor of smaller and smaller families. Population has always been a serious matter for Jews, particularly so in this generation after the Holocaust.[35]

Third, in view of the fact that most Jewish women also work, there has been an increasing fluidity of roles. Though there is a strong tradition of Jewish women in the workplace, for the past few generations these models were the exception, not the rule.[36] No one wants to rewrite the Mishnah to read, "And *his* duties are to cook, wash, and . . . weave in wool," but this greater sharing of breadwinning and nurturing roles is a fact of life. While the father was an ever-present figure in the medieval Jewish home, and Jewish fathers always instructed their children in a very personal way, the typical nurturing and homemaking tasks were left to the mother.[37] Today, like their non-Jewish counterparts, Jewish couples are restructuring these roles. Fourth, Jews are not much more immune to living in a divorce culture than are other modern Western people. Estimates of divorce among Jews are close to forty percent, compared with a nationwide rate of fifty percent. That represents a body-blow to the once-stable Jewish family. Feminism is surely not the cause, but its emphasis on independence has served as a catalyst—to wit, the great rise in female-initiated divorce.

Fifth, there is a call for equality in religious structures, including those relating to marriage. For example, in the traditional marriage ceremony the woman is a silent partner. In Reform and Conservative Judaism, this has been altered somewhat to incorporate a woman's voice. There are those among the Orthodox who have experimented in this area as well. Without changing the traditional ceremony, they have opened a place for women, such as a female guest reading the *ketubah* or the bride reciting several verses to the groom under the *chuppah*. Similarly, new strides are being made in an attempt to equalize men and women in Jewish divorce law; also, there is increased sharing of rituals in the home. These are relatively small changes, but they are not insignificant in the process of moving toward equality within traditional structures of marriage and divorce.

As the reader can see, it is a two-edged matter. Not all the winds of change blow kindly on Jewish marriages of today. Like all other people, Jews are affected by the erosive forces of contemporary values on marriage and family. And, like all other religious groups, those more anchored to tradition are most resistant to changes. Nevertheless, a balance must be found. Judaism must continue to search for ways to integrate into its permanent structures new rituals that represent the new equity in marriage today.

Intermarriage

From earliest times, intermarriage was explicitly forbidden to Jews (Dt. 7:3–4, Av. Zar. 36B). Because Judaism, the religion, took root in a family, the concept of Jewish peoplehood was perhaps more central than for most other religions. But it was not purity of bloodlines that was the issue; rather, it was the integrity of the faith community. One could join the faith community through conversion; one could not enter Judaism simply through the act of marrying a Jew. For Jews, exogamy represented not only a violation of halacha but also a dilution of Jewishness of the community. Throughout Jewish history, intermarriage was kept to a minimum. Several forces were at work. Aside from the laws—which were made very clear—the response that followed such a marriage was an even more powerful preventive. If a Jew married out of the faith, he or she knew what to expect: total banishment from family and community. Parents and siblings would rend their garments and observe the *shiva* mourning period, as if the child were no longer alive.

Non-Jewish society was similarly opposed to intermarriage. In almost every official Christian document on the Jews throughout the Middle Ages, the injunction against intermarriage with "perfidious Jews" appears.[38] Often, punishment for offenders was harsh. Nor was there any opportunity. In most Christian lands, Jews and non-Jews lived apart, either by mutual desire or by imposed ghettoes. Social interchange between Jews and Christians in nonbusiness settings was the exception, not the rule.

With the dawn of emancipation, the ghetto walls came tumbling down. Jews mainstreamed into the social, cultural, and political life of the nations in which they lived. Intermarriage grew as social contacts grew and as the stigma against Jews was muted. Equality and fraternity were the new and ideal ways to view the "other." Romantic love took precedence over issues of group survival and cohesion. Civil marriages were convenient, for one did not need to convert out of one's own religion to marry a person of another faith.

Until recently, the intermarriage rate for Jews was quite low in the United States. In the decades before 1960, the rate was only six percent. Between 1960 and 1966, it rose to seventeen percent, and by 1971 it had jumped to thirty-two percent.[39] Adjusting for conversion and other factors, the number of Jews in America in 1980 having non-Jewish spouses was approximately 300,000—or five percent of the total Jewish population. Today, it is thought to be considerably higher. The rise is due to many factors: a new age of pluralism and ecumenism has dawned, in which value is placed on every other faith, belief, and style of life; the new fascination with, and therefore desirability of, ethnics; the intense relationships constructed by humanists of all faiths in the course of sharing a crusade such as civil rights, anti-nuclear protest, or feminism; the vast number of Jews on the college campuses;[40] the loss of enforcement power in religious communities; the increasing number of Jews in all professions; and feminism and the increase of female out-marriage.[41] Finally, because of the sheer numbers, the old response of total cut-off is no longer a viable threat.

There are certain new facts about intermarriage that make it no less forbidden according to Jewish law but help to explain some of the contemporary communal responses: First, there is a not-insignificant phenomenon of conversion into Judaism—in fact, more into than out of Judaism. In approximately thirty percent of the intermarriages, the non-Jewish spouse converts to Judaism, either before or after the wedding.[42] Second, even where there is no conversion, intermarriage seems no longer to represent to exogamous Jews a negation of Judaism. Most of them continue to think of themselves as Jews—which they are—and to maintain family and community links.[43] Third, in a recent study of a sample of intermarried couples across the United States, it was found that twenty percent of the families with no conversion had, nonetheless, provided some formal Jewish education for their children, and thirty percent of them celebrated the bar or bat mitzvah rite as well.[44]

The denominational responses to intermarriage are varied. In 1983, the Reform rabbinate formally adopted the principle of Jewishness by patrilineal descent (as well as matrilineal). This is a *de jure* statement of what Reform Judaism has *de facto* accepted all along, but the statement represents a more aggressive outreach to and acceptance of intermarried couples. Conservative and Orthodox Jews have strongly objected to the patrilineal-descent motion on the basis of its halachic inadmissibility.

The leaders of Reform Judaism have also called for greater outreach in the area of conversion, and this would apply particularly to non-Jewish spouses. While conversion has always existed in Judaism—and many converts were given the appellation "righteous convert"—there was a controlling ambivalence in tradition and community. Today, Reform Judaism has deemed it the right moment to extend its hand to the "unchurched"; simultaneously, the Central Conference of American Rabbis (Reform) voted to censure rabbis performing a mixed marriage.

Conservative Jews have expanded their educational and rabbinic resources available to potential converts. The procedure for conversion involves several months of study preceding conversion rites, often on a one-to-one basis. This study takes a great deal of time in the life of a rabbi.

In the Orthodox community, the primary response has been one of prevention—to raise children more intensively Jewish, in educational and social settings not conducive to intermarriage, such as yeshiva high school, Jewish summer camps, etc. Orthodox Jews tend also to live in communities that are identifiably Jewish. With community reinforcing home environment, intermarriage remains extremely rare in the Orthodox community. Still, it occasionally does happen, for Orthodox Jews go to college and enter the workplace and are geographically mobile, like everyone else. Moreover, while converts rarely come via an Orthodox partner, many wish to convert through an Orthodox conversion, so that later there will be no questions of acceptability. And, for many, Orthodoxy signifies a certain authenticity of tradition. Once one makes the effort to convert, the rationale seems to be to want to do it in the most "kosher" manner.

Thus, there is today a less resistant attitude to conversion in the Orthodox community. The law has always been that an ulterior motive, such as marriage, is not a valid reason for conversion, which must be based on love of Judaism and not love of a particular Jew. Most rabbis, recognizing reality, have tended to work around that obstacle. More recently, there has been an attempt to point up halachic precedents that validate conversion for the sake of marriage.[45] On the personal level, traumatic though a mixed marriage is in an Orthodox family, the old response of sitting *shiva* and cutting off all ties no longer seems to be the common response. While an Orthodox Jew would not participate in or attend the wedding ceremony, and while the relationship is surely affected, the connections are often maintained—in part, to encourage that the grandchildren be raised as Jews (converted, if the mother is not Jewish); in part, hoping that with love, time, and familiarity with Judaism the non-Jew will eventually want to convert; in part, because we all live in a culture that places human values above all other claims.

Sexism in the Tradition

How is one to deal with sexism in sacred and quasi-sacred literature? One cannot rewrite the sources any more than one can rewrite history, nor does one like to tinker excessively with issues of divinity and ancient authority. But neither can one let sexist sources stand uncorrected. One answer lies in the hermeneutic: to teach the richness of tradition and the essence of its message—whether that be the sanctity of marriage or the caring nature of a relationship—yet apply the yardstick of equality. It is possible to point out where the sources, for reasons of sociology, culture, and timing, fall short. It is possible to explain certain texts in the context of those times—hierarchies in all pre-feminist cultures. And because of the conglomerate and cumulative nature of Jewish tradition, it is often possible to find a benign precedent or principle to substitute or counterbalance a sexist one. In other words, it is possible to engage a critical eye and a loving heart at one and the same moment.

Moreover, tradition has much of value to teach—including something important about role distinctiveness in male/female relationships. Role distinctiveness does not necessarily imply hierarchy or inequity. In the long run, role distinctions may be quite healthy for psyche and society. But there are limits to such a gentle and respectful approach. These are reached when outright discrimination affects the lives of real people. In such a case, the second answer lies in the area of politics and power. For example, Jewish divorce law is unequivocally sexist. However, before we explain the law, let us examine the procedure: After all attempts at reconciliation have failed, the couple appear before a *bet din,* a Jewish court of law, consisting of three rabbis who are experts in Jewish divorce law. A scribe and two witnesses are also present. At the instruction of the husband, the scribe hand-letters the *get* (the writ of divorce), filling in name, location, time, and the standard text of the divorce. The core of the *get* is the husband's attestation to divorcing his wife and setting her free to marry any other man. The witnesses sign the *get,* and the rabbi reads it aloud; then the man places it into the woman's hands, acclaiming, "This is your *get.* You are now free to marry another."

No divorce, no matter how amicable, is problem-free. To some extent, Jewish divorce action with its routine methodical procedures and its absence of interpersonal negotiations helps to keep the tensions low; the *get* proceedings lend a note of closure to the relationship, which speeds up the process of psychological closure as well. In that sense, a Jewish divorce proceeding has some positive impact. Moreover, the fact that divorce is permitted at all in Judaism is a fundamental recognition of the human right to happiness.

But, for all that, Jewish divorce law poses problems for women. First, only the husband can issue the *get*. There are cases where the recalcitrant husband, for reasons of spite or blackmail, withholds the *get* from his wife and thereby does not release her to marry. Her life is in limbo. In the closed society of former times, a Jewish court could compel him to authorize the *get* or could punish him under rules of internal autonomy until he acquiesced. Paradoxically, in an open society, the problem is exacerbated. A recalcitrant husband can slip through the cracks between civil and religious jurisdiction.[46] He can refuse to show up at the rabbinic court. A woman faithful to Jewish law is most vulnerable in this situation. Second, if a husband disappears and his whereabouts are not known, his wife remains an *agunah*—a woman anchored to an absentee husband. She, too, is in limbo and cannot get on with her life. While the rabbis in every generation bent their efforts to resolving the problem of an *agunah,* they could do so only within limits, and their efforts were not always successful. Third, the liberal branches of Judaism—Reform and secular Jews—operate by civil law. Jews of Orthodox and Conservative denominations cannot marry Jews divorced according to the laws of the secular state who lack a *get*.

The problems grow directly out of the original sources: "If a man finds something unseemly in his wife, he writes her a writ of divorce, delivers it into her hands, and sends her away" (Dt. 24:1). This principle, the absolute right of the husband in matters of divorce, has been diminished, circumscribed, and, through the process of rabbinic interpretation, whittled away from generation to generation. Moreover, the rights of women in family law have grown through the centuries. A woman may not be divorced against her will, and she has legitimate rights to sue in court for divorce. Yet, despite the many improvements instituted by the rabbis throughout the generations, and despite the new pre-nuptial clauses to prevent abuse, a woman still remains dependent on her husband's willingness to authorize and deliver a *get*. In fact, there are instances of blackmail on the part of recalcitrant husbands. And the community remains divided.

Until such time as Jewish law undergoes reinterpretation, in order to eliminate potential abuse and to incorporate the principle of equity into Jewish divorce proceedings, Jewish law will be subject to charges of sexism. Unless the potential for discrimination is rooted out, much of the credibility of what is good in the tradition, though sexist in language will be eroded.

Conclusion

Life in Western society as we approach the year 2000 has indeed changed. Whereas choosing a traditional marriage and family was the ultimate and only legitimate choice, today it is no longer the rule. With a divorce rate of fifty percent, a singles lifestyle, serial marriages—only twenty-four percent of all adult Americans fit the old model! But Judaism, while not forcing everyone into that mold, makes a strong case for traditional marriages: a long-term relationship characterized by love and the bonds of nurturing each other and children, and also bounded by traditional parameters of fidelity, mutual respect, and steadfastness.

I, for one, would hope that these teachings will never be muted, that this paradigm for living will never be replaced by another—no matter how modern or *au coruant* a view of adult relationships might prevail at any given moment in history. I believe a covenantal model works for human relationships. And—just as we have all learned from ecumenism that it is possible for people of different faiths to experience the gifts, the promises, and the covenant uniquely, and without putting down all others—so I believe it is possible to posit a primary model of marriage without demoting to the status of pariah all those who fall outside of it.

What does Judaism teach? That marriage is good, very good: that it is the Jewish way.

NOTES

1. See Ellen Willis' review of *The Second Stage* by Betty Friedan in *The Village Voice,* Literary Supplement, November, 1981, p. 1.

2. The traditional greeting for an infant girl differs: "May she be raised to the wedding canopy and to a life of good deeds." Now that women also study Torah, many modern Jews add that to the female greeting as well.

3. Yoreh Deah 240:25. Yoreh Deah is one of the four tractates of the *Shulkhan Arukh,* which is the most authoritative Jewish legal compendium of the Middle Ages.

4. Irving Greenberg, "The Jewish Family and the Covenant," unpublished manuscript, 1983.

5. For two different approaches to the issue of sexism in the sources, see Leonard Swidlet, *Women in Judaism* (Methuchen, NJ: Scarecrow Press, 1976); and Blu Greenberg, *On Women and Judaism* (Philadelphia: Jewish Publication Society, 1982).

6. There are several other nonnarrative references in the Torah to marriage; among them, see Ex. 21:7–10; Dt. 24:1, 5.

7. It is interesting to note that the Hebrew root, *davok,* to cleave, is the word used to describe the relationship between God and the Jewish people: "And you who cleave unto the Lord your God . . ." (Dt. 4:4).

8. Is. 61:10, 62:5; Ez. 16; Hos. 2:21; Song of Songs.

9. *Yevamot, Kiddushin, Ketubot, Gittin, Niddah, Sotah.*

10. See also Kid. 29B, 30A; Yev. 62b, 113a.

11. Bav. Bat. 110A, Bech 45B, Pes. 49a.

12. Kid. 70A.

13. See further Kid. 41A, Ket. 102B.

14. See also Sot. 2A on predestination of partners.

15. Cf. Yev. 63B; Ket. 63A; Sot. 4B. See also Swidler, *Women in Judaism,* pp. 111–113, for a fuller discussion of wives as a distraction to Torah study.

16. Oddly enough, the formal obligation, "Be fruitful and multiply" (Gen. 1:28), was interpreted by the rabbis to apply only to men, with the defense that a woman could not be commanded to do something that might cause her physical pain.

17. See Chaim Pearl, "Marriage Forms," in Peter Elman, ed., *Jewish Marriage* (London: Soncino, 1967), p. 17. See also Gen. 38 and Ruth 2–3.

18. On circumscribing and then forbidding polygyny, see Yev. 65b and Pes. 113A; also *Shulkhan Aruch*. Even Ha'ezer 1:9–11. On halitzah, see the fine article by Menachem Elon in *Encyclopedia Judaica* (Jerusalem: Keter Publishing, 1971), vol. 11, pp. 122–130. On marital rape, see Eruvin 100b.

19. See Yev. 63B; Git. 90B.

20. The Talmud consists of two layers: the earliest, the Mishnah, is largely a detailed exposition of the law; the second, the Gemara, is an elaboration and explication of the Mishnah, and includes law, history, theology, aphorisms, narrative, debate, etc. The Mishnah was closed in the year 250 C.E.; the Gemara, in 499 C.E.

21. Such a conclusion is derived from two facts: (a) the money specified was so minimal as to eliminate any possibility of constituting a financial transaction; (b) unlike other acquisitions, a wife could not be resold or transferred.

22. The language in rabbinic literature shifts from "he takes" or "he acquires" to "is required," teaching us that the woman has the power of consent or refusal.

23. This declaration is found in the Talmud and is most likely the oral version of what originally constituted "deed" in the Mishnaic reference.

24. Maurice Lamm, *The Jewish Way in Love and Marriage* (San Francisco: Harper & Row, 1982), pp. 213–215.

25. Though it seems primitive to Western sensibilities, the custom of examining the bride's linen after the first night of marriage for spots of blood as proof of her virginity is still practiced in some Oriental communities. See *Encyclopedia Judaica,* vol. 11, p. 1045.

26. See James Parkes, *The Conflict of the Church and the Synagogue* (New York: Atheneum, 1969), ch. 1.

27. See Jacob R. Marcus, *The Jew in the Medieval World* (New York: Harper and Row, 1938), particularly sections 1 and 2. See also Philip and Hanna Goodman, eds., *The Jewish Marriage Anthology* (Philadelphia: Jewish Publication Society, 1965), p. 171; Louis M. Epstein, *The Jewish Marriage Contract* (New York: Jewish Theology Seminary, 1927), pp. 281 ff.; and Zev Falk, *Jewish Matrimonial Law in the Middle Ages* (Oxford: Oxford University Press, 1966).

28. See also Rachel Biale, *Women and Jewish Law* (New York: Schocken Books, 1984), p. 66.

29. See Howard M. Sachar, *The Course of Modern Jewish History* (New York: Dell Publishing, 1957), p. 63.

30. See A. Mielziner, *The Jewish Law of Marriage and Divorce in Ancient and Modern Times and Its Relation to the Law of the State* (Cincinnati: Bloch Publishing, 1884), p. 94.

31. Ibid., chap. 16.

32. See Maimonides, *Mishneh Torah,* Book IV, Laws of Women.

33. In approximately 1025, Rabbenu Gershom, a leading rabbinic light of the Diaspora, formally banned polygamy.

34. It is interesting to note that the responsibility falls upon women; the condom is not permitted. See David M. Feldman, *Marital Relations, Birth Control, and Abortion in Jewish Law* (New York: Schocken Books, 1974).

35. It has been reliably estimated that 6,000,000 Jews were killed in the Holocaust. This represented one-third of world Jewish population. It is now forty years after the Holocaust, and the losses have nowhere been made up.

36. See Paula Hyman, "The Jewish Family: Looking for a Useable Past," in Susannah Heschel, ed., *On Being a Jewish Feminist* (New York: Schocken Books, 1983).

37. See Israel Abrahams, *Jewish Life in the Middle Ages* (New York: Atheneum, 1969), particularly chap. 7, "Monogamy and the Home."

38. See Marcus, *The Jew in the Medieval World,* pp. 4–5, 139.

39. The National Jewish Population Study, Council of Jewish Federations and Welfare Funds, 1972.

40. See Irving Greenberg, "Jewish Survival and the College Campus," in *Judaism* 17 (Summer, 1968): 259–281.

41. Prior to the 1970s, the Jewish male/female outmarriage ratio was three to one. As women have moved increasingly into professional settings, the female outmarriage rate has risen.

42. Egon Mayer, "Intermarriage among American Jews: Consequences, Prospects and Policies" (New York: National Jewish Resource Center, 1979).

43. Ibid., p. 3.

44. Ibid.

45. See Marc Angel, unpublished paper presented at Chevra, National Jewish Resource Center, Spring, 1983, New York City.

46. Normon Solomon, "Jewish Divorce Law in Contemporary Society," *Journal of Jewish Social Sciences,* vol. 25, no. 2 (December, 1983).

36.

Marriage in Islam

Lois Lamyā' Ibsen al Faruqi

There are a billion Muslims in the world today. Somewhere between two and three million of that number are in the United States. Islam, currently, is also the fastest-growing religion in the U.S. and is expected early in the next century to surpass Judaism as the third-largest religion in the country.

Stereotypes abound in the public perception of Islam in the West. The emergence of a revivalistic Islam strikes bewilderment and terror in many minds and hearts. Prejudice is fueled also by mass ignorance. The result is intolerance, misunderstanding, and chauvinism. This essay should go a long way toward alleviating misperceptions and positively affirming the wisdom of Islam in its marriage teachings and traditions.

The essay falls neatly into four major divisions: (1) the dominant influence of religion on Islamic marriage and the avowed purposes it serves; (2) specific requirements and codes necessary for its legitimacy; (3) mechanisms for the dissolution of marriage are detailed; and, (4) a brief look at the changes and challenges facing Islamic marriage in the future. The essay is a model of clarity and a reservoir of knowledge about our Muslim sisters and brothers.

Some questions suggested by the essay follow:

1. *What distinctive differences do you find between the Islamic and the Roman Catholic views of marriage? What common ground do they share?*
2. *Compare the purposes of marriage in Islam with the twofold purpose as stated in contemporary Roman Catholic theology.*
3. *Will the meeting of Islam and feminism be confrontation or cooperation in relation to marriage?*
4. *In what areas will Islamic marriage show resistance or/and assimilation to secular modernity?*

In order to explain how marriage is regarded and manifested in Islamic society, an organization of four major divisions has been utilized here. The first discusses

From *Journal of Ecumenical Studies* 22:1 (Winter 1985): 55–68. Reprinted with permission.

general characteristics and purposes of Islamic marriage; a second outlines the specific requirements which legitimize marriage; and a third details the mechanisms for its dissolution. A fourth and final section discusses briefly the issue of change and the future.

General Characteristics of Marriage in Islam

Religion as Dominant Factor

Probably the most dominant factor that influences or has influenced Islamic marriage is religion. This is due in part to the Islamic idea that religion (*dīn*) is not a body of ideas and practices which should be practiced or which should influence only that part of human life generally designated as dealing with the sacred. In Islamic culture there is little or no conception of a bifurcation between that which is sacred and that which is secular. Instead, since every aspect of life is the creation of Allah, it carries religious significance. It is the matériel with which humanity works to fulfill the will of God on earth, the ultimate human purpose in creation. Thus, for the Muslim, the ritual prayer conducted in any clean place is as equally valid and acceptable as that performed in the mosque; the commitment to political and social awareness and activity is as much a religious duty as the recitation of prayers; economic pursuits are equally regulated in accordance with religio-ethical pronouncements as the *zakāt* (Islamic levy for social welfare) and the *hajj* (pilgrimage to Makkah). Aesthetic products present not art for art's sake, but art for religion's sake: they are restatements and reminders of religious truth. In fact, every aspect of Islamic life is permeated with the effects of qur'ānic and religious teachings. Marriage is no less affected.

In Islamic society, however, marriage is not a religious sacrament. In other words, it is not a ceremony which necessitates the involvement of any clergy, presupposes a numinous or divine involvement, or, as a consequence, tends to be regarded as an indissoluble commitment.[1] In Islam there are no priests and little notion of sacredness in marriage which surpasses that of any other similarly recommended institution of the culture. Consequently, the religion has delineated accepted procedures for its dissolution.

Although marriage is not regarded as a religious sacrament, Islam recommends it for every Muslim. Commendations are to be found for it in the Qur'ān (4:1, 29; 7:107; 13:38; 24:32–33; 30:20), and the *ḥadīth* literature—recording sayings and events from the life of the Prophet—contains numerous passages in which celibacy is discouraged and marriage encouraged. For example, "Marriage is of my ways," the Prophet is reported to have taught, as well as, "When a man has married, he has completed one half of his religion," and "Whoever is able to marry, should marry."[2] Such stimulus from religious teachings has made marriage the goal of every Muslim and caused a considerable amount of social pressure to achieve that end for every member of the society.[3] Parents, relatives, and friends all feel committed to assist actively in the process, and few are the individuals who "escape the system."

Religion also guarantees certain rights and imposes certain responsibilities on the participants in marriage. Both men and women are regarded as deriving substantial benefits from the institution of marriage, but they are also bound to its obligations by actual qur'ānic prescriptives and by the legal elaboration and interpretation of the scriptural passages.[4] The religious laws of personal status are therefore crucial to any understanding of Islamic marriage. It is in these laws that the most detailed enunciation of both the woman's and the man's rights and obligations is to be found. Complying with the fulfillment of those mutual responsibilities is therefore regarded as a religious obligation for both parties. The acquiescence of the failure to comply is regarded as an act carrying divine reward or punishment, and therefore it is not a matter to be taken lightly.

Also important is the fact that matrimony in Islam is as much a joining of two families as it is a joining of two individuals. Given the level of interdependence in the Islamic family,[5] Muslims are especially likely to believe marriage necessitates a consideration of the welfare of the familial groups involved rather than merely the desires of the two individuals. For this reason a much larger participation in the choice of marriage partners is regarded as proper and beneficial than would be acceptable in a contemporary Western environment. Even after the marriage, the extended family organization reduces dependence on the single adult relationship of husband and wife and stimulates equally strong relationships between the wife or husband and other members of the family. Despite this heavily "familial" character of the Islamic marriage, the woman maintains her separate legal identity after marriage, her maiden name, her adherence to a particular school of law, and her right to separate ownership of her money, property, or financial holdings.

The Purposes of Marriage

The importance of marriage in Islamic society and its advocacy by the religious teachings rest on the avowed purposes it serves. First, Muslims regard marriage as providing a balance between individualistic needs and the welfare of the group to which the individual belongs. As such, it is regarded as a social and psychological necessity for every member of the community.

Second, marriage is a mechanism for the moral and mutually beneficial control of sexual behavior and procreation. Islam regards sexual activity as an important and perfectly healthy drive of both males and females. Thus, it is not shameful and should not be denied to members of either sex. Lack of sexual satisfaction is believed to cause personality maladjustments and to "endanger the mental health and efficiency of the society."[6] Islam, therefore, commends sex as natural and good but restricts it to participants of a union which insures responsibility for its consequences.[7]

A third purpose of marriage is its provision of a stable atmosphere for the rearing of children. Islam sees this purpose as inextricably tied to an extended family system. The extended family may vary in size, even in residential proximity, as is evidenced in different regions of the Muslim workd, but the cohesion of its members is inextricably bound to qur'ānic prescriptions and Islamic law. These explicity enunciate the rights and obligations of its members and the legal extent of those benefits and responsibilities.[8]

Fourth, marriage assures crucial economic benefits for women during their child-bearing and child-rearing years. Self-support during this period is difficult, if not impossible, for mothers who have no outside help. Even if sustained by the "supermom," of which we hear so much in recent times, the physical and emotional toll on such persons is beyond what most individuals can tolerate.

Fifth, the close companionship of the marital partners provides emotional gratification for both men and women. The importance of this purpose of marriage in Islam is evidenced by repeated references in the Qur'ān and *ḥadīth* literature to the quiescence (Qur'ān 30:21; 7:189) and protective nature (Qur'ān 9:71) of the bond between the husband and wife. The man and woman are considered to be so close that they are described as garments of one another (Qur'ān 2:187). The kindness, love, and consideration enjoined on the partners appears repeatedly in both religious and legal texts.

Specific Requirements for Marriage in Islam

Given its general characteristics and purposes as outlined above, Islamic marriage also entails certain specific features which are regarded necessary for its legitimacy. Let us first specify the criteria which must be met by the participants themselves.

Limitation of Participants

Applying to both parties are the Islamic boundaries of incestuous union. The list of acceptable persons whom one can marry rests on a firm ground of conformity since the basic exogamy/endogamy[9] patterns are fixed by the Qur'ān and *ḥadīth.*[10] Islam prescribes limits on marriages between certain blood (consanguine) relatives, between others closely related through marriage (affinal relatives), and between lactational relatives, that is, those who have been nursed by the same woman.[11] Since the institution of the wet nurse and the reciprocal nursing of their babies by women with close familial or affinal ties have often been widespread, the lactational limits have proved very influential for discouraging inbreeding. The jurists of the different schools have unanimously adhered to that prohibition and accepted the authenticity of the *ḥadīth*s on this matter,[12] although the details of how much nursing and how much milk constitutes a prohibiting amount, the ages of affected children, etc., were sometimes disputed.

The definition of allowable marriage partners is considered by Muslims to fulfill two major purposes: to prevent the biological effects of inbreeding, and to guard against excessive familiarity between sexual partners. Such familiarity is regarded as cause for sexual indifference in the partners. Therefore, marriage with someone as close as a mother, sister, daughter, or aunt would result, in most cases, in a denial of sexual gratification for the marriage partners. In the Muslim village, young people of the opposite sex are separated from the age of puberty or before. If they are to realize a sexually successful marriage in the village, the possibilities for familiarity must be limited and the aura of mystery and excitement engendered by marriage candidates of the opposite sex preserved. Whether consciously or unconsciously pursued by the Muslim peoples, this concern seems to be at the base of the preference—or, in some parts of the Muslim world, the demand—for segregation of the sexes. That is a much more logical underlying purpose for segregation than the need to curb sexual promiscuity. The latter is almost an impossibility in the close quarters and intensive interaction of the village.[13]

There are also religious affiliation boundaries for participation in an Islamic marriage. The male must be a Muslim; the female may be a Muslim, Jew, or Christian (Qur'ān 5:6). Sometimes the religious requirements for the female have been interpreted more widely to include anyone who is not idolatrous. The prohibition against a non-Muslim man's marrying a Muslim woman is, however, qur'ānic and has been unequivocally adhered to by all the legal schools.

Though a few early jusists rejected completely the idea of intermarriage, most Muslims have considered it permissible under certain conditions.[14] The qur'ānic and legal directives on intermarriage tended to be reinforced by the circumstances of the early centuries of Islamic history. As traders, warriors, missionaries, teachers, administrators, and religious or education-seeking pilgrims, many Muslim men traveled to live in different parts of the world. They often loved in predominantly non-Muslim societies where Muslim women were not available for marriage partners. Women, on the other hand, tended to remain in their predominantly Muslim societies and therefore had access to Muslim male partners.

There were no age limitations for marriage partners in Islamic culture until recent times. Since the individuals remained part of a larger family structure which did not call upon them to support themselves, to set up their own home, or to cope unaided with the problems of parenting children, Islam held a much more relaxed view of the prior preparedness for marriage. As the extended family of certain urban environments has been weakened in recent times, a greater emphasis has been placed on the readiness of the married couple to live a more isolated and self-sufficient existence. This has stimulated an appeal by certain individuals and groups in a number of Muslim countries to call for the setting of minimum age limits for marriage.[15] Since marriage in Islam means the signing of a contract, with consummation following perhaps at a much later time, marriage of minors did not raise the same sort of problems it might in another societal complex. This does not mean that the custom of child marriage was never abused; it does mean that this Islamic custom need not be detrimental if practiced in tandem with a properly functioning Islamic society. Not its use but its misuse in an Islamically inconsistent social complex has generated Muslim concern and the recent need for a minimum-age limitation for marriage partners.

Some writers have attributed the recent calls for minimum-age requirements to

Western influences on Islamic thinking and customs. In fact, that argument has been the main thrust of conservative opposition to the initiation of age restrictions on marriage. It seems much more likely that this so-called Westernization would not have taken root unless the misuse and/or imbalance within the system had not made some sort of change necessary.

Another requirement which relates to the qualifications of the marriage participants is that the Muslim woman must be unmarried. If she has formerly been married, she should not be pregnant or in the first three months following her previous marriage, a period known as *'iddah,* in which she may not be aware of a possible pregnancy. The former husband is obliged to support her during the *'iddah* or until the birth of her child, after which she may remarry. Such restrictions help verify paternity of a child resulting from the earlier marriage.

The male partner in marriage, however, is not limited to a single marriage. Islam can be described as permitting polygynous marriage, that is, it is a society in which plural marriages for males are possible. In Islamic society, however, only a small portion of the males practice polygyny. If those that do are known to have no valid reason for taking another wife,[16] or do not treat their wives with the complete equality commanded by the Qur'ān (4:3; 129), Muslims judge such instantiation of this form of marriage as unIslamic and religiously and morally reprehensible. As an excuse for sexual promiscuity, the practice is unconditionally condemned, but, if practiced according to Islamic moral exhortations and legal provisions, Muslims regard polygyny as a more equitable and humane solution to certain situations than the unconditional demand for monogamy.[17] In some schools of law, a woman who wishes to prevent her husband's future second marriage can ask that such a stipulation be written into the marriage agreement.

Mechanistic Requirements

Other specific requirements of Islamic marriage pertain to the actual execution of the marriage rather than to the qualities of its participants. These are designated here as "mechanistic requirements."

A WRITTEN CONTRACT: Marriage in Islam constitutes an agreement between a man and a woman which is embodied in a written contract. The marriage agreement includes specification of the dower (both an initial and delayed portion, see the section on Dower, below), signatures of the two participants and of their respective witnesses, and other terms agreed upon by the parties concerned. The contract is a legal document, which is filed with the local Islamic registry of the government and upholdable in a court of law.

The occasion for the signing of the contract is called *'aqd nikāḥ* ("marriage contract"). It usually takes place at the home of the bride's parents, in the presence of members of the family and close friends. The *'aqd nikāḥ* is accompanied by the surrender of the agreed-upon initial dower and the exchange of gifts by the marriage partners. It also marks the beginning of preparations for the consummation of the marriage. The actual marriage requirements are fulfilled in the *'aqd nikāḥ,* and the marriage is complete, but it is not until the *'urs* or actual wedding party that the marriage is consummated and the bride moves to the home of her husband. This may take place shortly after the contract signing or at a much later time, as the parties desire.

The *'urs* may be a simple party or an elaborate occasion. Refreshments, activities, and entertainment vary according to tastes, financial capabilities, and regional preferences. It may take place at the groom's home or at a public place reserved for the occasion. It is the common practice not only for gifts to be presented to the bride and groom by guests on this occasion but also for the couple to reciprocate with a token of their gratitude—usually a small dish or platter for carrying sweets or an item of clothing for closer relatives. These vary in extravagance to match the economic situation of the wedding participants.

The clothing of the bride and groom varies from one level of society to another and from one region to another. In the Arab Mashriq (near the eastern end of the Mediterranean Sea), for example, the wedding dress may be a white gown and headcovering similar to that common in Western societies or an elaborate example of the local style. Men wear Western-style suits or traditional dress. In the Indian subcontinent, brides are clothed in crimson saris elaborately decorated with gold, while grooms don the traditional "Nehru jacket" with its high collar and buttoned front. In Malaysia the typical attire for both bride and groom is a traditionally styled outfit made of the locally produced brocade woven with gold threads. If the wedding couple come from wealthy families, the clothing may be so heavily decorated with the precious metal that the pair have difficulty in moving. But this does not present too much of a problem since the function of the *'urs* is an elaborate reception at which the bride and groom and their families accept the congratulations and best wishes of friends and relatives.

TWO ADULT WITNESSES: Two witnesses—one representing the bride, the other representing the groom—are a necessary feature of the marriage-contract signing. Where possible, these are the two fathers of the couple, but any other adult Muslim could legitimately fill this role. No other intermediary is required for the performance of a marriage. Any person who is chosen by the parties may make and accept the marriage proposal. Often, however, a *qāḍī* or Muslim lawyer attends as a registrar of the marriage.

SU'ĀL AND *ĪJĀB* ("QUESTION AND CONSENT"): Another element of a legitimate marriage in Islamic society is an explicit request by the groom and his family or representatives and an explicit consent to marriage by the bride and her family or representatives, which may be either in writing or oral. The Qur'ān does not deal specifically with this matter, but there are a number of instances from the *ḥadīth* literature which pertain to this question.[18] The story is told of a woman in the time of Muhammad who demanded and was accorded by the Prophet the right to repudiate her marriage because she had not been asked and therefore had not given her consent to the bond.[19] Despite strongly authenticated instances in the *sunnah* ("example") of the Prophet Muḥammad, Muslim jurists have varied in their interpretations of the guardian's powers in arranging marriage for his wards or children. Some deem it a necessary condition for the father to seek the consent of his daughter before he gives her in marriage. Others argue that it is commendable rather than necessary. Still others limit the need for the consent of the bride to the mature woman who is a widow or divorcee.[20]

DOWER: A fourth requirement of the Islamic marriage is a marriage gift or dower to the bride by the groom. This gift may consist of anything deemed suitable by the participants—money, real estate, or other valuable items. In some cases it has entailed the transfer of great wealth to the bride; in others, it has been as modest as an iron ring or a token coin.[21] No maximum has been set by Islamic law, though some schools have specified minimums.[22] Such details are worked out by the representatives of the bride and groom prior to the *'aqd nikāḥ* and are specified in the written contract.

The dower may be immediate, that is, given at the time of signing the *'aqd nikāḥ*. More often, the parties agree that the dower be divided into two parts: one portion (the *mahr*) surrendered to the bride at the time of marriage, and another delayed portion (the *mu'akhkhar*) which falls due in case of death or divorce. If a man dies, Islamic law provides that his widow's *mu'akhkhar* settlement be paid before any other commitment on his estate is honored.

The immediate dower may be used for the trousseau and the household purchases needed for the newly married couple. At other times it is a kind of economic insurance for the future welfare of the bride, in which case she invests it and draws benefits from it. In the case of poor people, the amounts are so negligible that they can be viewed as little more than a symbol of the groom's willingness to take on financial responsibility for this future wife. Amounts of dower vary not only with the difference in the economic capabilities of different grooms and their families but also according to regional practices and social levels within a particular society.

It is clear that the Islamic marriage is not a "ceremony." Although it may be associated with a number of elaborate activities (procession of the bride to her new husband's home, beautification of the bride prior to marriage with henna decorations, gift-giving and well-wishing of friends and relatives, elaborate clothing, refreshments, and entertainment), marriage in Islam is essentially a legal agreement between two individuals and two families. While carrying the sanction and blessing of the religion, it cannot be considered a sacred ceremony. Marriage, like so many other aspects of Islamic culture, is neither wholly sacred nor wholly secular, neither religious nor nonreligious.

Dissolution of Marriage

Dissolution through Death

Dissolution of marriage occurs either by death of one of the parties or by divorce. In the case of the woman's death, there are inheritance laws which pertain to her wealth since her property remains separate from that of her husband. There are similarly specific requirements for the distribution of his wealth following the death of the husband. Other requirements pertain to the female survivor. These include payment of the *mu'akhkhar* dower to the widow and the illegality of the widow's remarriage before completion of the three-month *'iddah* period in which the possibility of pregnancy can be determined. During this time, the widow is financially supported by her late husband's estate. If she is pregnant, maintenance is guaranteed until birth of the child or the end of the nursing period, but she must not remarry until the birth of her child.

Divorce

More complicated is the dissolution of marriage through legal divorce. Although it is generally believed that dissolution of marriage takes place in Islam only by male repudiation of the wife—that is, by his pronouncing three times, "I divorce you"—the fact is that Islamic law provides various mechanisms and channels for ending a marriage. Despite the variety of means for divorce, it has remained a repugnant act in Islamic society,[23] to be invoked only when all methods of reconciliation have been exhausted. Some types of divorce are male-instigated, others female-instigated; still others are the result of mutual agreement or judicial process.

MALE-INSTIGATED DIVORCE: The most common form of divorce initiated by Muslim males is known as *ṭalāq* ("letting go free"), which involves a series of three statements by the husband that he divorces his wife. Contrary to common opinion, these repudiation statements cannot legally be rendered at a single time. In fact, very strict rules have been established in Islamic law to prevent misuse.[24] Unfortunately, these laws have not always been enforced.

Ṭalāq is to be pronounced with specific terms before two qualified witnesses. Each pronouncement must be made at a time when the wife is not incapacitated for sexual activity by menstrual flow. Having made the first statement of divorce, the man must wait to make the second statement until the woman completes her next monthly period. The third pronouncement must be similarly spaced. Only after the third repudiation is the divorce considered final. Each of the other two statements is revocable. The wife continues to live in her home, and she is provided full maintenance throughout the divorce proceedings. During this time, attempts are made to achieve reconciliation through the counseling and arbitration of family and friends. Only if this is not possible is the final pronouncement made and the marriage considered irrevocably broken. From that point, husband and wife are forbidden to live together or to remarry each other unless and until the woman has remarried someone else from whom she becomes widowed or divorced. Any Islamic divorce, like the dissolution of marriage by death, requires a three-months' waiting period in order to determine whether the divorcee is pregnant. She is not free to remarry until that period is completed or, in case of pregnancy, until she gives birth. As in the case of dissolution of marriage by death, maintenance of the wife is incumbent upon the husband for the *'iddah* the period of child-bearing.

Any *ṭalāq* divorce not made in accordance with these rules is considered to be an aberration, or *bid'ah*. Such practice is regarded as sinful, but unfortunately the actions of some Muslims and the positions of certain jurists have not always accorded with the ideal.[25]

FEMALE-INSTIGATED DIVORCE: In the Qur'ān it is stated that women have equitable rights in divorce to those of men (2:228). As in other Islamic matters, equitable does not mean equivalence or identity. Therefore, there are different procedures which apply to women in their initiation of divorce proceedings. The wife is entitled to originate dissolution of her marriage under four circumstances.

First, in a delegated divorce, the right of *ṭalāq,* or repudiation, by the wife may be agreed upon prior to marriage and stipulated as a condition of the marriage contract.[26] Second is a conditional divorce, a stipulation in the marriage contract that the wife will be free to divorce her husband if he does certain things contrary to his pre-marriage promises. This type of divorce is accepted by some jurists but rejected by others.[27] Third is a court divorce, in which freedom from the marriage bond is granted to a woman for any of the inadequacies of the husband which are generally regarded as legitimate causes for divorce: long absence or desertion, impotence, failure to provide adequate support, physical or mental mistreatment, serious physical or mental illness, apostasy, and proved debauchery. As noted earlier, a wife may also be granted a divorce if, upon reaching maturity, she rejects a marriage contracted by a guardian on her behalf while she was still a child. A fourth type of divorce instigated by the wife is known as *khul'*. It involves a release of the wife from the marriage contract on her agreement to pay compensation to the husband.

MUTUAL CONSENT OR *MUBĀRA'AH:* When a husband and wife reach mutual agreement to dissolve their marriage, it is called *mubāra'ah*. It differs from the *khul'* in being effected by mutual desire on the part of the marriage partners.[28]

JUDICIAL PROCESS: *Li'ān* or "double testimony" is the dissolution of marriage which results from the husband's accusation that his wife has committed adultery. If he proves his case (four eye-witnesses are necessary in Islamic law!), it it considered valid reason for divorce. If he has no witnesses other than himself, he must swear by God four times that his statement is true. The wife is called upon to admit her guilt or to testify in a similar way that her husband has lied about her. Both also invoke divine curses for false swearing. If no further proof either for or against the accusation can be substantiated, reconciliation between the two parties is deemed impossible, and the marriage is dissolved by judicial process.[29] Certain types of divorce instigated by women are also dealt with by the Islamic courts (see "court divorce," above).

Change and the Future

An increase in the number of women in the work force, increased education for women and men, the development of Islamic awareness and identity among

Muslims in all parts of the world, increased mobility, concentration of populations in urban centers, increased contact with alien cultures, as well as many other contemporary facts of life may require certain adjustments in marriage practice in any society. Whether such adjustments will prove disastrous to the institution of Islamic marriage as it is now known—and, therefore, to Muslim society—or merely productive of a new synthesis of twentieth- and twenty-first-century influences with the core premises depends on the ability of society to react with a strong and intelligent social conscience. Social change as such need not be unduly disruptive of Islamic marriage practices. In fact, the magnitude of the Muslim identity (nearly 1,000,000,000 persons) and the diversity of geographic, ethnic, linguistic, and cultural backgrounds out of which these people moved toward Islamization have resulted in an extremely rich and varied tradition in marriage as in all other aspects of the civilization.

Islam has been particularly tolerant of its new converts, and those converts have been particularly ingenious in adapting their local customs to basic Islamic premises wherever possible. At the same time, there were basic premises of the faith as expressed in the Qur'ān and exemplified in the teachings and example of Muḥammad which provided a religio-cultural core with which the more superficial variations could be sympathetically related. History has confirmed that, along with a maintenance of the core, Islamic society has been flexible enough to allow regional variations as well as changes accompanying the passage of time and the variation of circumstances. Each period has had to make its "peace" with those variations of circumstances, to take that which was acceptable and combinable with the Islamic core, to reject those which were culturally and religiously "indigestible," and to adapt—that is, to Islamize—still others before adopting them. This process must, of course, be carried on in the present day and into the future. To rule it out would be to kill the culture and the religion.

A society's proper reaction to change implies two prerequisites. First, it needs a study and awareness of the total societal complex, each of whose institutions and factors is integrally interwoven with and dependent upon the others. Any suggestion for change regarding marriage practices, therefore, should not be investigated in isolation. It should be studied in relation to the other factors of individual and group welfare which those factors may affect. It might be argued, for instance, that, because of increased participation of women in the work force in many Muslim countries, the Islamic custom of mandatory male support for women should be abolished; yet such a shortsighted view fails to take into account anything but the material aspects of the question. It might be counter-argued that much more important in this Islamic stipulation is the reinforcement it gives to the interdependence of the marriage partners; to break those bonds of rights for women and obligations for men would cut deeply into the strength of the marriage bond.

Similarly, some contemporaries might argue that Islam is old-fashioned in its unshakable condemnation of sex outside of marriage. They would cite contraceptive devices and new attitudes toward the sexual freedom of women as demanding a reappraisal of an old Islamic "fixation." But Islam's reason for rejecting greater sexual freedom was not that no adequate contraceptives were available in earlier times to prevent children born out of wedlock, nor did it insist on the importance of female virginity in order to discriminate against women or set a "double standard." Rather, Islam promoted this idea in order to strengthen the institutions of marriage and the family by making them carry benefits that could not be achieved elsewhere. To destroy the uniqueness of such marital benefits in an attempt to provide complete sexual equity would carry widespread and debilitating effects, not only for those institutions, but equally for the individuals who make up those institutions. We do not have to surmise about the effect on women that this innovation might have, for a living example is available in Western society. The consequences are already glaringly apparent. The increased sexual dispensability of the wife which this new promiscuity produces is one of the factors leading to the increased divorce rate. It also has drastically adverse effects on both the financial and emotional security of middle-aged and older women. Proper reaction of a society to change, therefore, demands a careful screening of the elements of change and their results for compatibility with the rest of the culture's goals and institutions.

Second, in order for Muslims to avoid rash acceptance of drastic and harmful changes in their marriage laws and customs, they must purge their society of the misuse of existent laws and customs. Often it is the widespread neglect or circumvention of the qur'ānic or legal prescriptives which are at the root of the problem, rather than the institutions and practices themselves. A case in point is the misuse of the institution of *ṭalāq,* many instances of which fail to comply with the regulations and restrictions which have been established for it. Adherence to those regulations and restrictions would obviate the need for drastic changes in the institution.

The imbalances caused by rapid social change are probably inevitable factors in any society. At this period of history, they are particularly challenging. Without careful reappraisal of the side-effects of contemplated innovation and its compatibility with other aspects of the religion and culture, and without a purging of misapplications of extant institutions, massive social disorientation and deleterious effects on the members of any society are inevitable. This is the contemporary challenge, not only for Muslim society, but also for every other society in the world.

NOTES

1. In Christianity, marriage was first recognized as a sacrament in the twelfth century, when Peter Lombard's "Sentences" (Book 4, dist. 1, num. 2) enumerated marriage as one of the seven sacraments. That work became the standard textbook of Catholic theology during the Middle Ages and was formally accepted by the Councils of Florence (1439) and Trent (1545–1563).

2. Muḥammad ibn Isma'il al Bukhārī, *Saḥiḥ Al-Bukhārī,* tr. Muhammad Muhsin Khān, Al-Medina al-Munauwara: Islamic University, 1974, Vol. 7, pp. 2–5, 8.

3. To this day, in the highly Westernized society of Beirut, Lebanon, unmarried young people are often embarrassed by the constant social pressures on them to marry.

4. The Muslim maintains that the Qur'ān is the word of God dictated verbatim to the

Prophet Muḥammad. Its basic ethical principles and prescriptive laws, therefore, carry the authenticity of divine provenance. These principles and laws are designated as the *sharī'ah* ("path"). The human elaboration and interpretation of the *sharī'ah,* i.e., its development into specific laws by the jurists of the five schools of law (four Sunnī and one Shī'ī *madhāhib,* s. *madhhab*), is the subject matter of Islamic jurisprudence or *fiqḥ*.

5. See Lois Lamyā' (Ibsen) al Faruqi, "An Extended Family Model from Islamic Culture," *Journal of Comparative Family Studies* 9 (Summer, 1978): 243–256; and "Islamic Traditions and the Feminist Movement: Confrontation or Cooperation?" *The Islamic Quarterly,* vol. 27, no. 3 (1983), pp. 132–139.

6. Hammūdah 'Abd al 'Atī, *The Family Structure in Islam* (Indianapolis: American Trust Publications, 1977), p. 50.

7. The Qur'ān and Islamic law, reflecting the practice of salvery which existed in the pre- and early Islamic periods, also condoned cohabitation of a master with his slave girls (Qur'ān 23:5–7; 70:29–31). Although most Muslims would try to rationalize the existence of the passages sanctioning this form of extramarital sex (see 'Abd al 'Aṭī, *Family Structures,* pp. 41–49) and draw attention to the companion mechanisms which legitimized the children of such unions, enjoined emancipation for the mothers, and generally controlled and regulated the practice, few, if any, Muslims today would regard sex as legitimate except within the bonds of marriage.

8. The Qur'ān contains not only repeated references to the rights of kin (17:23–26; 4:7–9; 8:41; etc.) but also inheritance and support provisions which are stipulated as reaching far beyond the nuclear family (2:180–182; 4:33, 176; eitc.). Dire punishment is threatened for those who ignore these measures for intra-family support (4:7–12).

While the commune dwellers in the 1960s in America realized rightly the benefits to their children of a family environment larger than that of the nuclear family, their experiments often resulted in frustration because of the weakness or nonexistence of a strong marriage bond between the procreative or adoptive parents. For various reasons, there were considerable instability and mobility among the adult members of the communes. Children, therefore, were constantly faced with the problems of separation from those "parents" whom they had come to know and love and of rebuilding substitue "parental" relationships with new arrivals to the group. This brought psychological problems for all members of the group, but especially for the children.

9. Exogamy is the custom of marrying only outside one's own tribe, clan, or family, i.e., outbreeding; endogamy is the opposite, i.e., inbreeding.

10.

> Prohibited to you
> (For marriage) are:
> Your mothers, daughters,
> Sisters; father's sisters,
> Mother's sisters; brother's daughters,
> Sister's daughters; foster-mothers;
> (Who gave you suck), foster-sisters;
> Your wives' mothers;
> Your step-daughters under your
> Guardianship, born of your wives
> To whom ye have gone in,
> No prohibition if ye have not gone in;
> (Those who have been)
> wives of your sons proceeding
> From your loins;

And two sisters in wedlock
At one and the same time,
Except for what is past;
For God is Oft-forgiving
Most Merciful. (Qur'ān 4:23)

See also al Bukhārī, *Ṣaḥīḥ Al-Bukhārī,* pp. 28–34.

11. The qur'ānic passage stating that it is unlawful for men to marry their "milk relatives" is to be found in 4:23. See note 10 above and al Bukhārī, *Ṣaḥīḥ Al-Bukhārī,* pp. 24–28.

12. 'Abd al 'Atī, *Family Structure,* p. 131; al Bukhārī, *Ṣaḥīḥ Al-Bukhārī,* pp. 28–34.

13. The ever-widening problems of impotence in contemporary males may in large part be a result of the excessive sexual freedom and familiarity which pertains in many of the societies of this century. See G. L. Ginsberg, et al., "The New Impotency," *Archives of General Psychology* 28 (1972): 218; and G. F. Gilder, *Sexual Suicide* (New York: Quadrangle/The New York Times Book Co., 1973).

14. See 'Abd al 'Atī, *Family Structure,* p. 139.

15. In Turkey and Pakistan, men and women can marry at eighteen and sixteen years of age, respectively; in Egypt, at nineteen and seventeen; in Jordan, at eighteen and seventeen; in Morocco and Iran, at eighteen and fifteen; and in Tunisia, at twenty for both sexes. Marriage of younger persons must have the consent of guardians and the permission of the court.

16. E.g., barrenness of the first wife, disbalance of male/female population, chronic illness of the wife, large numbers of helpless widows and orphans in the community, etc.

17. See Tanzil-ur-Rahman, *A Code of Muslim Personal Law* (Karachi: Hamdard Academy, 1978), pp. 94–101, for a summary of modern legislation pertaining to polygyny. The Tunisian Law of 1956 prohibits it outright, while other countries have placed various restrictions on having more than one wife—e.g., financial ability of the husband, just cause, consent of the first wife or wives, and/or permission of a court or *ad hoc* council.

18. Al Bukhārī, *Ṣaḥīḥ Al-Bukhārī,* pp. 51–53.

19. Tanzil-ur-Rahman, *Code,* pp. 51–52.

20. 'Abd al 'Atī, *Family Structure,* pp. 76–84. Tanzil-ur-Rahman, *Code,* chap. 3, see especially pp. 71–74, for modern legislation in various Muslim countries relating to consent to marriage.

21. Al Bukhārī, *Ṣaḥīḥ Al-Bukhārī,* pp. 51, 55, 59–61.

22. Asaf A. A. Fyzee, *Outlines of Muhammadan Law,* 2nd ed. (London: Oxford University Press, 1955), pp. 112–113; Tanzil-ur-Rahman, *Code,* pp. 218–221.

23. This paraphrases a *ḥadīth* from the life of the Prophet Muḥammad: "Of all the permitted acts the one disliked most by God is divorce" (S. Ameenul Hasan Rizvi, "Women and Marriage in Islam," *The Muslim World League Journal,* vol. 12, no. 1 (Muharram, 1404 A. H. [October, 1984], p. 26).

24. Tanzil-ur-Rahman, *Code,* pp. 313–316; Fyzee, *Outlines,* pp. 128–130.

25. Fyzee, *Outlines,* pp. 130–131.

26. Tanzil-ur-Rahman, *Code,* chap. 12, pp. 339ff.; Fyzee, *Outlines,* pp. 134–135.

27. Tanzil-ur-Rahman, *Code,* pp. 346–350.

28. Ibid., pp. 552ff.; Fyzee, *Outlines,* pp. 138–139; Alhaji A. D. Ajijola, *Introduction to Islamic Law* (Karachi: International Islamic Publishers for Ajijola Memorial Islamic Publishing Co. [Nigeria], 1981), pp. 172ff.

29. Tanzil-ur-Rahman, *Code,* pp. 504ff.; Ajijola, *Introduction,* pp. 176–177; Fyzee, *Outlines,* pp. 141–142.

37.

Marriage in the Hindu Religious Tradition

Arvind Sharma

This section concludes with a revealing examination of the Hindu marriage. Hinduism is the world's religious psychologist. Before the invention of psychology and psychoanalysis, Hinduism had discovered and practiced the essence of their insights. We can see this in its whole attitude toward marriage and sexuality.

Marriage may not be universal in Hinduism, but it is fundamental and surrounded by a sacramental aura. It is the foundation on which the edifice of Hinduism as a social system rests. Hinduism teaches that every able-bodied person should get married. Exceptions are allowed for the ascetic, but marriage has a pervasive social and eschatological value and purpose. Lifelong fidelity is the dominant image of marriage—symbolized in the couple's exchange of vows with fire as witness.

Most Hindu marriages are still arranged by parents, although the impact of modernity is increasing the availability of choice. The marriage rite is an elaborate (six-part) affair, culminating, it is said, in the fulfillment and completion of the married persons.

Questions for reflection follow:

1. The Hindu does not marry the woman he loves. He loves *the woman he marries. Compare and evaluate this cultural attitude from a Western point of view.*

2. Hindu womanhood has essentially remained conservative. Is this a strength or a liability today?

3. Compare the current status of women in Hinduism and Western Christianity.

4. Hinduism has a renowned sense of religious tolerance. Can this openness offer a model for interreligious marriage in our time?

From *Journal of Ecumenical Studies* 22:1 (Winter 1985): 69–80. Reprinted with permission.

I

The institution of marriage may not be universal in Hinduism, but it is fundamental. To see just how fundamental, it is helpful to recognize that, though Hinduism does not possess a church as such in the Christian sense, it sees all of society as one. Such a society, in Hindu thought, is organized around the "three great ideal structures": the four classes, the four stages of life, and the four ends of humankind.[1] The four classes consist of the four *varṇas,* or the hierarchy of the classes of priests (*brāhmaṇas*), warriors (*kṣatriyas*), merchants (*vaiśyas*), and laborers (*śūdras*) into which Hindu society has traditionally been divided. The four stages of life (*āśrama*) are those of student (*brahmacarya*), householder (*gṛhastha*), forest-dweller (*vānaprastha*), and renunciant (*sannyāsa*), each successive stage covering twenty-five years. The four ends of human beings are righteousness (*dharma*), prosperity (*artha*), sensuous pleasures (*kāma*), and spiritual liberation (*mokṣa*).

Such are the building blocks of the Hindu social structure, and, "while it is clear that the three idealized constructions were never fully realized within Indian society at any time, nevertheless they provided the background for social life, and still exercise a powerful appeal to all those who live within the Hindu milieu."[2] The ideological commitment of Hinduism to this social structure is so strong that Hinduism is itself sometimes defined as *varṇāśrama-dharma*—that religion distinguished from others by *varṇa* and *āśrama.*

To demonstrate the centrality of marriage to this whole structure, I must introduce the concept of *saṃskāras* or sacraments in Hinduism. These are variously listed, and one's eligibility for them differs according to one's class, but the two sacraments common to *all* Hindus seem to be those of marriage and obsequies. Thus, despite the differences in the sacramental duties of the various classes of Hinduism, marriage is common to all.

Let us now turn to the four stages of life. The standard view is that they should be undergone in that order. An option is allowed for the spiritually impatient to renounce the world after leading the life of a student. But the importance attached to marriage is seen in the claim that there is really only one *āśrama* or stage of life, "viz. that of the householder (*brahmacarya* being only preparation for it) and that the other *āśramas* are inferior to that of the householder."[3] It is true that there is an out—one can bypass the stage of married life, but to do so is exceptional and, in some texts, even forbidden today according to the Hindu scheme of *yugas* or ages. The remarkable verses of the most authoritative lawgiver in Hinduism, Manu, concerning the stage of the married householder deserve to be cited here:

> As all living creatures subsist by receiving support from air, even so (the members of) all orders subsist by receiving support from the householder.
>
> Because men of three (other) orders are daily supported by the householder with (gifts of) sacred knowledge and food, therefore (the order of) householder is the most excellent order.[4]

As for the four ends of humankind, according to Hindu axiology they are to be pursued harmoniously. As the famous Hindu text on marriage and erotics, the *Kāmasūtra,* says:

> Man, who could normally live up to a hundred years, must apportion his time and take to virtue, material gain, and pleasure in such a way that these are mutually integrated and do not harm each other.
>
> As a boy he must attend to accomplishments like learning; in youth he should enjoy himself; in later life he should pursue the ideals of virtue and spiritual liberation.[5]

The overall position in Hinduism, then, is that all four ends of life should be pursued by all the *varṇas,* who must all undergo the stage of the householder. Even the exceptional celibate monks depend for their subsistence on the householder. Marriage, therefore, is the foundation on which the edifice of Hinduism as a social system rests.

II

Hinduism teaches that every able-bodied individual should get married. There is an ascetic drop-out option, but the general teaching is very clear—all should marry. Even the Hindu gods are usually married.[6] Divorce was not permitted[7] for either party. A Hindu, it may be said, was born once, married once, and died once. This is not quite true, as remarriage and even polygamy[8] were possible, but the statement sums up the general spirit. In classical Hinduism widows were also expected not to remarry (and the more enthusiastic even concremated themselves with the dead husband: *satī*).[9] Men could remarry. Marriage, however, was "made obligatory for women but not for men"[10] by the classical lawgivers.[11]

The importance attached to marriage can further be seen from two stories which breathe of folklore. First is the story of a girl who practices severe penances till her old age. She was told that if she died unmarried she would not go to heaven, so she induced a man to marry her the day before she died. Second is the story of an ascetic who in a vision saw his ancestors suffering torment, since by not marrying and continuing the lineage he was going to deprive them of the traditional offerings to the *manes* which sustained them. After this experience the ascetic decided to get married. The value placed on marriage was both social and eschatological, the latter probably only a reflection of the former.

III

The dominant imagery of marriage is in a sense provided by the rite itself. The rite, to be analyzed later, consist essentially of the exchanging of vows of fidelity between the bride and the bridegroom with fire as witness. Hence, the dominant imagery of marriage is of lifelong (even postmortem!) fidelity. Such fidelity could be both deadly and death-defying, depending on how it was applied. One wife, Sāvitrī, much eulogized in the tradition, rescued her husband from the jaws of death by the power of her devotion.[12] The other example is provided by Kannagi, who remained faithful to her husband even when his affections had been alienated by a courtesan. He was wrongly executed through a miscarriage of justice, whereupon Kannagi reduced the whole city of Madurai to ashes by the imprecatory power born of chastity.[13]

Just as a wife who was totally devoted to her husband was known as a *pativratā,* a husband devoted to his wife was known as an *ekapatnīvratī,* though we hear less of him. The double standard applied. However, if Hinduism was legally polygamous, it was morally monogamous. The dominant model in this respect is supplied by the figure of the prince (later king) Rāma, the divine hero of the epic Rāmāyana. Although Rāma has been subjected to some criticism for his treatment of his wife Sītā, he never compromised his marriage vows, despite temptations and long periods of separation from her. Similarly, Sītā, who voluntarily accompanied him into exile and refused to yield to the blandishments of her demon-abductor, epitomizes the loyal wife. Finally rescued by Rāma, she underwent an ordeal by fire to prove her chastity. Rāma and Sītā thus supply the dominant image for the ideal Hindu couple and are normative figures of conjugal fidelity.[14] They embody the ideal of the law books: "Let mutual fidelity remain till death" (Manu IX:101). The marriage of Mahatma Gandhi and Kasturba in modern times seems to approximate this ideal closely.[15]

Hindu conjugal imagery emphasizes the role of the wife's devotion to her husband in the context of marital stability. This is symbolized by the figure of Arundhatī, the chaste and devoted wife of Vasiṣṭha, a sage. Her name is also the name of the star situated near the polar star. As part of the marriage ceremony, the couple look at the star and "make a vow they too will be like Vasiṣṭha and Arundhatī." Thus, the ritual is suggestive of "firmness in conjugal life."[16] Arundhatī as a *pativratā* epitomizes the virtues of wifely devotion, although an Indian judge declared recently: "Times have radically changed and modern husbands can no longer expect the forbearance, patience and complete effacement of personality shown by the legendary Arundhatī."[17]

IV

The marriage ceremony itself is a fairly elaborate affair,[18] though its essentials are fairly simple. The tradition by which most Hindu marriages were arranged by the parents continues. Increasingly, however, young men and women are choosing partners for themselves, difficult as this is in Hindu society, where women are relatively segregated in comarison to Western society—though perhaps less so than in traditional Islamic society. Most parents try to arrange the marriage within the same "caste," the commensal and endogamous groups or *jātis* into which Hindu society is divided. *Jātis* number 3,000, each notionally accommodated within a *varna.* There is an increasing tendency to marry within the larger *varna* rather than only within the *jāti;* classical Hindu lawgivers favored marriage within the *varna.* Marriages in which the man's caste was superior to the woman's (hypergamy) were grudgingly approved, but if the woman's rank was higher (hypogamy) the marriage was frowned upon as leading to downward mobility. These considerations are no longer as strong as they used to be, and even intercaste marriages have ceased to be an innovation for quite some time.

Once the couple is ready for marriage the ritual is performed under an auspicious astrological combination.[19] The rite itself consists of six main elements: *Madhuparka, Kanyādāna, Pāṇigrahaṇa, Aśmārohaṇa, Lājahoma,* and *Saptapadī.*

Madhuparka refers to the reception of the bridegroom at the bride's house. The word literally means the ceremony in which honey is poured, so it seems the Hindu wedding ceremony starts with a sort of honeymoon! The bridegroom is given a place of honor and attended to, along with the rest of his party. In some parts of India the bride and the bridegroom exchange garlands to show mutual acceptance.

Kanyādāna follows, after a break if astrological calculations make it necessary. In this ceremony, after the fire has been lit in the sacrificial pit, the father hands the bride over to the bridegroom. "The father of the girl says that the bridegroom should not prove false to the bride in dharma, artha and kāma and he responds with the words 'I shall not do so.'"[20]

In the next rite, *Pāṇigrahaṇa,* the bridegroom grasps the hand of the bride. The word itself is sometimes used as a synonym for marriage. In some parts of India a decorative cord called the *maṅgalasūtra* is tied around the neck of the girl at this time like a pendant, symbolizing her new status as wife.

The next major rite is *Aśmārohaṇa* or stepping on a stone. "The stone is placed to the north of the fire. The bridegroom makes her step on the stone with the right foot. Then he recites 'step on this stone and you be firm like a stone. Tread the foes down and turn away the enemies.'"[21]

The bride then offers fried grain to the fire thrice to the accompaniment of ritual formulae. This is called *Lājahoma.*

Then follows the more dramatic part of the ceremony, the *Saptapadī,* in which the bride and the bridegroom circumambulate the fire seven times with the ends of their garments tied together, usually with the bridegroom leading. The symbolism of these seven steps is variously interpreted. In one view each round symbolizes a decade of union. The more devout see the connection as lasting for seven births. A more direct significance is seen in the words uttered on the occasion:

> May you take one step for sap, second for juice (or vigour), third step for the thriving of wealth, fourth step for comfort, fifth step for offspring, sixth, step for seasons, may you be my friend with your seventh step! May you be devoted to me; let us have many sons, may they reach old age.[22]

The steps are also said to symbolize comradeship between the husband and wife in accord with the tradition that friendship among the virtuous is *sāptapadīna:* seven steps taken with a virtuous person suffice to create a lifelong bond of friendship. The taking of the seven steps together by the bride and the bridegroom is the crucial rite of the sacrament, as "the promise of a daughter and giving a daughter with water are not certain means of knowing wife-hood but *saptapadī* is known to be the completion of marriage. If any of the other ceremonies are wanting that would not vitiate the marriage."[23]

Of the other ceremonies which follow,[24] the most moving is *varagrha-prasthāna,* associated with the permanent departure of the bride from her house for the groom's house. Her family cries copiously, as does the bride, and the bridegroom feels almost like a heel for being responsible for this tragic drama. This occasion is filled with the tenderest sentiment, one which has not changed over the centuries, if we are to believe the words of a Hindu father of the seventh century, A.D., who reprobates

> this rule of law laid down by some one viz. that one's own children (daughters) sprung from one's body, fondled on one's knees and whom one would never forsake, are taken away all of a sudden by persons (husbands) who till then were quite unfamiliar. It is on account of this sorrow that although both (son and daughter) are one's own children the good feel sorrow when a daughter is born and who offer water in the form of tears to their daughters at the very time of their birth.[25]

The final rite, that of consummation, is not to be performed, according to the lawbooks, till three days after marriage; hence, it is called "the rite performed on the fourth day of marriage"—*caturthīkarma.* One of the texts captures the romance of the occasion in the husband's words:

> "I tie thee with the string of the Love; not to be released again." Then he embraces her with the verse "*mamānuvratā bhava* [be devoted to me] . . ." again imploring the bride to be the follower and friend of the husband. Then he kisses her with the words "Oh! honey, this is honey; O my tongue speaks honey [sweet words], give honey [obtained from] the bees up to my mouth; the concord is attained." Further he recites a verse referring to magical concord of the *cakravāka* birds [symbolic of romantic longing]. Here the author has become quite poetic about the first union of the bride and the groom.[26]

It may now be added that the principal purpose of marriage according to the tradition is threefold: the performance of *dharma* or religious observances; the begetting of children, especially sons; and sexual pleasure. As for the first, it is said that only after marriage does one become a complete person and that husband and wife must perform religious rituals together rather than separately. The same idea of wholeness and fulfilment also underlies the second purpose. As a text says: "The wife is indeed half of one's self; therefore as long as a man does not secure a wife so long he does not beget a son and so he is till then not complete (or whole); but when he secures a wife he gets progeny and then he becomes complete."[27] A folk etymology for the common word for son, "Putra," was created to provide an eschatological argument for having a son. A son, by performing one's last rites, saved one from falling into a hell called "Put."[28]

V

Religious structures controlled marriage and divorce within Hinduism till the Hindu Marriage Act of 1955. This meant, in effect, that the wife, according to the popular Hindu lawbook of Manu, was always to "present a smiling face, be alert and clever in her domestic duties, keep the domestic vessels clean and burnished and not be extravagant."[29] Just as it was her foremost duty to honor and serve her husband—and she was bound to stay with him—she had the right of residence and maintenance. The husband was bound not only to maintain the wife but also to cohabit with her. Not to do so without good cause was a sin. The wife similarly could not refuse herself to her husband without good cause. Lapses from fidelity created problems. Minor lapses apart, if the husband was morally degenerate (*patita*) some relief was available, as also if he became a renunciant or an outcaste, though the texts are not unanimous.[30] The wife was encouraged to look upon her husband as a deity.[31] She was always to be responsible to him, and obedience was emphasized.[32] Hindu texts abound in the discussion of details of her behavior, called *strīdharma*.

The husband was to respect and cherish his wife, and he could not easily abandon her. The "humane character of the legislation" in this respect is best exemplified by the remarkably restrained attitude of the lawmakers toward adultery. "The propositions about maintenance set out here are accepted as the modern Hindu Law by the courts in India."[33] It is perhaps of some interest to add that modern formulations of Hindu law, after first taking a harsh view of the matter, now seem to be closer to the traditional stance.[34] It may be added further that, although ritually husband and wife were identical, "this identity between husband and wife was not accepted . . . for secular or legal purposes."[35] Ordinarily they were *not* liable for each other's debts, nor did they exercise domain over each other's property.[36] Ordinarily a husband or a wife could not file a case against one another, though the king was expected to take cognizance of a grave violation.[37] During British Raj, however, Hindu law became justifiable. The situation changed radically with the enactment of the Hindu Marriage Act and the laws that followed it. A uniform body of law was created for Hindus by the Indian government, and it is to a consideration of this that we must now turn.

Here the relationships between religion and state and between religion and secular culture come into sharp focus. The state has seen fit to intervene in matters of personal religious law and has moved both to reform it and to secularize it. The reforms which were carried out were both prospectively opposed[38] and retrospectively criticized.[39] These reforms relate to the formal as well as to the ideological aspects of marriage. J. Duncan M. Derrett has remarked:

> If the Rishis [sages to whom the law books are attributed] have at the end of the day failed to persuade India to accept, at the official level, the need for pre-puberty marriages, the conditional and occasional necessity for male polygamy, the absence of remarriage for women, and the subordination of women to their male relations and protectors; if they have failed in their endeavour to spread amongst all classes the need for betrothal by gift of a virgin accompanied by "ornaments," rather than a large cash dowry; these failures apply only to the mechanical aspects of social organization.[40]

The Hindu Marriage Act of 1955 also provides for divorce,[41] and it is here that it breaks ideologically with the classical tradition. Critics have alleged that this kind of legislation is the work of a nuclear-family-oriented Westernized elite that is out of touch with the masses.[42] This charge may have substance in the sense that the legislation may be ahead of its time, but change in India often proceeds from the top downward.

In marriage legislation a watershed is represented by the Special Marriage Act of 1954 and the Hindu Marriage Act of 1955, the latter being the more "conservative."[43] Both the Hindu[44] and the secular[45] thrusts have been maintained, and Hindu legislation has been growing increasingly liberal. For example, just as divorce was introduced in 1955, divorce by mutual consent became possible under Hindu law in 1976.[46] The government is committed by the Constitution of 1950 to the "abolition of personal law and the establishment of territorial law,"[47] which in effect means that a uniform civil code for the nation will apply to all citizens irrespective of religious affiliation; events seem to be moving in the general direction of such a secular consummation of religious marriage laws.[48] There is, however, still a widespread sentiment in favor of the sacramental aura that marriage has in Hinduism, as was disclosed by a survey of Hindu youth about a decade ago:

> Three-fourths (74 percent) of all the students indicated a preference for a religious ritual at the time of marriage, while the remainder were about equally divided between preference for civil registration and lack of any strong opinion on the subject. The great majority of those who preferred a religious ritual held that it is a good tradition, makes a marriage sacred, and binds the couple together better than mere civil registration could do. Those stating a preference for civil registration emphasized the economic burden placed upon the family by the religious ritual and cited this as the reason for their choice of the inexpensive procedure of civil registration.[49]

The commitment to the integrity of marriage is strong to the point of being sacramental, and the "judiciary of India . . . will be united *to a man and to a woman* in the view that reconciliation is worth every possible effort and ingenuity."[50]

When the secular and the Hindu blend in India, no one can be quite sure of the result: Will Hinduism be secularized or secularism Hinduized?[51]

VI

The women's movement in India took its real start with women's widespread participation in the independence movement. After the achievement of Indian independence in 1947, the Indian Constitution of 1950 gave Hindu women, among others, equal political rights and wider access to employment. Socially, however, women's position changed only with the new legislation:

> The social code which took the place of the traditional personal laws of Hindus effected a revolutionary change in the position of women. Marriage was made a civil contract; polygamy was abolished. Divorce was permitted under certain well-defined conditions. Caste restrictions in respect of marriage were abolished. Also a

> uniform marriage law was enacted for all Hindus. The law of succession was changed so as to give women the right to inherit a share in the family property. Briefly, a far-reaching revolution in respect of women's rights had taken place.[52]

It seems, however, that what K. M. Panikkar said in 1963 still applies today: "Though the legal rights of women are now on par with those of the most advanced societies, the social temper has not changed sufficiently to reflect these trends."[53] Despite "new trends, Indian womanhood has remained conservative."[54]

VII

The Hindu stand on interfaith marriage is a reflection of its stand on intrafaith marriage. As noted above, Hindus are divided into castes and prefer to marry within them; hence, the attitude toward interfaith marriage in general is not very positive, although it is possible to exaggerate the Hindu's alleged instinctive horror of miscegenation, which even Gandhi is said to have shared.

Historically, interfaith marriages have been encouraged in India so as to develop solidarity among the various religions. The Muslim emperor Akbar encouraged such a policy, and in modern India also there is a tendency to encourage such marriages.[55] However, from a Hindu point of view, they have an adverse consequence: since Hinduism is not a proselytizing religion, an interfaith marriage often involves a change of faith on the part of the Hindu spouse, or at least of the children of the marriage. This situation may change somewhat as Hinduism becomes more self-assertive, but loss of Hindu identity upon marriage seems to be the general trend. An anecdote relating to Pandit Nehru's visit to Kenya is instructive on this point. When a Kenyan minister complained that the Indians in Kenya do not marry Kenyans, Pandit Nehru is said to have remarked: "We don't even marry among ourselves!" In this respect the Hindu experience is comparable to the Judaic. In times of crisis, religious and racial identities fuse. In Hinduism the situation has been complicated by the caste system. As these traditions lose their siege mentality and become more self-confident, their attitude toward interfaith marriage may change.[56]

Another aspect of the Hindu experience may be instructive. Although intercaste marriages have been frowned upon in India,[57] marriages between sects have not been a problem. A family whose household deity is Śiva, for example, sees no problem in marrying into one in which Visnu is worshipped. Such situations do occur, because caste and sect do not always coincide, but they seem not to pose any special difficulty, even within the family. If, as can plausibly be argued, marriages between sects rather than between castes are the true intra-Hindu analogue to interfaith marriages, then there is a message here—the same old Hindu one of religious tolerance. The evidence supplied by French cleric J. A. Dubois in this respect is interesting. Traveling through South India at the close of the eighteenth century, he noted the sectarian rivalry between worshippers of Visnu and Śiva, the major gods of Hinduism. However, he added:

In some parts a remarkable peculiarity is to be observed in reference to these two sects. Sometimes the husband is a Vishnavite and bears the *namam* on his forehead, while the wife is a follower of Śiva and wears the *lingam*. The former eats meat, but the latter may not touch it. This divergency of religious opinion, however, in no way destroys the peace of the household. Each observes the practices of his or her own particular creed, and worships his or her god in the way that seems best, without any interference from the other.[58]

NOTES

1. Ainslee T. Embree, ed., *The Hindu Tradition* (New York: Modern Library, 1966), p. 74.
2. Ibid.
3. P. V. Kane, *History of Dharmaśāstra,* vol. 2, part I (Poona, India: Bhandarkar Oriental Research Institute, 1974), p. 424.
4. Manusmrti III, 77–78; *The Laws of Manu,* tr. G. Bühler (Oxford: Clarendon Press, 1886), p. 89.
5. William Theodore de Bary, ed., *Sources of Indian Tradition,* vol. 1 (New York: Columbia University Press, 1956), p. 209. Although the language of this passage is androcentric, it applies, *mutatis mutandis,* to both men and women.
6. J. Gonda, *Viṣṇuism and Śivaism: A Comparison* (London: Athlone Press, 1970), pp. 127–131.
7. The general position in the *dharmaśāstras* or law books is stated thus by P. V. Kane: "Therefore divorce in the ordinary sense of the word (i.e. divorce *a vinculo matrimonii*) has been unknown to the *dharmaśāstras* and to Hindu society for about two thousand years (except on the ground of custom among the lower classes)" (*History,* vol. 2, part 1, p. 620; but also see pp. 619–622 and p. 611, n. 1436).
8. Ibid., pp. 611, 550 ff.
9. Non-Hindus find it baffling that wifely devotion should have been carried to such extremes. To get a feel for the Hindu attitude here, see Embree, *The Hindu Tradition,* pp. 98–100.
10. See A. S. Altekar, *The Position of Women in Hindu Civilization* (Delhi: Motilal Banarsidass, 1973), p. 35.
11. For possible explanations, see ibid.
12. See R. K. Narayan, *Gods, Demons, and Others* (New York: Viking Press, 1964), pp. 182–189.
13. Ibid., pp. 190–201.
14. For other normative models, see Prabhati Mukherjee, *Hindu Women: Normative Models* (New Delhi: Orient Longman Ltd., 1978).
15. *The Story of My Experiments with Truth,* tr. Mahadev Desai (Washington, DC: Public Affairs Press, 1948), pp. 46, 340.
16. This practice of pointing out the Arundhatī star must have been popular. It has entered the learned lore of Hinduism "in the form of the expression *Arundhatī-darsana-nyāna,* 'the method of spotting Arundhatī,' a tiny-looking star, which even today the Brahmana bride has to see at the time of marriage. The practice is first to show her a bright star somewhere near Arundhatī and tell her that it is Arundhatī. When she has seen it, she is told that that is not the star, but another near it. And this process is repeated till she comes to the real Arundhatī. We moderns would proba-

bly have chosen a different method and called a spade a spade'' (Haridas Bhattacharyya, ed., *The Cultural Heritage of India,* vol. 1 [Calcutta: The Ramakrishna Mission Institute of Culture, 1958], p. 354).

17. Quoted in J. Duncan M. Derrett, *The Death of a Marriage Law* (Durham, NC: North Carolina Academic Press, 1978), p. 154.

18. For details, see Raj Bali Pandey, *Hindu Saṁskāras* (Delhi: Motilal Banarsidass, 1969), chap. 8. The law books classify Hindu marriages into eight types, but the classification does not seem to possess contemporary relevance and is thus omitted (for details, see ibid., pp. 153–170).

19. In parts of India the standard way of indicating interest in a marriage alliance is to ask for the horoscope of the potential bride or groom. It has been suggested that to say ''the horoscopes didn't match'' is a culturally acceptable way of saying no, somewhat like ''I have a headache'' when one partner is not interested in achieving conjugal bliss at the time!

20. Kane, *History,* vol. 2, part 1, p. 533. For the meaning of *''dharma,'' ''artha,''* and *''kāma,''* see section I, above.

21. Usha M. Apte, *The Sacrament of Marriage in Hindu Society* (New Delhi: Ajanta Books International, 1978), p. 99.

22. Kane, *History,* vol. 2, part 1, p. 529.

23. Ibid., p. 539.

24. Panday, *Hindu Saṁskāras,* p. 222.

25. Quoted in Kane, *History,* vol. 2, part 1, p. 510.

26. Apte, *Sacrament of Marriage,* p. 116.

27. Kane, *History,* vol. 2, part 1, p. 428.

28. Manusmṛti IX, 138.

29. Kane, *History,* vol. 2, part 1, p. 563. In America it is said that marriage boils down ultimately to who does the dishes. In ancient India it apparently boiled down to how well they were done.

30. Ibid., pp. 620–621.

31. Ibid., pp. 558–559.

32. Ibid., p. 562.

33. Ibid., pp. 572–573.

34. Derrett, *Death of a Marriage Law,* p. 154.

35. Kane, *History,* vol. 2, part 1, p. 573.

36. Ibid.

37. Ibid.

38. Derrett, *Death of a Marriage Law,* pp. 20–22, 122.

39. Ibid., passim, especially pp. 125 ff.

40. Ibid., pp. 155–156.

41. Ibid., p. 26.

42. Ibid., pp. 141, 177.

43. For details, see ibid., pp. 22 ff.

44. Ibid.

45. Ibid., p. 121.

46. Ibid., p. 152.

47. Joseph W. Elder, ed., *Lectures in Indian Civilization* (Dubuque, IA: Kendall/Hunt Publishing Co., 1970), p. 262.

48. Derrett, *Death of a Marriage Law,* pp. 128 ff.

49. Philip H. Ashby, *Modern Trends in Hinduism* (New York: Columbia University Press, 1974), p. 60.

50. See Derrett, *Death of a Marriage Law,* p. 200.

51. Ibid., pp. 188, 190.

52. K. M. Panikkar, *The Foundations of New India* (London: George Allen & Unwin Ltd., 1963), p. 251. In relation to employment, the following figures may be of interest: "Indian women comprise 7.1 per cent of the doctors, 1.2 per cent of the lawyers and 10.9 per cent of the scientists, in spite of incredibly low literacy rates for the over-all female population (18.4 per cent of Indian women are literate)" (Doranne Jacobson and Susan S. Wadley, *Women in India: Two Perspectives* [Delhi: Manohar, 1977], p. 113).

53. Panikkar, *Foundations,* p. 251.

54. Ibid., p. 252.

55. John F. Hinnells and Eric J. Sharpe, eds., *Hinduism* (Newcastle-upon-Tyne: Oriel Press Ltd., 1972), p. 127.

56. Derrett, *Death of a Marriage Law,* p. ix.

57. The attitude toward interracial marriage would probably follow a similar pattern; see ibid.

58. J. A. Dubois, *Hindu Manners, Customs, and Ceremonies,* ed. and tr. Henry K. Beauchamp (Oxford: Clarendon Press, 1959), p. 119.

IX

Appendices
Church Documents

We conclude this book with a representative sample of three types of church documents that couples may find themselves working with, either in preparation for marriage or in moving toward an annulment. Policies, guidelines, and processes are clearly presented and dispel much of the confusion and misunderstanding surrounding Roman Catholic marital codes.

1. The first document, Marriage Preparation Policy (Appendix A), lays out the policies and guidelines for marriage preparation for the diocese of Paterson, New Jersey. These policies (or some equivalent form) are now common in Roman Catholic dioceses in the United States. They have regularized procedures for marriages and established a common set of expectations.

2. Catholic-Jewish Relationship Guidelines for the diocese of Brooklyn, New York (Appendix B) is the second document. It recognizes the growth of interreligious marriages (in this case, Catholic–Jewish) in the past twenty-five years, and it reflects the positive theological and ecumenical developments since Vatican II. The document is a model of sensitivity that structures a hospitable framework for Catholic–Jewish coupling. Dioceses lacking the large Jewish population of the Brooklyn area may not have guidelines for Jewish–Catholic relations at all or very brief ones.

3. The final document (Appendix C) outlines the Procedures for Annulments operative in the Diocese of Brooklyn. Again, the step-by-step process outlined has been standardized nationwide. This is not a divorce proceeding, as is sometimes misunderstood, but rather a task of assessing whether there ever was a true bond. The Procedures are offered here as one example of the kind of annulment process used in the Roman Catholic Church.

Appendix A

Marriage Preparation Policy, Diocese of Paterson, New Jersey

Christian marriage is a lifelong process of growth and the faithful commitment of the partners to a continuing supportive relationship in which each helps the other to develop as fully as possible. "Christian spouses have a special sacrament by which they are fortified and receive a kind of consecration in the duties and dignity of their state. By virtue of this sacrament, as spouses fulfill their conjugal and family obligations, they are penetrated with the Spirit of Christ . . . thus they increasingly advance their own perfection, as well as their mutual sanctification, and hence contribute jointly to the glory of God." (Gaudium et Spes, 48)

Marriage is primarily the work of God. It is a journey often fraught with pain, sorrow and tears but also with hope, love and ecstatic joy. And it is a dream—a dream to reflect the life and love of Jesus Christ through the chosen vocation of marriage. To live this dream is no easy task. Our society does not support couples in their efforts to live committed, intimate, and faithful lives.

Because the theology of the Second Vatican Council characterizes Christian marriage as an "Intimate Partnership," a "Conjugal Covenant" of life and love, it is evident that effective pastoral preparation for couples who express the desire to marry in the Catholic Church must be given special emphasis. The rate of divorce, the strong influence of mass media, the enormous pressures and challenges facing young couples today prompt us to affirm more clearly than ever the positive, hope-filled vision of Christian marriage offered by the Church. Serious concern for pre-marriage preparation has prompted the Diocese of Paterson to set forth the following policy on marriage preparation in this diocese:

1. Ordinarily, couples who want to marry in a Catholic ceremony within the Paterson Diocese are to begin preparation one year in advance of their marriage. Pastoral concerns must always be taken into consideration.
2. In line with the Province of New Jersey's Common Policy, preparation

of all couples for the Sacrament of Marriage will consist of a minimum of six sessions over a period of one year: Three formational sessions with a priest/deacon and a minimum of eight hours of instruction in a marriage preparation program.

3. A personality profile such as *Focus, Myers Briggs, PMI,* or *Prepare* may be offered to the engaged couple by the priest/deacon preparing the couple for marriage. If the minister chooses to offer such a personality inventory, the parish will be responsible for all costs and for securing a competent person to administer and interpret the inventory. The minister would then share the results of this inventory with the couple to help them discern the strengths and weaknesses of the relationship.
4. Upon completion of the first formational session, if the priest/deacon has any reservations about witnessing the marriage of the engaged couple, he will consult with one of the evaluators for exceptional cases. If a referral is needed, the policy and referral process will be explained to the engaged couple by the minister preparing them.
5. The eight hours instructional requirement for marriage preparation may be fulfilled by attending an Engaged Encounter weekend, a parish Pre-Cana or a Deanery Pre-Cana Conference, whichever is appropriate.
6. All parishes and marriage ministry centers are expected to follow the content for marriage preparation as contained in the Pre-Cana Leader's Guide issued by the Secretariat for Parish Life.
7. Marriage preparation is to meet the specific needs of the attending couples, e.g., older couples, previously married persons, and those entering an interfaith marriage.
8. Marriage preparation will always take into account the language and cultural differences of couples preparing for marriage.
9. A Marriage Ministry Center will exist in each deanery. Each center will be coordinated by a priest/deacon and lay couple, or priest/deacon and designated persons who are selected after consultation with the local dean, pastors, and the Secretariat for Parish Life. The deanery coordinators shall work with the Consultant for Adult and Family Ministry in the Secretariat for Parish Life. The deanery coordinators are accountable to the dean.
10. Marriage Ministry Centers are responsible for instructing all deanery marriage ministers functioning on a team in the areas of communication, theology/sacrament, sexuality, responsible family planning, personality and team leadership. The center is also responsible for scheduling, organizing, coordinating and implementing sufficient Pre-Cana Conferences within its deanery to accommodate any couple whose parish does not offer a Pre-Cana Conference, or who chooses not to attend an Engaged Encounter weekend.
11. All marriage ministers functioning on a team, in parishes, Marriage Ministry Centers, or Engaged Encounter weekends, are to be trained in marriage preparation at the appropriate Deanery Center.
12. Priests, deacons and lay persons involved in Marriage Ministry teams are

to update themselves with a minimum of two hours in-service every other year, e.g. lecture, book, tape, video, workshop, etc.

13. Each pastor will identify one couple to function as a contact person for Marriage Ministry in the parish. This couple will serve as the parish's liaison with the deanery Marriage Ministry Center.
14. All priests/deacons who witness marriages and counsel couples in marriage preparation, should be familiar with the Province of New Jersey's Common Policy concerning formation and instructional sessions with engaged couples. Further, they should be familiar with the specific requirements set forth in the Marriage Preparation Policy of the Diocese of Paterson, the Common Marriage Preparation Policy for the Province of New Jersey, and other pertinent diocesan guidelines dealing with marriage preparation.
15. Ordinarily, the local pastor is responsible for seeing that the pre-marital investigation is conducted, even if he does not do it personally. If someone outside the parish handles the investigation or any other aspect of marriage preparation, the proper pastor should be notified in writing.
16. The priest or deacon who works with the couple and witnesses the marriage has the right to make modifications in the diocesan guidelines to meet the special needs and circumstances of a particular couple. However, in all cases it is the parish priest/deacon in the church where the wedding takes place, who has the ultimate responsibility to certify that appropriate preparation in accordance with diocesan policy has been made.
17. The Secretariat for Parish Life will collaborate with appropriate institutions, agencies and offices, e.g., seminaries, Ministry to Priests, to ensure that all priests are familiar with the Common Policy and realize the importance of a thorough and effective marriage preparation for engaged couples.
18. Parishes that celebrate thirty or more marriages per year, will offer parish sponsored Pre-Cana programs and develop teams trained to conduct these programs in their parishes to meet the needs of those couples who do not participate in Engaged Encounter weekends.
19. Parishes that celebrate less than thirty marriages per year may choose to send their couples to an Engaged Encounter weekend or to a Deanery Marriage Ministry Center for Pre-Cana to fulfill their instructional requirement. The schedule for all Pre-Cana sessions will be made available to all priests and deacons.

Appendix B

Catholic–Jewish Relationship Guidelines, Diocese of Brooklyn, New York

Introduction

1. The term ''interreligious'' refers to our relationship as Christians with those who are not part of the Christian tradition. These relationships are developed theologically in Conciliar documents such as the *Declaration on the Relationship of the Church to Non-Christian Religions*. (*Nostra Aetate*)

2. One aim of interreligious dialogue is to foster a deeper understanding of, and respect for, the integrity of other peoples and faiths. Another aim is to identify the basic principles which are the common heritage of the world's great religions. However, it should be clear that interreligious dialogue does not promote homogenization, which may be defined as an attempt to create one religion out of the diverse traditions and sanctities of many religions.

3. In the document alluded to above, the Fathers of the Council made a momentous declaration regarding Judaism and Islam, Hinduism and Buddhism: we are urged to recognize these religions as positive forces with which the Church can and should enter into dialogue. The *Declaration* states: ''Let Christians, while witnessing to their own faith and way of life, acknowledge, preserve and encourage the spiritual and moral truths found among non-Christians, also their social life and culture. (*Nostra Aetate, N.2*)

Dialogue

4. Thus, the Church in dialogue is sensitive to the concrete forms in which man is seeking God. The quest for the Absolute and man's experience of the Absolute

are manifold. Through our sensitivity to this diverse richness we can also come to appreciate the special relation of Judaism and Islam to Christianity and the History of Salvation, based on the uniqueness of God's self-revelation through the Law, the Prophets, and His Son.

5. It is important for us to understand the meaning of dialogue. The responsibility to be involved in dialogue does not diminish the mandate of the Church to proclaim Jesus Christ to the world. "In virtue of her divine mission, and her very nature, the Church must preach Jesus Christ to the world." (VATICAN COMMISSION—GUIDELINES, 1974, I) Evangelization therefore is a primary responsibility of the Catholic Church.

6. Dialogue does mean that the Church recognizes her duty to foster unity and charity among individuals (*Nostra Aetate, N.I*) and mutual understanding between religious traditions. This means that the Church will show great sensitivity to the concerns of our Jewish brothers and sisters. "Lest the witness of Catholics to Jesus Christ should give offense to Jews, they must take care to live and spread their Christian faith while maintaining the strictest respect for religious liberty in line with the teachings of the Second Vatican Council." (VATICAN COMMISSION—GUIDELINES, 1974, I) Even outside the context of official dialogue, the Church respects and esteems non-Christian religions because they are living expressions of the soul of vast groups of people.

Catholic Concerns

7. In fairness to our Jewish brothers and sisters we should express to them areas of special sensitivity that flow from our Catholic tradition. There is a wide spectrum of concerns and sensitivities within the Catholic community due to various ethnic and national values but several are common to all:

a. Respect for life at all its stages. Life itself has value and is to be supported and defended. As Catholics we make no distinctions about respecting one mode of human life more than another.
b. Coupled with this we have a strong regard for the quality of life and a long social doctrine tradition that is now being expressed in terms of liberation from hunger, disease, oppression and ignorance.
c. Moral values and religious teaching is something that is a right of all persons to experience and know. Education is a concern of the family and the state has no monopoly over the children of society. The state should not penalize by double taxation the Catholic family that seeks to exercise its right to educate children in Catholic schools.
d. While the state has many rights it is not society. The demands of the common good limit the actions of government as do the clear rights of individual citizens.
e. A long tradition calls for the acceptance of all people into our community. The alien is to be regarded as not an enemy but as a subject who is to receive welcome and given hospitality.

f. Voluntary agencies by which love and concern is expressed to people in need are to be fostered and assisted by individual charity and also by government assistance.

Jewish Concerns

8. We should be aware of the special concerns and sensitivities of the Jewish Community. Although there is a very wide spectrum of opinion within the Jewish community, the following concerns are shared by all:

a. *The Holocaust:* The wanton murder of six million Jews by the Nazis weighs very heavily and painfully upon all Jews.
b. *Anti-Semitism:* On the basis of centuries of Jewish suffering and martyrdom at the hands of some who styled themselves as Christians, many Jews cannot escape the fear, either conscious or unconscious, that anti-Semitism, which is so deeply embedded in Western culture, may yet erupt in lesser or greater degree. In dialogue Jews and Christians should be encouraged to probe their deepest feelings toward each other. Sources of interreligious animosity should be identified with openness and candor, and an earnest effort should be made to reach a mature understanding of each other's convictions. Hatred of our neighbors is a grievous sin against the God in Whom both Christians and Jews believe.
c. *State of Israel:* American Jews feel a very strong bond of kinship to Jews throughout the world, and especially to Jews in the State of Israel. Though American Jews have no political allegiance to Israel, they take very great pride in its accomplishments, in the holy city of Jerusalem, in the new life Israel has made possible for the survivors of the Holocaust and for those Jews who have obtained permission to emigrate from the Soviet Union. To many Jews, indeed as to many Christians, the establishment of the State of Israel represents the fulfillment of the Divine promises set forth in Scripture. American Jews will do whatever is in their power to ensure the security of Israel. Almost all of them contribute to the welfare of the people of Israel through the United Jewish Appeal and other philanthropic agencies.
d. *Proselytizing:* Jews are always highly sensitive to activities that, to them, appear to encourage conversion. It must be remembered that Jews are mindful of the time when overly zealous churchmen used to compel them to attend services in which Judaism was belittled and condemned; in the past forced conversions were not infrequent. Therefore, from the Jewish perspective, efforts to convert Jews to Christianity, whether overt or subtle, or the implication that Judaism is an incomplete faith are unacceptable and destructive of dialogue. Our conversations with one another should be regarded as forums wherein each side freely expresses its views in the hope of attaining mutual knowledge and understanding. The motivation of dialogue is not conversion.

e. *Interfaith Marriage:* Jews are greatly concerned with preserving the Jewish people and Judaism as its way of life, and experience has demonstrated that intermarriage will inevitably lead to the diminution of the Jewish community.

While the foregoing five topics are not intended to be all-inclusive, they call for careful preparation. As Catholics we ought to be aware that these issues are likely to emerge in every dialogue with members of the Jewish community.

Opportunities for Dialogue

9. a. *Priest-Rabbi Dialogue:* In these dialogues, priests and rabbis will explore areas of mutual concern. Since we share a common scriptural heritage through the Hebrew Bible, the study of Scripture by scholars and clergy of both faiths is highly encouraged. The Second Vatican Council reminds us that "since Christians and Jews have such a common spiritual heritage, this sacred Council wishes to encourage and further mutual understanding and appreciation. This can be obtained, especially, by way of biblical and theological enquiry and through friendly discussions. (*Nostra Aetate N.4*)

b. *Seminaries:* Seminaries are ideal centers for student exploration of our heritage. In addition to homiletic and liturgical instruction, care should be given to implementing the 1974 Guidelines of the Vatican Commission for Religious Relations with Jews. Such instruction could counteract a sometimes anti-Semitic Gospel interpretation and do much to develop mutual richness in music, festival, and symbol. "With respect to liturgical reading, care will be taken to see that homilies based on them will not distort their meaning, especially when it is a question of passages which seem to show the Jewish people as such in an unfavorable light.

Efforts will be made to instruct the Christian people that they will understand the true interpretation of all the texts and their meaning for the contemporary believer." (VATICAN COMMISSION GUIDELINES, 1974, II)

c. *Marriage and Family:* Since marriage and family life, justice and morality are of extreme concern in both communities, joint study, understanding and action might well prove of mutual benefit. It is important that priests and rabbis participate in dialogues about these issues on their own level as a preparation for congregational dialogue.

d. *Education:* The well-developed parochial and secondary religious education system of the Jewish and Catholic communities in Brooklyn and Queens offers a unique opportunity to continue the positive achievements of the Jewish–Catholic dialogue. By exposing students to the religious traditions and by elaborating on the sources common to both, today's dialogue will ensure future good.

e. *Parish and Synagogue Interaction:* If the dialogue between Catholics and Jews is to be thorough and significant, it must also include a grassroots exchange between Catholic and Jewish congregations. However, as in any other dialogue, great care, patience, and understanding are required.

In exploring such parish–synagogue interactions, priests and rabbis might find it beneficial to seek suggestions from the Diocesan Ecumenical Commission and the Catholic–Jewish Relations Committee.

Catholic–Jewish Marriages

Pastoral Guidance

10. When a Catholic and a Jew decide to enter into marriage, the priest who is helping them prepare a marriage ceremony should be sensitive to the religious conviction and customs of both parties. Neither party to the marriage should be asked to violate the integrity of his or her faith.

11. The priest should advise the couple that neither the Catholic Church nor the Synagogue encourages mixed marriages; indeed, both the Church and the Synagogue greatly desire that Catholics marry Catholics and that Jews marry Jews. In counselling the interfaith couple the priest should remind them of the likelihood that the extended family of each party may be reluctant to accept their child's or sibling's spouse, and that tensions frequently arise as family ties are stretched to the breaking point.

Dispensations

In mixed marriages where the priest is to officiate at the ceremony, he may proceed as follows:

12. a. A priest or deacon may officiate at the wedding of a Jew and a Catholic, with a dispensation from the impediment of disparity of worship, in the sanctuary or other part of the Catholic Church, or in any suitable building on the parish grounds. For good reason, permission can be given for such a wedding to take place in a reception hall, private home or other suitable place.

In cases where the priest or deacon is officiating, a rabbi may be invited to participate in the marriage ceremony and he may offer prayers for the couple and invoke God's blessing on them. He may not share in the marriage ritual as such and is not an official witness to the ceremony. He should not request the vows of either party or lead in the recitation of the vows or co-sign a license.

b. In cases where very serious reasons exist which preclude the possibility of a marriage ceremony before a Catholic priest or deacon, a dispensation from the Catholic form of marriage may be sought so that the Catholic party may marry in a religious ceremony before a rabbi in a sacred or private place.

In this case, a priest or deacon invited to participate in the marriage ceremony may offer prayers for the couple and invoke God's blessing on them. He may not share in the marriage ritual as such and is not an official witness to the ceremony. He should not request the vows or co-sign the license.

c. If for very serious reasons, it is not possible to have either a Catholic or Jewish ceremony, the Church will reluctantly permit a public ceremony recognized in civil law provided the Catholic party has obtained a dispensation from the canonical form of marriage and provided all other requirements are fulfilled.

d. The priest should be aware that Orthodox and Conservative Rabbis are forbidden to officiate at mixed marriages. Some members of the Reform Rabbinate will officiate at mixed marriages. Of those Reform Rabbis who will officiate at mixed marriages, only a very small group will officiate together with a priest.

13. Jews may be admitted as witnesses and attendants at a marriage ceremony in a Catholic Church. Catholics may act as witnesses and attendants at the wedding of friends who are Jews, provided that the wedding to be witnessed will be lawful and valid and takes place within the Diocese of Brooklyn.

14. When a priest or deacon is invited to a marriage ceremony of two Jews conducted by a Rabbi, he may wear, if appropriate, the proper liturgical vestments, offer prayers for the couple and invoke God's blessing on them. It is understood that he will participate in this manner only if in the eyes of the Catholic Church the wedding will be considered lawful and valid and takes place in the Diocese of Brooklyn.

15. When a rabbi is invited to a marriage ceremony of two Catholics conducted by a priest, the rabbi should be offered a place of honor in the sanctuary. He may wear appropriate robes, offer prayers for the couple and invoke God's blessing on them.

Conclusion

It is the hope of the Catholic–Jewish Relations Committee that these guidelines will set a continued tone for constructive dialogue and creative cooperation among the Catholic and Jewish communities of our Diocese.

These guidelines are to be considered in conjunction with the decrees issued by the Holy See as well as the National Conference of Catholic Bishops.

In light of the "spiritual bond linking the people of the New Convenant with Abraham's stock" (*Nostra Aetate, 4*), the unique importance of these guidelines is evident on a socio-spiritual level because of the large number of Catholics and Jews that reside in the boroughs of Brooklyn and Queens.

The members of the Catholic-Jewish Relations Committee welcome the words of Pope John Paul II at a recent audience with Presidents and Representatives of the Jewish World Organizations:

> The Church cannot forget that she received the revelation of the Old Testament through the people with whom God, in His inexpressible mercy, deigned to establish the ancient covenant. It is on the basis of all this that we recognize with utmost clarity that the path along which we should proceed with the Jewish religious community is one of fraternal dialogue and fruitful collaboration.
>
> According to this solemn mandate, The Holy See has sought to provide the instruments for such dialogue and collaboration, and to foster their realization both here at the center and elsewhere throughout the church. John Paul II, March 12, 1979, Rome

Appendix C

Diocese of Brooklyn "Procedures for Annulments"

Church Teaching on Marriage

The constant teaching of the Church is that marriage is indissoluble. It is a sacred covenant, something more than a mere contract, whereby the parties truly and totally commit themselves to each other as persons. Marriage is a sacrament which reflects the loving union between Christ and His Church and is, moreover, a participation in that model union. By far the majority of marriages are happy in just this way. A church tribunal upholds this biblical principle of indissolubility. When a marriage case is presented, efforts are made to ascertain whether reconciliation is possible. If not, the tribunal must scrupulously strive to determine whether there was some defect in the marital consent, or in some other vital element of the marital union, which the parties may have been unaware of, but which, nonetheless, was real and verifiable. This is what an annulment process seeks to do. A tribunal does not esteem marriage lightly. It reveres and honors it. It does not allow itself to be the victim of any currently fashionable view that marriage can be terminated at will or on demand; that is, the selfish philosophy of a society without God, conscience or morals. A tribunal has the duty though to establish whether in a particular case a true marriage ever came into being in the first place.

The Definition of an Annulment

An ecclesiastical annulment is a formal statement of a church tribunal that a particular marriage was never a marriage in the true and full sense of the word according to the teachings of the Catholic Church. It differs, therefore, from a divorce in that a divorce action "breaks" a marriage bond; an annulment is a declaration that there never was a true bond to begin with. In church law, it should be noted, an annulment does not mean that any children that may have been born of the union are illegitimate. In this country an ecclesiastical decision has no civil effect in law. Therefore, a civil annulment or divorce must also be granted by a competent civil court.

A marriage is declared null and void if it is proved that one of the four necessary requirements for a valid marriage was not present on the day of the union. In general, the necessary requirements are:

1. A validly celebrated marriage ceremony;
2. Freedom of the parties to marry each other;
3. Each intending to accept God's plan for marriage as taught in sacred scripture;
4. Psychological maturity of each partner.

The Annulment Process

The annulment request goes through the following steps:

A. An investigation is made concerning each of the four general requirements. The partner applying for the annulment will discuss with a priest trained in Canon Law the facts surrounding the union. The priest will normally ask for a written report. The written report of the marriage helps the priest attorney to understand all the factors in the marital history. It also becomes part of the permanent record in the case file. Writing the report at home allows the person ample time and freedom to recall, at his own pace, the events of the courtship and marriage, especially the circumstances which led to the break up of the union. The matters to be treated in the report are found in detail later on in this document.

Upon receipt of this report the other party to the marriage will then be invited to come in for an interview. The statements of the parties alone do not constitute proof. The facts must be corroborated by witnesses. The witnesses may be close relatives or friends. Those who agree to act as witnesses are also required to write a report and usually will be called to appear in Court on the day set for formal hearing. The more preliminary witnesses' reports submitted, the better. In this way, the attorney can select those witnesses who are most knowledgeable. Guide lines to help witnesses prepare these reports will be provided by your priest attorney.

Tribunals in recent years have entertained petitions for annulments on the grounds of psychological unpreparedness, using the insights developed by psychology. Their findings have helped clarify what is required to enter marriage validly.

It is accepted in church tribunals that if one or both of the parties lacked the maturity to understand the obligations of marriage, the freedom to commit themselves to marriage, or the ability to fulfill the demands of marriage, no true union came into existence. Such a state of affairs must be proved. Such proof is secured from medical records pertaining to either one of the parties, or most usually, through the help of a psychiatrist present on the date of the formal hearing.

B. By reason of present church law, the Brooklyn Tribunal may only adjudicate cases if:

a. The petitioner has a REAL RESIDENCE in Kings or Queens County; OR
b. The other party has a REAL RESIDENCE in Kings or Queens County; OR
c. The marriage ceremony in question took place in Kings or Queens County.

If the Brooklyn Diocese has competence, the trial may be conducted by the Brooklyn Tribunal.

C. The parties appoint a Tribunal Advocate to represent them. The same Advocate can represent both: but two are necessary, one for each, if the parties disagree as to whether they want an annulment. (The other party in the marriage can elect to be a "copetitioner" if he desires to cooperate fully.) The Priest Lawyer for the Petitioner(s) introduces the "Libel," the formal petition for the annulment. This document, which is only presented when the parties' statements (and medical statements, if any) have been submitted, alleges the marriage is null and void for a particular reason and offers the names of witnesses.

A date on the court calendar is then assigned by the Secretary of the Tribunal who will advise both parties of the time of the formal hearing of the case. The parties and their witnesses appear separately before the Court. Such privacy gives each person the opportunity to be completely candid. Although all usually will appear on the same day, the time is arranged so that the parties purposely do not meet. This prevents any unnecessary recriminations.

A court-appointed psychiatrist is necessary in each case involving maturity. His task is to review all the testimony taken during the trial and to supply the Court with an independent evaluation of the psychological factors at work at the time of the marriage.

After all parties have been heard, the Judges will call for argumentation from the Priest Lawyers and the Defender of the Bond. The latter is a priest appointed by the Bishop to act in defense of the marriage bond. The Defender of the Bond is summoned for all the important acts of the trial. He is free to question the parties, as are the Judge and Priest Lawyer. The Court, a panel of three Judges, subsequently deliberates the facts of the case once all the possible testimonies are on hand and arguments are completed. The Court then gives its decision.

D. *A review procedure* normally is required in annulment cases. If the Court has found in favor of nullity and the Defender of the Bond agrees that the decision given was just, he will ask the Bishop to apply for special permission to dispense from this review. The Bishop applies for this dispensation from the National Bishops' Office in Washington. Upon the granting of this dispensation the decision becomes final.

Expenses

The expenses incurred by the Tribunal in a case heard in formal trial are approximately $1,200. The Diocese is willing to bear half this sum; the petitioner is asked to assume the remainder. From persons living in the Brooklyn Diocese (and who are presumed, therefore, to be supporting their parish and the Diocese) a fee of $600 is expected. From a person now living elsewhere and whose former spouse also lives elsewhere, but where the Brooklyn Tribunal has competence only because the marriage in question was performed in Kings or Queens County, a fee of $850 is expected. This payment includes ecclesiastical attorneys' fees, fees for psychiatric and psychological experts, and operating office expenses such as rent, secretarial help, ordinary telephone calls, etc. Nothing else will be asked for except possible extraordinary long-distance telephone calls. Should payment be truly impossible for an individual, some accommodation can be made. It is suggested that the petitioner begin payment once the case is introduced. Checks should be made payable to the "Tribunal—Diocese of Brooklyn."

Additional Information

It is impossible to estimate the length of time for any case. Each case is unique. There can be delays, for instance, in securing the statements of the parties themselves and their witnesses, or securing an appointment with the medical expert. There is always a crowded Court calendar.

UNDER NO CIRCUMSTANCES SHOULD A DATE FOR A NEW MARRIAGE BE ARRANGED BEFORE RECEIVING THE FINAL DECREE.

Documents to be Supplied

Copy of the marriage record

Copy of the civil annulment or divorce; if any

A recent copy of both baptismal records

Names and addresses of psychiatrists or marriage counselors possibly consulted before or during the marriage; approximate date of visits

Written Reports to be Supplied

The report from each of the marriage partners is to be typed, if possible, dated, signed and marked "Privileged and Confidential."

A. Using the following outline, please provide the following information concerning the husband in the case:

1. Childhood; adolescence; school history; personality of parents, which was dominant; relationship with brothers and sisters; friends. Medical history. Is anyone known to have been eccentric or afflicted by emotional or mental illness? Ever under psychiatric care? If in military service, give dates and service history.
2. Please give us a personality picture considering the following: unusual

fears in childhood or later; selfishness; self-confidence; suspiciousness of others; misinterpretation of other person's thoughts, words or actions; nervousness; moodiness, ability to make and maintain friends; inappropriate anger, tantrums, silliness or outbreaks of crying; stubbornness; weakness in character such as lying, stealing, etc. (Please illustrate comments by examples as far as practicable.)

3. State life goals. Please explain any problems with gambling, drugs, alcohol, handling of money. Any history of arrests or imprisonments? Job history.

B. Using the following outline, please provide the following information concerning the wife in the case:

1. Childhood; adolescence, school history; personality of parents, which was dominant; relationship with brothers and sisters; friends. Medical history. Is anyone known to have been eccentric or afflicted by emotional or mental illness? Ever under psychiatric care?
2. Please give us a personality picture considering the following: unusual fears in childhood or later; selfishness; self-confidence; suspiciousness of others; misinterpretation of other person's thoughts, words or actions; nervousness; moodiness; inclination towards anger or tantrums; stubbornness; ability to make and maintain friends; inappropriate silliness or outbreaks of crying; weakness of character such as lying, stealing, etc. (Please illustrate by examples as far as practicable).
3. State life goals. Please explain any problems with gambling, drugs, alcohol, handling of money. Any history of arrests or imprisonments? Job history.

C. Information about the courtship covering: how you met; length of courtship; source of attraction; dating problems; breakups (if so, why you came back together); attitudes of each towards children; hesitancy or reluctance to wed; part each played in wedding preparations; any opposition by either family towards this marriage. Was there any unusual pressure for you to marry? Any prior serious courtships? If so, why were they terminated?

D. Did anything unusual happen at the Church or reception on the wedding day or during the honeymoon?

E. Please detail the problems in the marriage, e.g., physical or emotional mistreatment; misuse of money, alcohol; gambling; inability to face responsibility of work, household, etc.; communication; role of in-laws; attitude of each towards sex, having and raising children. When did problems arise? Did you seek professional help? Were there any temporary separations and, if so, what brought you back together? Names and ages of children, if any, and present status; any emotional or school problems with the children?

F. *Final Separation,* detailing what caused it and describe, with incidents, the history of each party since the break up.

G. Has either party remarried? If so, details: name, place, date.

The Reports of the Witnesses

The witnesses are to send their reports directly to the Priest Attorney. They are not to be sent by way of the principals in the case. Witnesses are to be advised that all information provided by them will be held in the strictest confidence. They should preface their statements ''Privileged and Confidential'' and sign them in the presence of a Notary Public. As mentioned above, guidelines to help witnesses prepare their reports will be provided by the Priest Attorney.

The Trial Day and Decision

The Tribunal Secretary will inform the parties by mail of the date set for the formal hearing of the case. You are asked to reply as to whether you can be present. Please take care to answer. Inform your witnesses of the date and inform them to come on time.

Once the formal hearings have been completed, it normally takes several months before a final decision is reached. You will be informed by letter of that decision. Should you remarry, the letter should be shown to the priest arranging the new marriage; a new marriage date cannot be set *until that decision is in his hand.*